D0856283

THE LOEB CLASSICAL LIBRARY

FOUNDED BY JAMES LOEB, LL.D.

EDITED BY
G. P. GOOLD, PH.D.

HIPPOCRATES

VOL. V

HIPPOCRATES

VOL. V

WITH AN ENGLISH TRANSLATION BY

PAUL POTTER

UNIVERSITY OF WESTERN ONTARIO

CAMBRIDGE, MASSACHUSETTS
HARVARD UNIVERSITY PRESS

LONDON
WILLIAM HEINEMANN LTD

MCMLXXXVIII

American ISBN 0–674–99520–1
British ISBN 0 434 99472 3

First published 1988

Printed in Great Britain by
Thomson Litho Ltd, East Kilbride, Scotland

CONTENTS

Note: Tables of Weights and Measures, an Index of Symptoms and Diseases and an Index of Foods and Drugs are to be found at the end of volume VI.

PREFACE TO VOLUMES V AND VI

In his preface to volume IV (1931), W. H. S. Jones writes: "This book completes the Loeb translation of Hippocrates," offering no explanation why the rest of the Collection is to be ignored, unless it is implied in his next sentence: "The work of preparing the volume has taken all my leisure for over five years . . ."

Whatever Jones' reasons for stopping may have been, the lack of a complete English translation has been noted and regretted by classicists and historians of medicine alike. A plan to continue the Loeb *Hippocrates* has now existed in America for several decades, and it is chiefly due to the untiring efforts of Dr. Saul Jarcho and Mr. Richard J. Wolfe that volume V sees the light of day.

The cost of preparing and publishing volumes V and VI has been met by NIH Grant LM 02813 from the National Library of Medicine, and the examination of Hippocratic manuscripts in Florence, Paris, Rome, Venice and Vienna made possible by grants generously provided by the Jason A. Hannah Institute for the History of Medicine.

PREFACE

Work on volumes V and VI was greatly facili-
tated by the use of computer texts and indexes
kindly furnished by Prof. Gilles Maloney and his
team at the *Laboratoire de recherches hippocra-
tiques* in Quebec.

Finally, it is my pleasant duty to thank Prof.
M. P. Goold, Associate Editor of the series, Prof.
Dr. Fridolf Kudlien, Prof. Wesley D. Smith, William
B. Spaulding M.D., F.R.C.P.(C.), and Lynn Wilson
Ph.D., all of whom read the volumes in various
stages of their preparation, for their manifold help-
ful comments.

Rome, November 1983   Paul Potter

INTRODUCTION TO VOLS V AND VI[1]

These volumes contain the most important Hippocratic works on the pathology of internal diseases. Presumably in consequence of their common purpose, these six treatises tend to share the same general structure: independent chapters of constant form each devoted to one specific nosological entity.[2]

About the treatises' interdependencies, authors, and relative dates of composition, nothing can be said with any degree of certainty. There is neither

[1] This introduction deals only with the treatises in volumes V and VI; for an orientation to Hippocrates and the Hippocratic Collection in general, the reader is referred to W. H. S. Jones' "General Introduction" (Loeb *Hippocrates* I. ix-lxix) and "Introductory Essays" (Loeb *Hippocrates* II. ix-lxvi). Useful guides to Hippocratic scholarship since Jones are Ludwig Edelstein's article "Hippokrates" in *Paulys Real-Encyclopädie der classischen Altertumswissenschaft*, Supplement VI, Stuttgart, 1935, cols. 1290–1345, H. Flashar (ed.), *Antike Medizin*, Darmstadt, 1971, Robert Joly's article "Hippocrates of Cos" in the *Dictionary of Scientific Biography*, vol. VI, New York, 1972, 418–31, and G. Maloney and R. Savoie, *Cinq cent ans de bibliographie hippocratique*, Quebec, 1982.

[2] The individual works are analysed in more detail in their particular introductions.

INTRODUCTION

Fig. 1

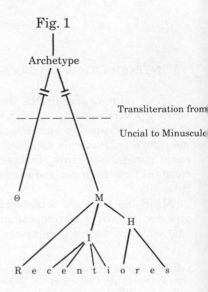

Archetype

Transliteration from
Uncial to Minuscule

10th c.

12th c.

14th c.

16th c.

Θ¹ = Vindobonensis Medicus Graecus 4
M = Marcianus Venetus Graecus 269
H = Parisinus Graecus 2142
I = Parisinus Graecus 2140
Recentiores = approximately twenty manuscripts

¹ Littré (VI. 139) assigned the siglum θ to this manuscript, but several later editors and translators, to whose number I belong, prefer Θ in order to avoid possible confusion with a lost manuscript.

any evidence that would confirm, nor any evidence that would call into doubt, their traditional time of origin about 400 B.C.

In the first century A.D. Erotian knew *Diseases I* and *III* and *Regimen in Acute Diseases (Appendix)*, and Galen (129–199) makes reference, in addition, to *Affections*, *Diseases II* and *Internal Affections*.

MANUSCRIPT TRADITION

Five of the six works in these volumes (*Affections*, *Diseases I–III* and *Internal Affections*) share a transmission that can be represented by the *stemma codicum* that appears as Fig. 1 (p. x).

The transmission of the sixth work, *Regimen in Acute Diseases (Appendix)*, is more complex both because of the existence of a commentary by Galen, which provides a fertile source of variant readings, and also because it was translated into Latin at an early date.[1] The *stemma codicum* that appears as Fig. 2 (p. xii) indicates the relationships among the Greek manuscripts upon which the critical editions, including this one, are based.

Furthermore a papyrus (Rylands Greek Papyrus 56)[2] of the first half of the second century A.D. con-

---

[1] See Hermann A. Diels, *Die Handschriften der antiken Ärzte*, Berlin, 1905–1907, pp. 8 f. and Supplement p. 25.

[2] Edited by A. S. Hunt in *Catalogue of the Greek Papyri in the John Rylands Library at Manchester*, vol. I, Manchester, 1911, 181 f.

Fig. 2

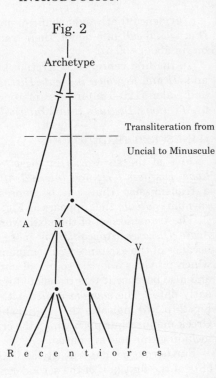

Archetype

Transliteration from

Uncial to Minuscule

10th c.

A   M

12th c.

V

14th c.

R e c e n t i o r e s

16th c.

A = Parisinus Graecus 2253
M = Marcianus Venetus Graecus 269
V = Vaticanus Graecus 276
Recentiores = approximately twenty manuscripts

taining two fragments[1] of the text of *Regimen in Acute Diseases (Appendix)* makes a limited but significant contribution to the establishment of the text.

TEXT AND TRANSLATION

For *Diseases I* and *III* and *Regimen in Acute Diseases (Appendix)* I have generally relied on the collations given in the critical editions.[2] For the other three works, which lack critical editions since Littré, collations of Θ and M have been made from microfilms and supplemented by inspection of the actual manuscripts.

In establishing the Greek text and making the English translation, I have consulted many earlier texts, translations and commentaries, among which the most important are:

Hippocratis Coi ... opera ... per M. Fabium [Calvum] ... Latinitate donata ..., Basel, 1526. (Calvus)
Hippocratis Coi ... libri omnes, ad vetustos Codices ... collati & restaurati [per Ianum Cornarium], Basel, 1538. (= Cornarius)

---

[1] Chapter 24 φιλέει τῷ τοιῷδε — (25) ὑγρὰ διαχωρήσῃ καὶ and Chapter 26 τὸ ἕτερον παρὰ τὸ ἕτερον — (27) Τοὺς τοιούσδε.
[2] See below p. 97, and vol. VI pp. 5 and 261.

INTRODUCTION

Hippocratis Coi ... opera ... omnia. Per Ianum Cornarium ... Latina lingua conscripta, Lyons, 1554. (= Carnarius/Latin)

Magni Hippocratis ... opera omnia ... latina interpretatione & Annotationibus illustrata Anutio Foesio ... Oeconomia Hippocratis alphabeti serie distincta, Geneva, 1657–1662. (= Foes)

Magni Hippocratis Coi Opera omnia edita ... industria & diligentia Joan. A. Vander Linden, Leiden, 1665. (= Vander Linden)

Hippokrates Werke aus dem Griechischen ... von J. F. C. Grimm. Revidiert ... von L. Lilienhain, Glogau, 1837–1838. (= Grimm)

E. Littré, *Oeuvres complètes d'Hippocrate*, Paris, 1839–1861. (= Littré)

F. Z. Ermerins, *Hippocratis ... reliquiae*, Utrecht, 1859–1864. (= Ermerins)

Car. H. Th. Reinhold, ΊΠΠΟΚΡΑΤΗΣ Κομιδῇ, Athens, 1865–1867. (= Reinhold)

H. Kuehlewein, *Hippocratis Opera omnia*, Leipzig, 1894–1902. (= Kuehlewein)

R. Fuchs, *Hippokrates, sämmtliche Werke. Ins Deutsche übersetzt ...*, Munich, 1895–1900. (= Fuchs)

The English translation attempts to be as close to the original as possible while still remaining readable. In matters of vocabulary, I have taken the *Shorter Oxford English Dictionary* and *Dorland's Illustrated Medical Dictionary* as a basis.

INTRODUCTION

Bibliographical Note: Supplementary bibliographical information is to be found in the introductions to individual treatises and in notes to the Indexes, which are printed at the end of volume VI.

AFFECTIONS

INTRODUCTION

Galen, alone of the ancients, mentions a Hippocratic *Affections*[1] and, as the following two passages show, means the same treatise that our manuscripts know under this title:

> For in the *Affections* of Hippocrates, whether in fact the book is by Hippocrates himself or by his pupil Polybus, the following is written about lientery: "foods pass off undigested and watery; no pain is present; patients become lean of body."[2]

> τελεῖν : to consume, as he says in *Affections*.[3]

Affections consists of two parts: a nosological part (1–38) and a dietetic part (39–61).

The nosological part has the following plan:

1: Proem[4]
2,4,5: Diseases of the Head; 3: Importance of Early Treatment

---

[1] Besides the two instances quoted here see also C. G. Kühn, *Cl. Galeni Opera omnia*, Leipzig, 1821–33, XV. 587 = Corpus Medicorum Gracorum V 9, 1 p. 198 and Kühn XVIII(1). 11.

[2] Kühn XVIII(1). 8; the reference is to *Affections* 24.

[3] Kühn XIX. 145; the reference is to *Affections* 43 and 44.

[4] The proem reveals, through its studied architecture and polished style, an unmistakable dependence on the

2

AFFECTIONS

The individual disease descriptions, which make up most of this part, vary greatly in completeness and emphasis, but do, in principle, all follow the same format: name; symptoms and course; treatment; aetiology.

The dietetic part of *Affections* lacks any very apparent order. Chapters 39, 44–47, 50–51, 55 and

rhetorical art. After winning his reader's attention by an appeal to "any man who is intelligent", the author states and justifies the purpose of his work, expounds his medical theory, explicitly defines the limits to which a layman's knowledge must extend, and concludes with a sentence on how he intends to proceed. The proem, and by implication the work that follows, is addressed not to the physician, but to the intelligent layman.

3

AFFECTIONS

59–61 are devoted mainly to more general theoretical considerations, the rest to rules for the use of specific dietetic agents:

40: Gruels and Drinks
41: Diet after taking Medications
42: Anointing
43: Foods to Moisten, Dry, Restore
48: Wines
49: Meats
52: Cereals, Wines, Meats, Fish
53: Hot Bath
54: Vegetables
56: Boiled Vegetables
57: Melons
58: Honey

Affections has long been regarded,[1] no doubt on account of its first chapter, as a book of popular medicine, but this view is mistaken, as the following two points prove. First, in every single instance in which the reader is addressed in the second person, the context dictates that it must be the physician, and not the layman, that is meant.[2] Second, the general level of technical sophistication evident in *Affections* is no different from that of other Hippocratic works universally held to be addressed to

[1] See e.g. Littré VI. 206.
[2] E.g. 3: you administer; 14: Give ... whatever you think suitable to drink; 25: no one will blame your understanding; 37: When you come to a patient; 43: When you wish to moisten a patient's cavity; 44: If you wish to give; 47: If you make your administrations to patients.

AFFECTIONS

physicians.[1] Thus, we must suppose that the first chapter of the work, together perhaps with one or two other sentences,[2] represents a frame, into which a two-part medical treatise has been set.

In modern times, besides finding a place in all the standard collected editions and translations, *Affections* has been the subject of two special studies:

Jean de Varanda, *Opera Omnia*, Lyons, 1658: "Explicatio Libri Hippocratis Περὶ Παθῶν De Gravissimis Morbis", pp. 828–834.[3]

Jürgen Wittenzellner, *Untersuchungen zu der pseudo-hippokratischen Schrift* Περὶ Παθῶν, Diss. Erlangen-Nürnberg, 1969.

A recent interpretative work on the Hippocratic Collection contains a newly edited text for several chapters of *Affections*:

Jacques Jouanna, *Hippocrate. Pour une archéologie de l'école de Cnide*, Paris, 1974. (= Jouanna)

[1] Cf. e.g.
Affections 19: *Diseases II* 71: *Internal Affections* 21
Affections 21: *Diseases III* 14
Affections 29: *Internal Affections* 51
Affections 52: *Regimen II* 42, 44, 46, 48: *Regimen in Acute Diseases (Appendix)* 49, 50
Affections 54–56: *Regimen II* 54: *Regimen in Acute Diseases (Appendix)* 45–47.

[2] E.g. chapter 33: "Through understanding these things, a layman will be less likely to fall into incurable diseases. . . ."

[3] This commentary ends abruptly with a *Reliqua desunt* after discussing the first sentence of *Affections* as far as the word ὠφελέεσθαι.

ΠΕΡΙ ΠΑΘΩΝ

VI 208
Littré

1. Ἄνδρα χρή, ὅστις ἐστὶ συνετός, λογισάμενον ὅτι τοῖσιν ἀνθρώποισι πλείστου ἄξιόν ἐστιν ἡ ὑγιείη, ἐπίστασθαι ἀπὸ τῆς ἑωυτοῦ γνώμης ἐν τῇσι νούσοισιν ὠφελέεσθαι·[1] ἐπίστασθαι δὲ τὰ ὑπὸ τῶν ἰητρῶν καὶ λεγόμενα καὶ προσφερόμενα πρὸς τὸ σῶμα ἑαυτοῦ καὶ διαγινώσκειν· ἐπίστασθαι δὲ τούτων ἕκαστα ἐς ὅσον εἰκὸς ἰδιώτην.

Ταῦτ' οὖν ἐπίσταιτο ἄν τις μάλιστα εἰδὼς καὶ ἐπιτηδεύων τάδε· νοσήματα τοῖσιν ἀνθρώποις ἅπαντα γίνεται ὑπὸ χολῆς καὶ φλέγματος. ἡ δὲ χολὴ καὶ τὸ φλέγμα τὰς νούσους παρέχει ὅταν ἐν τῷ σώματι ὑπερυγραίνηται ἢ ὑπερξηραίνηται ἢ ὑπερθερμαίνηται ἢ ὑπερψύχηται· πάσχει δὲ ταῦτα τὸ φλέγμα καὶ ἡ χολὴ καὶ ἀπὸ σίτων καὶ ποτῶν, καὶ ἀπὸ πόνων καὶ τρωμάτων, καὶ ἀπὸ ὀσμῆς καὶ ἀκοῆς καὶ ὄψιος καὶ λαγνείης, καὶ ἀπὸ τοῦ θερμοῦ τε καὶ ψυχροῦ· πάσχει δέ, ὅταν τούτων ἕκαστα τῶν εἰρημένων ἢ μὴ ἐν τῷ δέοντι προσφέρηται τῷ σώματι, ἢ μὴ τὰ εἰωθότα, ἢ πλείω τε καὶ ἰσχυρότερα, ἢ ἐλάσσω τε καὶ ἀσθενέστερα.

Τὰ μὲν οὖν νοσήματα γίνεται τοῖσιν ἀνθρώποι-

AFFECTIONS

1. Any man who is intelligent must, on considering that health is of the utmost value to human beings, have the personal understanding necessary to help himself in diseases, and be able to understand and to judge what physicians say and what they administer to his body, being versed in each of these matters to a degree reasonable for a layman.

Now a person would best be able to understand such things by knowing and applying the following: all human diseases arise from bile and phlegm; the bile and phlegm produce diseases when, inside the body, one of them becomes too moist, too dry, too hot, or too cold; they become this way from foods and drinks, from exertions and wounds, from smell, sound, sight, and venery, and from heat and cold; this happens when any of the things mentioned are applied to the body at the wrong time, against custom, in too great amount and too strong, or in insufficient amount and too weak.

All diseases in men, then, arise from these

---

[1] The passage Ἄνδρα ... ὠφελέεσθαι recurs in *Regimen in Health* 9 (Loeb vol. IV. 58).

ΠΕΡΙ ΠΑΘΩΝ

σιν ἅπαντα ἀπὸ τούτων· δεῖ δὲ πρὸς ταῦτα τὸν
ἰδιώτην ἐπίστασθαι ὅσα εἰκὸς ἰδιώτῃ. ὅσα δὲ
τοὺς χειροτέχνας εἰκὸς ἐπίστασθαι καὶ προσφέρειν
καὶ διαχειρίζειν, περὶ δὲ τούτων καὶ τῶν λεγομέ-
νων καὶ τῶν ποιουμένων οἷόν τ' εἶναι τὸν ἰδιώτην
γνώμῃ τινὶ συμβάλλεσθαι.

Ἤδη οὖν ὁπόθεν τούτων ἕκαστα δεῖ τὸν ἰδιώτην ἐπίστασθαι ἐγὼ φράσω.

210   2. Ἢν ἐς τὴν κεφαλὴν ὀδύναι ἐμπέσωσι, τούτου τὴν κεφαλὴν συμφέρει διαθερμαίνειν λούοντα
πολλῷ καὶ θερμῷ, καὶ πταρμὸν ποιεῦντα φλέγμα
καὶ μύξας ὑπεξάγειν. καὶ ἢν μὲν πρὸς ταῦτα
ἀπαλλάσσηται τῆς ὀδύνης, ἀρκεῖ ταῦτα. ἢν δὲ
μὴ ἀπαλλάσσηται, κάθηραι τὴν κεφαλὴν
φλέγμα, διαιτᾶν δὲ ῥοφήματι καὶ ποτῷ ὕδατι,
οἶνον δὲ μὴ προσφέρειν, ἔστ' ἂν[1] ἡ περιωδυνίη
παύσηται· τὸν γὰρ οἶνον ὅταν θερμὴ ἡ κεφαλὴ
ἐοῦσα σπάσῃ, ἡ περιωδυνίη ἰσχυροτέρη γίνεται.

Τὰ δὲ ἀλγήματα ἐσπίπτει ὑπὸ φλέγματος,
ὅταν ἐν τῇ κεφαλῇ κινηθὲν ἀθροισθῇ.

Ἢν δὲ ἄλλοτε καὶ ἄλλοτε ὀδύνη καὶ σκοτοδινίη ἐμπίπτῃ ἐς τήν κεφαλήν, ὠφελέει μὲν καὶ
ταῦτα προσφερόμενα· ὠφελέει δέ, κἢν αἷμα
ἀφαιρεθῇ ἀπὸ τῶν μυκτήρων, ἢ ἀπὸ τῆς φλεβὸς
τῆς ἐν τῷ μετώπῳ. ἢν δὲ πολυχρόνιον καὶ ἰσχυρὸν τὸ νόσημα ἐν τῇ κεφαλῇ γίνηται, καὶ μὴ
ἀπαλλάσσηται καθαρθείσης τῆς κεφαλῆς, ἢ σχά-

8

things. The layman must understand as much about them as befits a layman; and what it is fitting for the expert to understand, to administer, and to manage, about these matters, both what is said and what is done, let the layman be able to contribute an opinion with a certain amount of judgement.

So now, from the point whence the layman must comprehend each of these things, let me proceed to explain them.

2. If pains befall the head, it benefits the patient to warm his head by washing it with copious hot water, and to carry off phlegm and mucus by having him sneeze. If, with these measures, he is relieved of his pain, that suffices; but if he is not relieved, clean his head of phlegm, and prescribe a regimen of gruel and drinking water. Do not give wine until the intense pains stop; for when the head, in its warmness, draws wine to itself, its pain becomes even more intense.

These pains attack as the result of phlegm, when, having been set in motion, it collects in the head.

If, from time to time, pain and dizziness befall the head, the above administrations are also of benefit; it helps, too, if blood is let from the nostrils or from the vessel between the eyes. If the disease in the head is protracted and intense, and does not go away when the head is cleaned out, you must

---

[1] Θ adds μή.

σαι δεῖ τούτου τὴν κεφαλήν, ἢ τὰς φλέβας κύκλῳ
ἀποκαῦσαι. τῶν γὰρ λοιπῶν ἀπὸ τούτων μόνων
ἐλπὶς ὑγιέα γενέσθαι.

3. Τοὺς νοσέοντας χρὴ σκοπεῖν εὐθὺς ἀρχομέ-
νους ἐν τῇ καταστάσει τῶν νοσημάτων, ὅτου ἂν
δέωνται, καὶ οἵους τε ὄντας καὶ φαρμακευθῆναι
καὶ ἄλλο ὅ τι ἂν τις θέλῃ προσενέγκαι. ἢν δέ,
τὴν ἀρχὴν παρείς, τελευτώσῃς τῆς νόσου προσφέρῃς ὃ[1] ἀπειρηκότι ἤδη τῷ σώματι δεῖ[2] ἐνισχύον
τι προσενέγκαι, κίνδυνος ἁμαρτάνειν μᾶλλον ἢ
ἐπιτυγχάνειν.

4. Ἢν ἐς τὰ ὦτα ὀδύνη ἐμπέσῃ, λούειν συμφέρει πολλῷ καὶ θερμῷ, καὶ πυριᾶν τὰ ὦτα. καὶ
ἢν μὲν πρὸς ταῦτα περιίστηται τὸ φλέγμα λεπτυνόμενον ἀπὸ τῆς κεφαλῆς, καὶ ἡ ὀδύνη ἀπο-
212 λείπῃ, ἀρκεῖ | ταῦτα. εἰ δὲ μή, τῶν λοιπῶν ἄρι-
στον φάρμακον πῖσαι ἄνω ὅ τι φλέγμα ἄγει,[3] ἢ
τὴν κεφαλὴν καθῆραι φλέγμα. τὸ δὲ ἄλγημα καὶ
τοῦτο γίνεται, ὅταν ἔσωθεν πρὸς τὴν ἀκοὴν
φλέγμα ἀπὸ τῆς κεφαλῆς προσπέσῃ.

Ἢν δὲ τὰ παρὰ τὴν φάρυγγα φλεγμαίνῃ, ἀνα-
γαργαρίστοις χρῆσθαι· γίνεται δὲ καὶ ταῦτα ὑπὸ
φλέγματος.

Ἢν δὲ τὰ οὖλα ἢ τῶν ὑπὸ τὴν γλῶσσάν τι
φλεγμαίνῃ, διαμασητοῖσι χρῆσθαι· ἀπὸ φλέγματος δὲ καὶ ταῦτα γίνεται.

either incise the patient's head, or cauterize the vessels all around it. For, of the possible measures that remain, only these offer a hope of recovery.

3. It is immediately upon their becoming ill that you must examine patients for what they require, when they are still able to take a medication and whatever else you might wish to administer. If, however, having neglected the beginning, when the disease is approaching its end you administer the sort of strengthening remedy necessary for a body that is already sinking, there is the danger of failing more often than succeeding.

4. If pain befalls the ears, it helps to wash with copious hot water and to administer a vapour-bath to the ears. If, with this, the phlegm is thinned and devolves from the head, and the pain stops, that suffices. If not, the best of the possible measures that remain is to have the patient drink a medication that draws phlegm upwards, or to clean his head of phlegm. This pain, too, is due to phlegm, when from the head it invades the ear internally.

If the area along the throat swells up, give gargles; this, also, arises because of phlegm.

If the gums or any of the parts beneath the tongue swell up, give medications that are chewed; this too arises from phlegm.

---

[1] Potter: ἐν ΘM. [2] Potter: δὲ δι Θ: δὲ δεῖ M. [3] Θ: καθαίρει M.

ΠΕΡΙ ΠΑΘΩΝ

Ἢν δὲ ἡ σταφυλὴ κατακρεμασθῇ καὶ πνίγῃ—ἔνιοι δὲ τοῦτο καλοῦσι γαργαρεῶνα—
παραχρῆμα μὲν τοῖσιν ἀναγαργαρίστοισι χρῆσθαι,
σκευάζων ὡς γέγραπται ἐν τοῖς Φαρμάκοις. ἢν
δὲ πρὸς ταῦτα ἰσχνὴ μὴ γίνηται, ὄπισθεν ξυρή-
σαντα τὴν κεφαλήν, σικύας προσβάλλειν δύο, καὶ
τοῦ αἵματος ἀφαιρέειν ὡς πλεῖστον, καὶ ἀνασπά-
σαι ὀπίσω τὸ ῥεῦμα τοῦ φλέγματος. ἢν δὲ μὴ
τούτοισι καθίστηται, σχάσαντα μαχαιρίῳ τὸ ὕδωρ
ἐξιέναι, σχάζειν δ᾽ ὅταν τὸ ἄκρον ὑπέρυθρον γένηται. ἢν δὲ μὴ τοιοῦτον τμηθῇ, φλεγμαίνειν
ἐθέλει, καὶ ἔστιν ὅτε ἄπνουν ἔπνιξε. γίνεται δὲ
τοῦτο ὑπὸ φλέγματος, ὅταν ἐκ τῆς κεφαλῆς θαλ-
φθείσης[1] ἀθρόον καταρρυῇ.

Ὅσα δὲ περὶ ὀδόντας γίνεται ἀλγήματα· ἢν
μὲν βεβρωμένος ᾖ καὶ κινέηται, ἐξαιρέειν· ἢν δὲ
μὴ βέβρωται ἢ μὴ κινέηται, ὀδύνην δὲ παρέχῃ,
καύσαντα ἀποξηρῆναι· ὠφελέει δὲ καὶ τὰ διαμα-
σήματα. αἱ δὲ ὀδύναι γίνονται, ὅταν φλέγμα
ὑπέλθῃ ὑπὸ τὰς ῥίζας τῶν ὀδόντων· ἐσθίονται δὲ
οἱ μὲν ὑπὸ φλέγματος, οἱ δὲ ὑπὸ σίτων, ἢν φύσει
ἀσθενεῖς ἔωσι, καὶ κοιλίην ἔχοντες, καὶ πεπηγό-
τες ἐν τοῖσιν οὔλοισι κακῶς.

214   5. Ἢν δὲ ἐν τῇ ῥινὶ πώλυπος γένηται, οἷον γαργαρεὼν γίνεται[2] καὶ ἀπογκέει ἐκ τοῦ μυκτῆρος

___

[1] Θ adds ὅταν.   [2] γαργαρεὼν γίνεται Potter (cf. *Diseases II*
33): πρῆγμα πνίγεταί τε Θ: πρῆγμα πνέεταί τε Μ: πρῆσμα γίγνεταί τε
Ermerins.

AFFECTIONS

If the swollen uvula hangs down and chokes the patient—some call this condition *gargareon*—at once give gargles, preparing them as recorded in the *Medications*.[1] If, with these, the swelling does not go down, shave the back of the head, apply two cups, and remove as much blood as possible, in order to draw the flux of phlegm back up again. If, with this, the uvula still does not return to normal, incise it with a knife and discharge the fluid; incise when its extremity becomes reddish. If the uvula is not incised when it is in this state, it is inclined to swell up, and on occasion it has choked a patient to the point of suffocation. This condition arises because of phlegm, when it pours down in quantity out of the head which has become heated.

Pains that arise about the teeth: if the tooth is decayed and loose, remove it; if it is not decayed or loose, but produces pain, dry it out by cautery; medications that are chewed are useful as well. These pains occur when phlegm invades beneath the roots of the teeth; some teeth are decayed by phlegm, others by foods, when they are weak by nature, have caries, and are poorly fixed in the gums.

5. If a polyp forms in the nose, it is like a swollen uvula and protrudes out of the nostril towards

[1] This work, referred to as *Medications* (Τὰ φάρμακα) or *Medication Book* (Φαρμακῖτις), would seem to be lost. For a discussion of its possible relationship with the Hippocratic fragment *On Medications* (Περὶ φαρμάκων) see Hermann Schöne, "Hippokrates ΠΕΡΙ ΦΑΡΜΑΚΩΝ" in *Rheinisches Museum* (N.F.) 73, 1920–24, 434–48.

ἐς τὸ πλάγιον. ἐξαιρέεται δὲ βρόχῳ διελκόμενος
ἐς τὸ στόμα ἐκ τῆς ῥινός· οἱ δὲ καὶ φαρμάκοις
ἐκσήπονται. φύεται δὲ ὑπὸ φλέγματος.

Ταῦτα μὲν ὅσα ἀπὸ τῆς κεφαλῆς φύεται
νοσήματα, πλὴν ὀφθαλμῶν· ταῦτα δὲ χωρὶς
γεγράψεται.

6. Περὶ δὲ τῶν κατὰ κοιλίην νοσημάτων ἐνθυμέεσθαι χρὴ τάδε· πλευρῖτις, περιπλευμονίη,
καῦσος, φρενῖτις· αὗται καλεῦνται ὀξεῖαι, καὶ
γίνονται μὲν μάλιστα καὶ ἰσχυρόταται τοῦ χειμῶ-
νος· γίνονται δὲ καὶ τοῦ θέρεος, ἧσσον δὲ καὶ
μαλακώτεραι. ἢν δὲ παρατυγχάνῃς, τάδ' ἂν καὶ
ποιέων καὶ συμβουλεύων τυγχάνοις μάλιστα.

7. Πλευρῖτις· πυρετὸς ἔχει, καὶ τοῦ πλευροῦ
ὀδύνη, καὶ ὀρθοπνοίη, καὶ βήξ. καὶ τὸ σίελον
κατ' ἀρχὰς μὲν ὑπόχολον πτύει, ἐπειδὰν δὲ πεμπταῖος γένηται ἢ ἑκταῖος, καὶ ὑπόπυον.

Τούτῳ τοῦ μὲν πλευροῦ τῆς ὀδύνης διδόναι ὅ
τι ἀποστήσει ἀπὸ τοῦ πλευροῦ τό τε φλέγμα καὶ
τὴν χολήν· ἡ γὰρ ὀδύνη οὕτως ἂν εἴη μαλακωτά-
τη. τὴν δὲ κοιλίην ὑπάγειν θεραπεύοντα[1] καὶ
ψύχοντα κλύσματι· οὕτω γὰρ τῇ νούσῳ τῇ συμ-

[1] ὑ. θ. Θ: θεραπεύειν ὑπάγοντα Μ.

---

[1] The only work in the Hippocratic Collection this
remark could refer to is *Sight* (Littré IX. 122–161).

[2] I have translated κοιλίη throughout with the general
term "cavity" because it represents an anatomico-
physiological concept incommensurable with modern

14

the side. It is removed with a snare, by drawing it from the nose through into the mouth. Otherwise, it is made putrid with medications. It arises because of phlegm.

Such are the diseases that arise from the head, except for those of the eyes, which will be handled separately.[1]

6. With regard to diseases in the cavity, you must consider the following: pleurisy, pneumonia, ardent fever, and phrenitis. These are called "acute", and occur most frequently and violently in winter; they occur in summer as well, but less frequently and more mildly. If you meet them, you will be most successful by acting and counselling as follows.

7. Pleurisy: there are fever, pain in the side, orthopnoea and coughing. At the beginning the patient expectorates sputum that is slightly bilious, but then by the fifth or sixth day also somewhat purulent.

Against the pain in his side, give this patient a medication to remove phlegm and bile from the side, for if you do this the pain will be mildest. Clean the cavity[2] downwards by giving a medication and cooling it with an enema; this is very beneficial throughout the whole course of the

terminology. Generally the "cavity" is the thorax and/or abdomen, or, more frequently, the gastro-intestinal tract, but defined less by anatomical criteria than according to subjective and functional phenomena. Anything a person feels to be "high up" or that involves nausea or vomiting is located in the "upper cavity" (ἡ ἄνω κοιλίη), anything felt to be "low down" or that has a relation to defecation is in the "lower cavity" (ἡ κάτω κοιλίη).

πάσῃ συμφορώτατα. προσφέρειν δὲ ποτὸν καὶ
ῥύφημα, καὶ τὰ πώματα διδόναι ὀξύτερον,[1] ὡς τὸ
σίελον ἀνακαθαίρηται ἀπὸ τοῦ πλευροῦ. ὅταν δὲ
καθαίρεσθαι ἄρξηται τὸ πύον, θερμαίνοντα συμφέρει τὸ πλευρὸν ἔξωθεν πεπαίνειν τὰ πρὸς τῷ
πλευρῷ· πρόσθεν δὲ οὐ ξυμφέρει· ξηραίνεται γάρ.

Γίνεται δὲ ἡ νοῦσος αὕτη μάλιστα μὲν ἐκ
ποσίων, ὅταν τις, ὑγράζοντος τοῦ σώματος, ἢ
μεθύων ἢ νήφων ῥιγώσῃ· γίνεται δὲ καὶ ἄλλως.
216 κρίνεται δὲ ἡ νοῦσος, ἡ μὲν | βραχυτάτη ἑβδόμῃ,
ἡ δὲ μακροτάτη τετάρτῃ καὶ δεκάτῃ. κἢν μὲν ἐν
ταύτῃ πτυσθῇ καὶ καθαρθῇ τὸ πύον ἀπὸ τοῦ
πλευροῦ, ὑγιὴς γίνεται· ἢν δὲ μὴ πτυσθῇ, ἔμπυος
γίνεται, καὶ ἡ νοῦσος μακρή.

8. Κρίνεσθαι δέ ἐστιν ἐν τῇσι νούσοισιν, ὅταν
αὔξωνται αἱ νοῦσοι, ἢ μαραίνωνται, ἢ μετα-
πίπτωσιν εἰς ἕτερον νόσημα, ἢ τελευτῶσι.

9. Περιπλευμονίη· πυρετὸς ἴσχει καὶ βήξ· καὶ
ἀποχρέμπτεται τὸ μὲν πρῶτον φλέγμα παχὺ καὶ
καθαρόν, ἕκτῃ δὲ καὶ ἑβδόμῃ ὑπόχολον καὶ ὑποπέλιον, ὀγδόῃ δὲ καὶ ἐνάτῃ ὑπόπυον.

Τούτῳ ἢν μὲν ὀδύνη ἐγγίνηται ἢ τοῦ νώτου ἢ
τῶν πλευρέων, διδόναι ὅπερ ἐν τῇ πλευρίτιδι τοῦ
πλευροῦ τῆς ὀδύνης ἐν τῇ Φαρμακίτιδι γέγραπται. ποτοῖσι δὲ καὶ ῥυφήμασι καὶ τῆς κοιλίης
ἐς τὴν ὑποχώρησιν καὶ ψύξιν κατὰ ταὐτὰ θερα-

[1] Θ. -τερα Μ.

disease. Administer drinks and gruel: give these draughts quite quickly in order that the sputum will be cleaned thoroughly from the side. When the side begins to be cleaned, it is beneficial to bring the material lying against the chest wall to maturity by warming the side from the outside; earlier this is not useful, as the material only becomes dry.

Pleurisy generally arises from drinks, when a person in a moist state of body, either drunk or sober, has a chill; it also arises in other ways. The condition has its crisis, when it is shortest, on the seventh day, when it is longest, on the fourteenth day. If the pus is expectorated and cleaned from the side within this period, the patient recovers; but if it is not expectorated, he suppurates internally, and the disease becomes long.

8. To be judged[1] in diseases is when they increase, diminish, change into another disease, or end.

9. Pneumonia: there are fever and a cough; at first the patient coughs up thick clear phlegm, on the sixth and seventh day sputum that is slightly bilious and livid, and on the eighth and ninth day somewhat purulent sputum.

If pain in the back or side supervenes in this patient, administer the remedy recorded in the *Medication Book* for pleuritic pain. Treat just as in pleurisy with drinks and gruels to bring about an evacuation and cooling of the cavity. In order that

[1] I.e. by the physician; cf. chapter 37 below.

πεύειν τῇ πλευρίτιδι. ὅπως δὲ τὸ σίαλον ἐκ τοῦ πλεύμονος ἀνακαθαίρηται καὶ τὸ πύον, διδόναι φάρμακα ποτά, οἷσιν ὁ πλεύμων ὑγραίνεται, καὶ καθαίρεται[1] τὸ πύον ἄνω.

Ἡ δὲ νοῦσος αὕτη γίνεται, ὅταν ἐκ τῆς κεφαλῆς φλέγμα ἀθροῦν ῥυῇ ἐς τὸν πλεύμονα· ἔστι δὲ ὅτε καὶ ἐκ πλευρίτιδος μεθίσταται ἐς περιπλευμονίην, καὶ ἐκ καύσου. κρίνεται δὲ ἐν ἡμέρῃσιν, ἡ μὲν βραχυτάτη ἐν τεσσερεσκαίδεκα, ἡ δὲ μακροτάτη ἐν δυοῖν δεούσαιν εἴκοσι· διαφεύγουσι δὲ ταύτην ὀλίγοι. γίνονται δὲ ἔμπυοι καὶ ἐκ ταύτης τῆς νόσου, ἢν μὴ ἐν ταῖσι κυρίαισιν ὁ πλεύμων καθαρθῇ.

10. Φρενῖτις ὅταν λάβῃ, πυρετὸς ἴσχει βληχρὸς τὸ πρῶτον, καὶ ὀδύνη πρὸς τὰ ὑποχόνδρια, μᾶλλον δὲ πρὸς[2] τὰ δεξιὰ πρὸς τὸ ἧπαρ. ὅταν δὲ τεταρταῖος γένηται καὶ πεμπταῖος, ὅ τε πυρετὸς ἰσχυρότερος γίνεται καὶ αἱ ὀδύναι, καὶ τὸ χρῶμα ὑπόχολον γίνεται, καὶ τοῦ νοῦ παρακόπτει.

Τούτῳ τῆς μὲν ὀδύνης ἅπερ ἐν τῇ πλευρίτιδι διδό|ναι, καὶ χλιαίνειν ἢν ὀδύνη ἔχῃ· τὴν κοιλίην δὲ θεραπεύειν, καὶ τἆλλα τοῖς αὐτοῖς, πλὴν τοῦ ποτοῦ· ποτῷ δὲ χρῆσθαι πλὴν οἴνου τῶν ἄλλων ὅτῳ ἂν θέλῃς, ἢ ὄξος καὶ μέλι καὶ ὕδωρ διδόναι, ἢ ὕδωρ· οἶνος δὲ οὐ συμφέρει τοῦ νοῦ παρακόπτοντος, οὔτ᾽ ἐν ταύτῃ τῇ νούσῳ, οὔτ᾽ ἐν τῇσιν

---

[1] M: κ., καὶ ὑ. Θ. [2] Θ: ἐς M.

the sputum may be thoroughly cleaned from the lung and the pus with it, give medicinal drinks by which the lung will be moistened and pus cleaned upwards.

Pneumonia arises when phlegm flows in quantity from the head into the lung; also, sometimes there is a change from pleurisy to pneumonia, or from ardent fever. The crisis arrives, when the pneumonia is shortest, in fourteen days, when it is longest, in eighteen days; few patients survive. Patients also suppurate internally after this disease, if the lung is not cleaned out on the critical days.

10. In phrenitis, at first there are mild fever and pain over the hypochondrium, more on the right towards the liver. When the fourth or fifth day arrives, the fever becomes more intense, as do the pains, the colour becomes somewhat bilious, and the patient's mind becomes deranged.

For the pain, treat this patient with the same measures as in pleurisy; when pain is present, warm him. Give a medication for the cavity, and conduct the rest of the treatment along the same lines, except with regard to drink: as drink give any one you wish except wine; give vinegar, honey, and water, or water alone; wine, however, does not benefit a deranged mind in either this disease or

ἄλλῃσι. λούειν δὲ πολλῷ καὶ θερμῷ κατὰ
κεφαλῆς ἐν ταύτῃ τῇ νούσῳ συμφέρει·
μαλασσομένου γὰρ τοῦ σώματος, καὶ ἱδρὼς
μᾶλλον γίνεται καὶ ἡ κοιλία καὶ τὸ οὖρον
διαχωρεῖ καὶ αὐτὸς αὑτοῦ ἐγκρατέστερος γίνεται.

Ἡ δὲ νοῦσος γίνεται ὑπὸ χολῆς, ὅταν κινηθεῖσα πρὸς τὰ σπλάγχνα καὶ τὰς φρένας προσίζῃ.
κρίνεται δὲ ἡ μὲν βραχυτάτη ἑβδόμῃ, ἡ δὲ μακροτάτη ἑνδεκάτῃ. διαφεύγουσι δὲ καὶ ταύτην
ὀλίγοι, μεθίσταται δὲ καὶ αὕτη ἐς περιπλευμονίην·
καὶ ἢν μεταστῇ, ὀλίγοι διέφυγον.

11. Καῦσος δὲ ὅταν ἔχῃ, πυρετὸς ἴσχει καὶ
δίψα ἰσχυρή. καὶ ἡ γλῶσσα τρηχέη καὶ μέλαινα
γίνεται ὑπὸ θερμότητος τοῦ πνεύματος, καὶ τὸ
χρῶμα ὑπόχολον γίνεται, καὶ τὰ ὑπιόντα χολώ-
δη· καὶ τὰ μὲν ἔξω ψυχρὸς γίνεται, τὰ δ' ἔσω
θερμός.

Τούτῳ συμφέρει ψύγματα προσφέρειν, καὶ
πρὸς τὴν κοιλίην, καὶ ἔξωθεν πρὸς τὸ σῶμα,
φυλασσόμενον μὴ φρίξῃ. καὶ τὰ πώματα καὶ τὰ
ῥυφήματα διδόναι πυκνὰ κατ' ὀλίγον ὡς ψυχρό-
τατα. τὴν δὲ κοιλίην θεραπεύειν· ἢν μὲν μὴ
ὑποχωρέῃ τὰ ἐνεόντα, κλύσαι· ψύχειν δὲ κλύσ-
μασιν ὡς ψυχροτάτοις ἢ ὁσημέραι ἢ διὰ τρίτης.

Ἡ δὲ νοῦσος αὕτη γίνεται ὑπὸ χολῆς, ὅταν κινηθεῖσα ἐντὸς τοῦ σώματος καταστηρίξῃ. φιλέει
δὲ καὶ ἐς περιπλευμονίην μεθίστασθαι. κρίνεται

any other one. It is of benefit in this disease to wash with copious hot water from the head downwards, for, as the body is softened, sweating increases, the cavity discharges, urine passes, and the patient gains more control over himself.

Phrenitis arises from bile, when, having been set in motion, it settles against the inward parts and the diaphragm. It has its crisis, when it is shortest, on the seventh day, when it is longest, on the eleventh day. Few escape this disease, either, and it too sometimes changes into pneumonia; in the cases where it has, few have escaped.

11. In ardent fever there are fever and a violent thirst; the tongue becomes rough and black because of the heat of the breath, the patient's colour becomes somewhat bilious, and his stools are full of bile; on the outside he becomes cold, but inside he is hot.

It benefits this patient to administer cooling agents both to the cavity and externally on the surface of the body, but taking care that he does not suffer a chill. Give drinks and gruel often, a little at a time, and as cold as possible. Give a medication for the cavity; if its contents do not pass down, administer an enema; also, cool with very cold enemas every day or every other day.

Ardent fever arises from bile, when, having been set in motion, it is deposited inside the body; it, too, is liable to change into pneumonia. The disease

ΠΕΡΙ ΠΑΘΩΝ

220 δὲ ἡ μὲν βραχυτάτη | ἐνάτῃ,[1] ἡ δὲ μακροτάτη τεσσαρεσκαιδεκάτῃ. καὶ ἢν μὲν μεταστῇ ἐς περιπλευμονίην, ὀλίγοι διαφεύγουσιν· ἢν δὲ μὴ μεταστῇ, διαφεύγουσι πολλοί.

Αὗται μὲν οὖν ὀξεῖαι καλέονται, καὶ δεῖ ταύτας οὕτω θεραπεύειν.

12. Ὅσοι δὲ ἄλλοι τοῦ χειμῶνος πυρετοὶ γίνονται, εἴτ᾿ ἐξ οἴνου, εἴτ᾿ ἐκ κόπου, εἴτε ἄλλου τινός, φυλάσσεσθαι χρή· μεθίσταται γὰρ ἐνίοτε ἐς τὰς ὀξείας νούσους. ἡ δὲ μετάστασις αὐτῶν τοιαύτη γίνεται· ὅταν, κεκινημένων φλέγματός τε καὶ χολῆς, μὴ τὰ συμφέροντα προσφέρηται[2] τῷ σώματι, συστρεφόμενα αὐτὰ πρὸς ἑαυτὰ τό τε φλέγμα καὶ ἡ χολὴ προσπίπτει τοῦ σώματος ᾗ ἂν τύχῃ, καὶ γίνεται ἢ πλευρῖτις, ἢ φρενῖτις, ἢ περιπλευμονίη. φυλάσσεσθαι οὖν χρὴ τοὺς πυρετοὺς τοὺς ἐν τῷ χειμῶνι· ἡ δὲ φυλακὴ ἔστω ἡσυχίη καὶ ἰσχνασίη καὶ τῆς κοιλίης κένωσις. ῥυφήμασι δὲ καὶ ποτήμασι διάγειν, ἕως ἂν ὁ πυρετὸς μειωθῇ.

13. Τῶν νούσων σχεδόν τι μάλιστα αἱ ὀξεῖαι καὶ ἀποκτείνουσι καὶ ἐπιπονώταταί εἰσι, καὶ δεῖ πρὸς αὐτὰς φυλακῆς τε πλείστης καὶ θεραπείης ἀκριβεστάτης. καὶ ἀπὸ τοῦ θεραπεύοντος κακὸν μὲν μηδὲν προσγίνεσθαι, ἀλλὰ ἀρκέειν τὰ ἀπ᾿ αὐτῶν τῶν νοσημάτων ὑπάρχοντα, ἀγαθὸν δὲ ὅ τι οἷός τε ἂν ᾖ. καὶ ἢν μέν, ὀρθῶς θεραπεύοντος τοῦ

has its crisis, when it is shortest, on the ninth day, when it is longest, on the fourteenth day. If it changes into pneumonia, few patients escape; if it does not change, many escape.

These diseases, then, are called "acute", and you must treat them thus.

12. With the other fevers of winter, which arise either from wine or weariness or anything else, you must take care, for sometimes they change into acute diseases. The change comes about in the following way: when, with phlegm and bile set in motion, what is beneficial is not administered to the patient's body, the phlegm and bile collect together and fall upon some chance part of the body, and pleurisy or phrenitis or pneumonia results. Thus, you must pay heed with winter fevers. The precautions required are quiet, leanness, and emptiness of the cavity; continue with gruels and drinks until the fever diminishes.

13. Generally speaking, it is the acute diseases that cause the most deaths and that are the most painful, and with these the greatest care and the strictest treatment are necessary. Let nothing bad be added by the person treating—rather let the evils resulting from the diseases themselves suffice—but only whatever good he is capable of. If, when the physician treats correctly, the patient is

---

[1] M adds ἢ δεκάτη.   [2] M: -φέρῃ Θ.

ἰητροῦ, ὑπὸ μεγέθους τῆς νούσου κρατέηται ὁ
κάμνων, οὐχὶ τοῦ ἰητροῦ αὕτη ἡ ἁμαρτίη ἐστίν.
ἐὰν δὲ μὴ θεραπεύοντος ὀρθῶς ἢ μὴ γιγνώσκον-
τος ὑπὸ τῆς νούσου κρατέηται, τοῦ ἰητροῦ.

14. Τοῦ δὲ θέρεος τάδε γίνεται· πυρετὸς ἴσχει
ἰσχυρὸς καὶ δίψα, καὶ ἔνιοι ἐμοῦσι χολήν· ἐνίοισι
δὲ καὶ κάτω διαχωρέει. τούτοισι πίνειν διδόναι ὅ
τι ἄν σοι δοκέῃ ἐπιτήδειον εἶναι, καὶ ρυφεῖν. [ἢν
δὲ προσίστηται πρὸς τὴν καρδίην χολὴ ἢ
φλέγμα, ἐπιπίποντες ὕδωρ ψυχρὸν ἢ μελίκρητον,
ἐμούντων.][1] ἢν δὲ ἡ γαστὴρ μὴ ὑποχωρέῃ, κλύσματι χρῆσθαι ἢ βαλάνῳ. ἡ δὲ νοῦσος γίνεται ὑπὸ
χολῆς. ἀπαλλάσσονται δὲ μάλιστα ἑβδομαῖοι ἢ
ἐναταῖοι.

Ἢν δὲ τοῦ πυρετοῦ ἔχοντος μὴ καθαίρωνται
μήτε κάτω μήτε ἄνω, πόνος δὲ ἐνῇ καθ' ἅπαν τὸ
σῶμα, ὅταν ᾖ τριταῖος ἢ τεταρταῖος, ἢ φαρμάκῳ
ὑποκαθῆραι ἐλαφρῷ κάτω, ἢ πώματι, ἀπὸ δὲ
κέγχρου ποιέειν ἢ τοῦ ἀλήτου ρύφημα· καὶ πώμασι τοῖς αὐτοῖς θεραπεύειν. πάσχουσι δὲ καὶ
ταῦτα ὑπὸ χολῆς.

Ἢν δὲ τὰ μὲν ἔξω μὴ πυρώδης ᾖ σφόδρα, τὰ
δ' ἔσω, καί ἡ γλῶσσα τρηχεία καὶ μέλαινα
γίνηται, καὶ οἱ πόδες καὶ αἱ χεῖρες ἄκραι ψυχραί,
τούτῳ φάρμακον μὲν μὴ διδόναι, θεραπεύειν
δὲ προσφέρων ψύγματα καὶ πρὸς τὴν κοιλίην

[1] This sentence disturbs the sense of the paragraph; I
suspect that it is a marginal note referring to ch. 15.

overcome by the magnitude of his disease, this is not the physician's fault. But if, when the physician treats either incorrectly or out of ignorance, the patient is overcome, it is his fault.

14. During summer the following occur: violent fever and thirst set in, some patients vomit bile, and in some bile passes off below, as well. Give these patients whatever you think suitable to drink and to take as gruel. [If bile or phlegm invades the cardia, let the patients drink cold water or melicrat, and then vomit.] If the belly does not pass anything, administer an enema or suppository. This disease arises from bile. Patients usually recover on the seventh or ninth day.

If, while the fever is present, patients are cleaned neither downwards nor upwards, and pain is present throughout the whole body, on the third or fourth day clean downwards with either a gentle medication or a potion and make a gruel from millet or flour. Treat with the same drinks. Patients suffer these things, too, from bile.

If the patient is not very feverish externally, but is so internally, and if his tongue becomes rough and black, and his feet and hands are cold at the extremities, do not give him a medication, but treat by administering cooling agents both through the

καὶ πρὸς τὸ ἄλλο σῶμα. καλεῖται δὲ καυσώδης ὁ πυρετὸς οὗτος· κρίνεται δὲ[1] μάλιστα δεκαταῖος καὶ ἑνδεκαταῖος.[2]

Ἢν δὲ τὸ πῦρ λαμβάνῃ καὶ μεθίῃ, τοῦ δὲ σώματος αὐτὸν βάρος ἔχῃ, τούτον, ἕως μὲν τὸ πῦρ ἔχῃ, ῥυφήμασι καὶ πόμασι θεραπεύειν· ὅταν δὲ μὴ ἔχῃ, διδόναι καὶ σιτία. καθῆραι δὲ ὡς τάχιστα[3] φαρμάκῳ, ἤν τε ἄνω σοι δοκέῃ δεῖσθαι, ἤν τε κάτω.

(15.) Ἢν δὲ πυρετὸς μὲν μὴ ἔχῃ, τὸ δὲ στόμα πικρὸν ἔχῃ, καὶ τὸ σῶμα βαρύνηται, καὶ ἀσιτέῃ, φάρμακον διδόναι. πάσχει δὲ ταῦτα ὑπὸ χολῆς, ὅταν ἐς τὰς φλέβας καὶ τὰ ἄρθρα καταστηρίξῃ.

Ὁπόσαι δὲ ἄλλαι ὀδύναι ἐν τῷ θέρει κατὰ <τὴν>[4] κοιλίην γίνονται, ὅσαι μὲν πρὸς τὰ ὑποχόνδρια καὶ τὴν καρδίην· μελίκρατον ὑδαρὲς ποιῶν, ὅσον | τρεῖς κοτύλας, ὄξος παραχέας, δὸς πιεῖν χλιερόν· καὶ ἐπισχὼν ὀλίγον χρόνον, καὶ συνθαλφθεὶς πυρὶ καὶ ἱματίοισιν ἐμείτω. ἢν δὲ ἀπεμέσαντι αὖτις προσιστῆται καὶ πνίγηται,[5] αὖτις ἔμετον ποιείσθω, ἢ λούσας αὐτὸν πολλῷ καὶ θερμῷ, ὑποκλύσαι· καὶ χλιάσματα προστιθέ-

224

---

[1] In Θ this δὲ precedes πυρετὸς οὗτος.   [2] M adds καὶ τεσσαρεσκαιδεκαταῖος.   [3] Θ: μάλιστα M.   [4] Added by a later ms.   [5] Θ: πνίγῃ M.

---

[1] Compare chapter 11 above; this awkward repetition

cavity and to the rest of the body. This fever is called "ardent"[1]; it usually has its crisis on the tenth or eleventh day.

If the fever attacks and then remits, and a heaviness invades the body, treat this patient, as long as the fever is present, with gruels and drinks; when it is not present, give foods as well. Clean as soon as possible with a medication either upwards or downwards, whichever you think is required.

(15.)[2] If fever is not present, but the patient's mouth has a pungent taste, his body is weighed down, and he has no appetite, give a medication. He suffers these things because of bile, when it is deposited in the vessels and the joints.

Other pains that occur in the cavity in summer: pains that attack the hypochondrium and the cardia: make dilute melicrat in the amount of three cotylai, add vinegar, and give warm to drink; then, waiting a short while and having the patient warmed by a fire and blankets, let him vomit. If, after he has vomited, pains attack again and he is choked, let him induce vomiting again; or, after washing him in copious hot water, administer an enema; also apply fomentations if pain is present.

of the same disease is evidence of the author's compilatory method of composition.

[2] This paragraph continues the series of conditions that began in chapter 14, and is closely connected to the preceding paragraph. Thus, I have returned to Vander Linden's division and reduced Littré's 2 chapters to 1.

ναι ἐὰν ὀδύνη ἔχῃ. πάσχουσι δὲ ταῦτα μάλιστα
ὑπὸ φλέγματος, ὅταν κινηθὲν προσπέσῃ πρὸς τὴν
καρδίην. διδόναι δὲ τοῖσι τὰ τοιαῦτα ἀλγήματα
ἀλγέουσι καὶ τῶν φαρμάκων ἃ γέγραπται τῆς
ὀδύνης παύοντα ἐν τῇ Φαρμακίτιδι.

Ἢν δὲ μεθιστῆται ἡ ὀδύνη ἄλλοτε ἄλλῃ τῆς
κοιλίης, λούειν πολλῷ καὶ θερμῷ ἢν ᾖ ἄπυρος,
καὶ πίνειν διδόναι τῆς ὀδύνης εἵνεκα, ὅπερ ἐν τῇ
Φαρμακίτιδι[1] γέγραπται, ἢ τῶν ἄλλων ὅ τι ἂν
σοι δοκέῃ. ἢν δὲ μὴ ἀπαλλάσσηται τῆς ὀδύνης,
ὑποκαθῆραι φαρμάκῳ κάτω, σίτων δὲ ἀπέχεσθαι,
ἕως ἂν ἡ ὀδύνη ἔχῃ. τὰ δὲ τοιαῦτα ἀλγήματα
ὅσα πλανᾶται ὑπὸ χολῆς γίνεται.

Ὅσαι δὲ κάτωθεν τοῦ ὀμφαλοῦ ὀδύναι γίνον-
ται· ὑποκλύσαι μαλακῶ· ἢν δὲ μὴ παύηται, φάρμακον δοῦναι κάτω.

16. Ὅσαι δ' ὀδύναι ἐξαπίνης γίνονται ἐν τῷ
σώματι ἄνευ πυρετοῦ· συμφέρει λούειν πολλῷ καὶ
θερμῷ, καὶ χλιαίνειν. τὸ γὰρ φλέγμα καὶ ἡ χολὴ
συνεστηκότα μὲν ἰσχυρά ἐστι, καὶ κρατεῖ καθ'
ὁποῖον ἂν τοῦ σώματος στῇ, καὶ πόνον τε καὶ
ὀδύνην ἰσχυρὴν παρέχει· διακεχυμένα δὲ ἀσθε-
νέστερά[2] ἐστι καθ' ὃ ἂν ἔνδηλα[3] ᾖ τοῦ σώματος.

17. Τὰ δὲ νοσήματα, ὅσα τοῦ θέρεος γίνεται,
εἴωθε γίνεσθαι οὕτως· ὅταν τὸ σῶμα ὑπὸ τοῦ
226 ἡλίου θαλφθῇ, ὑγραίνεται· ὑγραινό|μενον δὲ νοσέει,

---

[1] Potter: πλευρίτιδι ΘΜ.   [2] Θ adds τε.   [3] Μ: -λον Θ.

AFFECTIONS

Patients generally suffer these pains because of phlegm, when, being set in motion, it falls upon the cardia. To those suffering pains of this kind give also the medications recorded in the *Medication Book* as stopping pain.

If the pain moves so that it is at one time in one part of the cavity and at another time in another part, wash with copious hot water when the patient is without fever, and for the pain have him drink what is recorded in the *Medication Book*, or whatever else you think suitable. If the pain is not relieved, clean downwards with a medication, and withhold foods as long as the pain is present. Pains of this kind that wander arise because of bile.

Pains that occur below the navel: apply a gentle enema; if the pain does not go away, give a medication that acts downwards.

16. Pains that suddenly arise in the body without fever: it benefits to wash with copious hot water, and to warm. For phlegm and bile, when gathered, are powerful and have dominance in whichever part of the body they occupy, and they produce suffering and violent pain; but dispersed, they are weaker in any part of the body in which they appear.

17. The diseases that occur in summer tend to arise thus: when the body is warmed by the sun, it becomes moist; on becoming moist, it becomes ill,

ἢ πᾶν, ἢ ἐς ὅ τι ἂν καταστηρίξῃ τὸ φλέγμα καὶ ἡ
χολή. ἢν μὲν οὖν τις αὐτὰ ἀρχόμενα θεραπεύῃ,
οὔτε μακρὰ γίνεται, οὔτε ἐπικίνδυνα· ἢν δὲ ἢ[1] μὴ
θεραπευθῇ[2] ἢ κακῶς θεραπευθῇ,[2] φιλέει καὶ
μακρότερα γίνεσθαι, καὶ ἔστιν ὅτε καὶ κτείνει.

18. Καὶ τριταῖοι δὲ καὶ τεταρταῖοι πυρετοὶ ἐκ
τῶν αὐτῶν γίνεσθαι πεφύκασιν. αὕτη ἡ κατά-
στασις τῶν νοσημάτων μάλιστα μὲν τοῦ θέρεος
γίνεται· γίνεται δὲ ἐνίοισι[3] καὶ τοῦ χειμῶνος.
τριταῖος δὲ πυρετὸς ὅταν ἔχῃ, ἢν μέν σοι δοκέῃ
ἀκάθαρτος εἶναι, τετάρτῃ φάρμακον δοῦναι. ἢν δὲ
μή σοι δοκῇ φαρμάκου δεῖσθαι, διδόναι φάρμακα
ποτά, οἷσι μεταστήσεται ὁ πυρετὸς ἢ ἀπολείψει·
διδόναι δ' ὥσπερ γέγραπται ἐν τοῖς Φαρμάκοις.
καὶ τῇ μὲν λήψει ῥυφήμασι καὶ ποτῷ διαιτᾶν,
ταῖς δὲ διὰ μέσου, σιτίοισι διαχωρητικοῖσι. καὶ[4]
λαμβάνει δὲ ὡς ἐπὶ τὸ πολὺ οὐκ ἐπὶ πλεῖστον·
ἢν δὲ μὴ θεραπεύηται, ἐθέλει μεθίστασθαι ἐς
τεταρταῖον καὶ γίνεσθαι[5] πολυχρόνιος.

Ἢν[6] τεταρταῖος λαμβάνῃ, ἢν μὲν ἀκάθαρτος
ᾖ, καθαίρειν πρῶτον μὲν τὴν κεφαλήν· καὶ διαλιπὼν τρεῖς ἢ τέσσερας ἡμέρας, φάρμακον δοῦναι
ἄνω αὐτῇ τῇ λήψει· διαλιπὼν δέ, ἕτερον κάτω
αὐτῇ τῇ λήψει.[7] ἢν δὲ μὴ πρὸς ταῦτα παύηται,

---

[1] ἢ om. M.   [2] Θ: θεραπεύῃ M.   [3] Jouanna (p. 276): γίνεται
δὲ ἐν τοῖσι Θ: ἐνίοισι δὲ M.   [4] καὶ om. M.   [5] M: γίνεται Θ.
[6] M adds δὲ.   [7] διαλιπὼν δέ . . . λήψει om. Θ.

either wholly, or in the part where phlegm and bile are deposited. Now, if someone treats these diseases at the beginning, they are neither long nor dangerous; but if they are either left untreated or are badly treated, they are likely to become longer, and sometimes they even kill.

18. Both tertian and quartan fevers are naturally disposed to arise from the same factors; this order of diseases usually occurs in summer, although in some instances it is seen in winter, too. When a tertian fever is present, if the patient seems to you to be in an unclean state, on the fourth day give him a medication. If you do not think he needs one, give medicinal drinks that will make the fever change or remit; administer these as described in the *Medications*. At the accession, prescribe gruels and drinks, on the days between, laxative foods. Generally, this disease does not attack with particular severity, but, if left untreated, it is likely to change into a quartan and to become chronic.

If a quartan attacks while a person is in an unclean state, first clean out his head; then, leaving an interval of three or four days, give a medication to act upwards just during the accession; then, leaving another interval, give a medication to act downwards at the next accession. If, with this treatment, the fever does not go away, leave

διαλιπών, λούσας πολλῷ καὶ θερμῷ, δοῦναι τῶν
φαρμάκων ἃ γέγραπται. ποτοῖσι δὲ καὶ ῥυφήμασι καὶ τῇ ἄλλῃ διαίτῃ χρῆσθαι, ὥσπερ ἐπὶ τοῦ
τριταίου. λαμβάνει δὲ οὗτος ὁ πυρετὸς τοὺς μὲν
πλείστους πολὺν χρόνον, τοὺς δὲ καὶ ὀλίγον.

Καὶ γίνεται μὲν ὁ τριταῖος καὶ ὁ τεταρταῖος
228 ὑπὸ χολῆς καὶ φλέγματος· διότι δὲ τρι|ταῖος καὶ
τεταρταῖος ἑτέρωθί μοι γέγραπται.

Δύναμιν ἔχει δὲ τούτων τῶν πυρετῶν τὰ φάρμακα πινόμενα, ὥστε τὰ σώματα[1] κατὰ χώρην
εἶναι ἐν τῇ ἑωθυίῃ θερμότητι καὶ ψυχρότητι, καὶ
μήτε θερμαίνεσθαι παρὰ φύσιν μήτε ψύχεσθαι·
διδόναι δὲ ὡς ἐν τῇ Φαρμακίτιδι γέγραπται.

19. Φλέγμα λευκὸν ὅταν ἔχῃ, τὸ σῶμα οἰδέει
πᾶν λευκῷ οἰδήματι, καὶ τῆς αὐτῆς ἡμέρης τοτὲ
μὲν δοκέει ῥᾴων εἶναι, τοτὲ δὲ φλαυρότερος· καὶ
τὸ οἴδημα ἄλλοτε ἄλλῃ τοῦ σώματος μεῖζόν τε
καὶ ἔλασσον γίνεται.

Τούτῳ φάρμακα[2] διδόναι κάτω ὑφ᾽ ὧν ὕδωρ ἢ
φλέγμα καθαίρεται· διαιτᾶν δὲ σιτίοισι καὶ ποτοῖσι καὶ πόνοισιν ὑφ᾽ ὧν ὡς ξηρότατος ἔσται καὶ
ἰσχνότατος.

Ἡ δὲ νοῦσος αὕτη γίνεται ὑπὸ φλέγματος,
ὅταν τις ἐκ πυρετῶν πολυχρονίων φλεγματώδης
ὢν ἀκάθαρτος γένηται, τρέπηταί τε τὸ φλέγμα

---

[1] Θ: τὸ σῶμα Μ.   [2] Μ: -κον Θ.

another interval, wash with copious hot water, and give one of the medications mentioned. Prescribe drinks, gruels, and the rest of the regimen as in a tertian fever. This fever attacks most patients for a long time, but others for a short time.

The tertian and the quartan fevers, too, arise because of bile and phlegm. Why the fever comes every third or fourth day I have explained elsewhere.[1]

Medications drunk in these fevers act in such a way that bodies remain undisturbed in their accustomed heat and cold, being neither abnormally heated nor abnormally chilled; give them as recorded in the *Medication Book*.

19. In white phlegm, the whole body swells up with a white swelling, and on one and the same day the patient seems sometimes better, at other times worse; the swelling increases and decreases at different times in the different parts of his body.

Give this patient a medication that will clean water and phlegm downwards; prescribe a regimen of foods, drinks and exercises as the result of which he will become as dry and lean as possible.

This disease arises because of phlegm, when a person that is phlegmatic after chronic fevers becomes unclean, and the phlegm turns into his

[1] This is the interpretation of Cornarius, Vander Linden, Ermerins and Fuchs. Littré, following Foes and Grimm, offers: *c'est pourquoi j'ai traité ailleurs de ces deux fièvres.*

αὐτοῦ ἀνὰ τὰς σάρκας. καὶ λευκότερον μὲν οὐδὲν
τοῦτο τοῦ ἄλλου [ἢ]¹ φλέγματος, ὁ δὲ χρὼς φαίνεται λευκότερος· τὸ γὰρ αἷμα ὑπὸ πλήθους τοῦ
φλέγματος ὑδαρέστερον γίνεται, καὶ οὐκ ἔνι ἐν
αὐτῷ ὁμοίως τὸ εὔχρουν· καὶ διὰ τοῦτο λευκότεροί
τε φαίνονται, καὶ καλεῖται ἡ νόσος φλέγμα
λευκόν. ἢν μὲν οὖν θεραπευθῇ ἀρχομένης τῆς
νούσου, ὑγιὴς γίνεται· ἢν δὲ μὴ θεραπευθῇ, ἐς
ὕδρωπα μεθίσταται ἡ νοῦσος, καὶ διέφθειρε τὸν
ἄνθρωπον.

20. Ὁπόσοι δὲ σπλῆνα ἔχουσι μέγαν· ὅσοι
μέν εἰσι χολώδεες, κακόχροοί τε γίνονται καὶ
καχελκέες καὶ δυσώδεες ἐκ τοῦ στόματος καὶ
λεπτοί· καὶ ὁ σπλὴν σκληρός, καὶ ἀεὶ παρα-
πλήσιος τὸ μέγεθος· καὶ τὰ σιτία οὐ διαχωρέει.
230 ὅσοι δὲ φλεγματίαι, ταῦτά τε ἧσσον | πάσχουσι,
καὶ ὁ σπλὴν² μέζων γίνεται, ἄλλοτε δὲ ἐλάσσων.

Τούτοισι δὲ συμφέρει, ἢν μὲν ἀκάθαρτοι
φαίνωνται, καθαίρειν τὴν κεφαλὴν καὶ τὸ ἄλλο
σῶμα· ἢν δὲ μὴ δέωνται φαρμακείης διαιτᾶν.
ὅσοι μὲν φλεγματώδεες ξηραίνονται τὸ σῶμα
καὶ ἰσχναίνονται³ σιτίοις καὶ ποτοῖς καὶ ἐμέτοις
καὶ γυμνασίοις ὡς πλείστοις καὶ περιπάτοις· καὶ
τοῦ ἦρος ἐλλεβόρῳ καθαίρειν ἄνω. ὅσοι δὲ χολώδεες συμφέρει διυγραίνοντα τῇ διαίτῃ ὑπάγειν τὴν

¹ Del. Littré.   ² M adds ἄλλοτε.   ³ Θ: ξηραίνοντα . . .
ἰσχναίνοντα Μ.

34

tissues. This phlegm is no whiter than any other, but the patient's skin appears whiter because his blood, in consequence of the large amount of phlegm, becomes more watery than normal, so that the usual healthy colour is no longer present in it to the degree that it was before; thus, patients appear whiter and the disease is called white phlegm. Now if the patient is treated at the onset of the disease, he recovers; if he is not, the disease changes to dropsy, and has actually killed the person.

20. Persons that have a large spleen: those who are bilious take on a poor colour, suffer from malignant ulcers, smell foully from the mouth, and become thin; their spleen is hard and always about the same size; foods do not pass off below. Those who are phlegmatic suffer these things less, and their spleen sometimes increases in size, sometimes decreases.

It benefits these patients, if they appear to be in an unclean state, to be cleaned from both the head and the rest of the body; if they do not require the use of medications, then clean by means of regimen. Let the phlegmatic patients have their bodies thoroughly dried and made lean by foods, drinks, vomiting, as many exercises as possible, and walks; in spring, clean upwards with hellebore. In the bilious patients, it benefits to moisten by means of regimen, and to evacuate the cavity and the

κοιλίην καὶ τὴν κύστιν, καὶ τὴν φλέβα τὴν
σπληνῖτιν ἀφεῖναι[1] πυκνά· καὶ τοῖσι διουρητικοῖσι
φαρμάκοις χρῆσθαι, ἃ γέγραπται τὸν σπλῆνα
μαλθάσσοντα· καὶ καθαίρειν ἔτεος ὥρῃ, καὶ τοῦτο
χολήν.

Ἔνιοι δὲ τῶν σπληνιώντων ὑπὸ μὲν τῶν φαρμάκων πίνοντες οὐκ ὠφελέονται, οὐδ' ὑπὸ τῆς
ἄλλης θεραπείας· οὐδ' ἰσχνότερος οὐδὲν γίνεται ὁ
σπλὴν αὐτῶν· ἀλλὰ κρατέεται τὰ προσφερόμενα
ὑπὸ μεγέθεος τῆς νούσου. προϊόντος δὲ τοῦ χρόνου ἐνίοισι μὲν ἐς ὕδρωπα περιίσταται ἡ νοῦσος,
καὶ δι' οὖν ἐφθάρησαν· ἐνίοισι δὲ ἐκπυΐσκεται, καὶ
καυθέντες ὑγιέες ἐγένοντο·[2] ἐνίοισι δὲ καὶ συγκα-
ταγηρᾷ σκληρός τε ἐὼν καὶ μέγας.

Τὸ δὲ νόσημα γίνεται, ὅταν ἐκ πυρετῶν καὶ
κακοθεραπείης χολὴ ἢ φλέγμα ἢ ἀμφότερα ἐς
τὸν σπλῆνα καταστηρίξῃ· καὶ πολυχρόνιον μέν
ἐστι, θανατῶδες δὲ οὔ. τῶν φαρμάκων ὅσα δίδοται τοῦ σπληνός, τὰ μὲν [καὶ][3] διὰ τῆς κύστιος
καθαίρει καὶ ποιεῖ λαπαρώτερον, τὰ δὲ καθαίρει
μὲν οὔτε διὰ τῆς κύστιος οὐδὲν ὅ τι καὶ φανερὸν[4]
οὔτ' ἄλλῃ οὐδαμῇ, λαπάσσει δὲ τὸν σπλῆνα.

21. Εἰλεὸς ὅταν λάβῃ, ἡ γαστὴρ σκληρὴ
γίνεται, καὶ διαχωρέει οὐδέν· καὶ ὀδύνη κατὰ[5]

---

[1] Potter: ἀφῆναι Θ: ἀφιέναι Μ.   [2] Θ: γίνονται Μ.   [3] Del.
Vander Linden.   [4] Θ: -ώτερον Μ.   [5] Later mss: κάτω
ΘΜ.

bladder, and also to let blood frequently from the splenic vessel. Also, give the diuretic medications said to soften the spleen; clean these patients of bile when it is the season.

Some splenic patients are not helped by drinking medications or by any other treatment; nor does their spleen become at all thinner; instead, what is administered is overcome by the magnitude of their disease. In some, as time advances the disease turns into dropsy, and they have actually died; in others, the spleen suppurates, and on being cauterized patients have recovered; in yet others, the disease grows old with the patient, the spleen remaining large and hard.

This disease arises when, from fevers and faulty therapy, phlegm or bile or both are deposited in the spleen; it is chronic, but seldom mortal. Of the medications given for the spleen, some clean through the bladder to make the spleen softer, while others do not clean in any visible way at all, either through the bladder or along any other path, but still soften the spleen.

21. In ileus the belly becomes hard and no longer passes anything; pain is present through the

πᾶσαν τὴν κοιλίην ἔχει, καὶ πυρετός, καὶ δίψα·
ἐνίοτε δὲ ὑπὸ πόνου καὶ ἐμέει χολήν.

232 Τοῦτον χρὴ διυγραίνειν | καὶ ἔξωθεν καὶ ἔσω-
θεν· καὶ λούειν πολλῷ καὶ θερμῷ, καὶ πίνειν ὅσα
τήν τε κοιλίην κινεῖ καὶ τὸ οὖρον ὑπάγει, καὶ
ὑποκλύζειν ἢν δέχηται. ἢν δὲ μὴ δέχηται τὸ
κλύσμα, αὐλίσκον προσδήσας πρὸς ποδεῶνα
ἀσκίου, φυσήσας, ἐνεῖναι τὴν φῦσαν πολλήν· καὶ
ἐπειδὰν ἀρθῇ τὸ ἔντερον ὑπὸ τῆς φύσης καὶ ἡ
γαστήρ, ἐξελὼν τὸν αὐλίσκον, ἐνεῖναι παραχρῆμα
κλύσμα· καὶ ἢν δέξηται, ὑποχωρήσει καὶ ὑγιὴς
ἔσται· ἢν δὲ μηδ' οὕτω δέξηται τὸ κλύσμα, ἀποθνῄσκει ἑβδομαῖος μάλιστα.

Ἡ δὲ νοῦσος γίνεται, ὅταν τῆς κόπρου
συγκαυθῇ ἀθρόον ἐν τῷ ἐντέρῳ· περὶ τοῦτο περιίσταται φλέγμα, καὶ τὸ ἔντερον, ἅτε τούτων
ἀθρόων ἐνεσκληκότων, περιοιδεῖ· καὶ οὔτε τῶν
ἄνωθεν πινομένων φαρμάκων δέχεται, ἀλλ'
ἀπεμεῖ, οὔτε τῶν κάτωθεν προσφερομένων κλυσ-
μάτων δέχεται. ἔστι δὲ τὸ νόσημα ὀξὺ καὶ ἐπι-
κίνδυνον.

22. Ὕδερος δὲ γίνεται τὰ μὲν πλεῖστα, ὅταν
τις ἐκ νούσου μακρῆς ἀκάθαρτος διαφέρηται πολὺν χρόνον· φθείρονται γὰρ αἱ σάρκες καὶ τήκονται καὶ γίνονται ὕδωρ· γίνεται δὲ ὕδρωψ καὶ ἀπὸ
τοῦ σπληνός, ὅταν νοσήσῃ, καὶ ἀπὸ τοῦ ἥπατος,
καὶ ἀπὸ λευκοῦ φλέγματος, καὶ ἀπὸ δυσεντερίης

entire cavity, and there are fever and thirst; sometimes, from the stress, the patient also vomits bile.

You must moisten this patient from both the outside and the inside: wash him with copious hot water, have him drink potions that will set the cavity in motion and evacuate urine, and, if he will accept it, administer an enema. If he will not accept the enema, bind a tube to the mouth of a small wine skin, inflate it, and blow into the patient's anus a good amount of air; then, when the intestine is distended by the air, and the belly too, remove the tube and immediately introduce an enema. If the patient accepts it, he will evacuate downwards and recover; if he does not accept the enema in this way, either, he usually dies on the seventh day.

This disease occurs when a thick mass of faeces is burnt together in the intestine; around this gathers phlegm, and the intestine, inasmuch as these masses become hardened, swells around them. The patient accepts neither the medications drunk from above, vomiting them up instead, nor enemas administered from below. The disease is acute and dangerous.

22. Dropsy arises, in most cases, when a person continues for a considerable time after a lengthy illness in an unclean state; for the tissues become corrupted, melt, and turn to water; it can also take its origin from the spleen becoming diseased, from the liver, from white phlegm, or from dysentery or

καὶ λειεντερίης. κἢν μὲν ἐξ ἀκαθαρσίης γένηται
ὕδρωψ, ἡ μὲν γαστὴρ ὕδατος πίμπλαται, οἱ δὲ
πόδες καὶ αἱ κνῆμαι ἐπαίρονται, οἱ δὲ ὦμοι καὶ αἱ
κληῗδες καὶ τὰ στήθεα καὶ οἱ μηροὶ τήκονται.

Τοῦτον ἢν ἀρχόμενον λάβῃς πρὶν ὑπέρυδρον
γενέσθαι, φάρμακα πιπίσκειν κάτω, ὑφ' ὧν ὕδωρ
ἢ φλέγμα καθαίρεται, χολὴν δὲ μὴ κινεῖν· σιτίοισι δὲ καὶ ποτοῖσι καὶ πόνοισι καὶ περιπάτοισι
διαιτᾶν, ὑφ' ὧν ἰσχνὸς καὶ ξηρὸς ἔσται, καὶ αἱ
σάρκες[1] ἰσχυρόταται. ἡ δὲ νοῦσος θανατώδης,
234 ἄλλως τε κἢν φθῇ ἡ γαστὴρ | μεστωθεῖσα
ὕδατος.

Ὅταν δὲ ἀπὸ σπληνός, ἢ ἥπατος, ἢ λευκοῦ
φλέγματος, ἢ δυσεντερίης ἐς ὕδρωπα μεταστῇ,
θεραπεύειν μὲν τοῖς αὐτοῖς συμφέρει, διαφεύγουσι
δὲ οὐ μάλα· τῶν γὰρ νοσημάτων ὅ τι ἂν ἕτερον
ἐφ' ἑτέρῳ γένηται ὡς τὰ πολλὰ ἀποκτείνει.
ὅταν γὰρ ἐν ἀσθενεῖ τῷ σώματι ὄντι[2] ὑπὸ τῆς
παρούσης νούσου ἑτέρα νοῦσος ἐπιγένηται, προ-
απόλλυται ἀπὸ[3] ἀσθενείης, πρὶν τὴν νοῦσον τὴν
ὑστέρην γενομένην τελευτῆσαι.

Τὸ δὲ ὕδωρ γίνεται οὕτως· ἐπειδὰν αἱ σάρκες
ὑπὸ φλέγματος καὶ χρόνου καὶ νόσου καὶ
ἀκαθαρσίης καὶ κακοθεραπείης καὶ πυρετῶν δια-
φθαρῶσι, τήκονται καὶ γίνονται ὕδωρ· καὶ ἡ μὲν

---

[1] M adds ὡς.   [2] Later mss: τῷ σ. M: ἀσθενείη τῷ σ. Θ.
[3] Θ: ὑπὸ M.

lientery. If dropsy arises from uncleanness, the belly becomes filled with water, the feet and the legs below the knees swell up, and the shoulders, regions about the collar-bones, chest and thighs melt away.

If you take on this patient at the beginning, before he becomes very dropsical, have him drink a medication that will clean water and phlegm downwards, but not set bile in motion; prescribe a regimen of foods, drinks, exercises, and walks from which he will become lean and dry and his tissues will be strengthened as much as possible. The disease is often mortal, especially if the belly has already swollen up with water.

When dropsy develops out of a disease of the spleen or liver, from white phlegm, or from dysentery, it helps to employ the same treatment. In this case, patients do not survive very well, since any disease that develops out of another one is usually fatal; for when a second disease befalls the body weakened by a disease already present, the patient perishes from weakness before the second disease reaches its end.

The water in dropsy arises as follows: when the tissues become corrupted as the result of phlegm, the passage of time, disease, uncleanness, faulty therapy, and fevers, they melt and turn to water;

41

ΠΕΡΙ ΠΑΘΩΝ

κοιλίη οὐ διαδιδοῖ τὸ ὕδωρ ἐς αὐτήν, κύκλῳ δὲ περὶ αὐτὴν γίνεται.

Ἢν μὲν οὖν ὑπὸ τῶν φαρμάκων καὶ τῆς ἄλλης διαίτης ὠφελέηται, καὶ ἡ γαστὴρ λαπάσσεται αὐτοῦ. εἰ δὲ μή, ταμὼν ἀφεῖναι τοῦ ὕδατος· τάμνεται δὲ ἢ παρὰ τὸν ὀμφαλόν, ἢ ὄπισθεν κατὰ τὴν λαγόνα. διαφεύγουσι δὲ καὶ ἐντεῦθεν ὀλίγοι.

23. Δυσεντερίη ὅταν ἔχῃ, ὀδύνη ἔχει κατὰ πᾶσαν τὴν κοιλίην, καὶ στρόφος, καὶ διαχωρέει χολήν τε καὶ φλέγμα καὶ αἷμα συγκεκαυμένον.

Τούτου καθήρας τὴν κεφαλήν, φάρμακον πῖσαι ἄνω, ὅ τι φλέγμα καθαίρει· καὶ τὴν κοιλίην γάλακτι ἑφθῷ διανίψας, τὸ ἄλλο σῶμα θεραπεύειν. κἢν μὲν ἄπυρος ᾖ, τὴν μὲν κοιλίην λιπαροῖς καὶ πίοσι καὶ γλυκέσι καὶ ὑγροῖσιν ὑπάγειν[1] τὰ ἐνεόντα, καὶ λούειν θερμῷ τὰ κάτω τοῦ ὀμφαλοῦ ἢν ὀδύνη ἔχῃ· τὰ δὲ πώματα καὶ ῥυφήματα καὶ τὰ σιτία προσφέρειν κατὰ τὰ γεγραμμένα ἐν τῇ Φαρμακίτιδι.

Ἡ δὲ νοῦσος γίνεται ἐπειδὰν χολὴ καὶ φλέγμα καταστηρίξῃ ἐς τὰς φλέβας καὶ τὴν κοιλίην· νοσέει μὲν τὸ αἷμα καὶ διαχωρέει ἐφθαρμένον, νοσέει δὲ τὸ ἔντερον καὶ ξύεται καὶ ἑλκοῦται. γίνεται δ' αὕτη ἡ νοῦσος καὶ μακρὴ καὶ πολύπονος καὶ θανατώδης· κἢν μὲν ἔτι τοῦ σώματος

[1] M adds αἰεί.

42

the cavity does not transmit this water into itself, but instead it forms in the region round the cavity.[1]

Now, if the patient is helped by the medications and the rest of the regimen, the swelling in his belly goes down too. If not, incise and draw off water; make the incision either beside the navel or at the back in the region of the flank. Few escape from this disease, either.

23. In dysentery pain and colic are present throughout the whole cavity, and the patient passes bile, phlegm, and burnt-up blood.

Clean out this patient's head, have him drink a medication that will clean bile upwards, and, after you have washed out his cavity well with boiled milk, treat the rest of his body. If the patient is without fever, evacuate his cavity of its contents by means of rich fat sweet moist substances, and, if pain is present, wash the area below his navel with warm water; administer drinks, gruels and foods according to what is written in the *Medication Book*.

This disease arises when bile and phlegm are deposited in the vessels and the cavity; the blood ails and passes off corrupted in the stools, and the intestine becomes diseased, dried, and ulcerated. The disease is long, painful, and usually mortal; if

[1] I.e. the fluid does not enter the intestine, but remains free in the abdomen.

ΠΕΡΙ ΠΑΘΩΝ

236 ἰσχύοντος θεραπεύηται, ἐλπὶς | διαφυγεῖν· ἢν δὲ
ἤδη ἐκτετηκότος καὶ τῆς κοιλίης παντάπασιν
ἡλκωμένης,[1] οὐδεμία ἐλπίς.

24. Λειεντερίη· τὰ σιτία διαχωρέει ἄσηπτα,
ὑγρά· ὀδύνη δὲ οὐκ ἔνι· λεπτύνονται δὲ τὸ σῶμα.
τοῦτον θεραπεύειν τοῖσιν αὐτοῖσι, οἷσι τοὺς ὑπὸ
δυσεντερίης ἐχομένους.

Ἡ δὲ νοῦσος γίνεται, ὅταν ἐκ τῆς κεφαλῆς καὶ
τῆς ἄνω κοιλίης κατάρροος γένηται τοῦ φλέγματος ἐς τὴν κάτω κοιλίην· ὅταν δὲ τοῦτο ᾖ, ὑπ᾽
αὐτοῦ τὰ σιτία ψύχεται, καὶ ὑγραίνεται, καὶ ἡ
ἄφοδος αὐτῶν ἄσηπτος ἐν τάχει[2] γίνεται· καὶ τὸ
σῶμα τήκεται,[3] ἅμα μὲν οὐ πεσσομένων τῶν σιτίων ἐν τῇ κοιλίῃ χρόνον ἱκανόν, ἅμα δὲ ὑπὸ τῆς
κοιλίης θερμῆς ἐούσης παρὰ φύσιν θερμαινόμενον.

25. Διάρροια δὲ μακρή ὅταν ἔχῃ, διαχωρέει
πρῶτον μὲν τὰ ἐσιόντα ὑγρά, ἔπειτα δὲ φλέγμα·
καὶ ἐσθίει μὲν ἐπιεικῶς, ὑπὸ δὲ τῆς διαχωρήσιος
ἀσθενὴς καὶ λεπτὸς γίνεται.

Τούτου τὰ ἄνω ἀποξηραίνειν ἐλλέβορον πιπίσκων καὶ τὴν κεφαλὴν καθαίρων φλέγμα· καὶ
τὴν κοιλίην διανίψαι γάλακτι ἑφθῷ· ἔπειτα τά
τε ἄλλα σιτίοισι καὶ ποτοῖσι θεραπεύειν, ὑφ᾽ ὧν
ξηρανεῖται ἥ τε κοιλίη καὶ τὸ σῶμα πᾶν.

Ἡ δὲ νοῦσος ἀπὸ τῶν αὐτῶν γίνεται, ἀφ᾽ ὧν

---

[1] Later mss: εἱλκωμένης ΘΜ. [2] Θ: ἀσήπτων ταχείη Μ.
[3] Θ adds καί.

44

it is treated while the body is still strong, there is hope of recovery; but if the body is already melted away and the cavity altogether ulcerated, there is none.

24. Lientery: foods pass off undigested and watery; no pain is present; patients become lean of body. Treat this patient with the same measures employed in dysentery.

The disease arises when a defluxion of phlegm occurs from the head and the upper cavity into the lower cavity; when this happens, the foods are chilled by it and become moist, and the excretions pass off quickly in an undigested state; the body is melted partly because the food is not digested for an adequate length of time in the cavity, and partly because it is abnormally warmed by the cavity's heat.

25. In long-standing diarrhoea, first watery ingesta pass off, then phlegm. The patient eats a reasonable amount but, because of his excretions, becomes weak and thin.

Dry out this patient's upper regions by having him drink hellebore, and by cleaning his head of phlegm; wash out his cavity well with boiled milk; then, for the rest, treat with foods and drinks that will dry the cavity and the body as a whole.

This disease arises from the same factors as

καὶ ἡ λειεντερίη. αὗται αἱ νοῦσοι, ἥ τε λειεντερίη
καὶ ἡ δυσεντερίη, παραπλήσιοί εἰσι, καὶ δεῖ αὐτὰς
οὕτως ἰᾶσθαι· τὸν κατάρρουν ἀπολαμβάνειν τὸν
ἀπὸ τῆς κεφαλῆς καὶ τῆς ἄνω κοιλίης, ἢ ἀπο-
τρέπειν· τοῦ γὰρ νοσήματος ἡ φύσις ἐντεῦθε
γίνεται, καὶ οὐδείς[1] σου μέμψεται τὴν διάνοιαν.

Σχεδὸν δὲ καὶ τἆλλα νοσήματα ὧδε δεῖ σκοπεῖν, ὁπόθεν ἑκάστῳ ἡ φύσις γίνεται· καὶ οὕτως
σκοπῶν καὶ λαμβάνων τὴν ἀρχὴν τῶν νοσημάτων ἥκιστ᾽ ἂν ἁμαρτάνοις.

238   26. Τεινεσμὸς ὅταν λάβῃ, διαχωρέει αἷμα
μέλαν καὶ μύξα, καὶ πόνος ἐν τῇ κάτω κοιλίῃ
ἐγγίνεται, καὶ μάλιστα ὅταν ἐς ἄφοδον ἵζῃ.

Τούτου συμφέρει τὴν κοιλίην διυγραίνειν καὶ
λιπαίνειν καὶ ἀλεαίνειν, καὶ ὑπάγειν τὰ ἐνεόντα,
καὶ λούειν θερμῷ, πλὴν τῆς κεφαλῆς.

Φιλέει δὲ ἡ νοῦσος αὕτη τὰ σιτία πλείω
ποιέειν·[2] οἱ γὰρ στρόφοι κενουμένης τῆς κοιλίης
ὑπὸ τοῦ αἵματος διεξιόντος καὶ τῆς μύξης καὶ
προσπιπτόντων πρὸς τὸ ἔντερον γίνονται·
ἐνόντων δὲ τῶν σιτίων, ἧσσον δῆξιν παρέχει τῷ
ἐντέρῳ.

Καὶ γίνεται μὲν ἀπὸ αὐτῶν, ὧν καὶ ἡ δυσεντε-
ρίη· ἀσθενεστέρη δὲ καὶ ὀλιγοχρόνιος καὶ οὐ θανατώδης.

---

[1] Potter: οὐδέν ΘΜ: οὐδεὶς οὐδέν later mss.   [2] Θ: τελέειν Μ.

lientery. These diseases, lientery and dysentery, are similar, and you must treat them as follows: cut off the defluxion from the head and the upper cavity, or turn it aside; for the origin of the disease is from this, and no one will blame your understanding.

In general, you must investigate other diseases too in the same way, looking to see whence each takes its origin; by investigating in this way and seizing upon the beginning of diseases, you will err least.

26. In tenesmus, dark blood and mucus pass in the stools, and pain is felt in the lower cavity, most especially when the patient is sitting at stool.

It is of benefit to moisten thoroughly, oil and warm this patient's cavity, to evacuate its contents downwards, and to wash him with hot water, except for his head.

This disease tends to increase the amount of food consumed, for, when the cavity is in an empty state, colic arises from blood being evacuated together with mucus and from these coming into contact with the intestine; but when foods are present, there is less gnawing in the intestine.

Tenesmus arises from the same things as dysentery; it is milder, of short duration, and not mortal.

27. Ὅταν δὲ ἐξ οἴνου ἢ εὐωχίης χολέρη λάβῃ
ἢ διάρροια, τῇ μὲν διαρροίῃ συμφέρει διανη-
στεύειν, καὶ ἢν δίψος ἔχῃ, διδόναι οἶνον γλυκὺν
καὶ¹ στέμφυλα γλυκέα, ἐς ἑσπέρην δὲ διδόναι
ταῦτα, ἃ καὶ τοῖς ὑπὸ φαρμάκου κεκαθαρμένοις.
ἢν δὲ μὴ παύηται, ἐθέλῃς δὲ παῦσαι, ἔμετον ἀπὸ
σίτου² ἢ φακίου ποιῆσαι· καὶ παραχρῆμα ἀνέσπα-
σται ἄνω ἡ κάτω ἄφοδος. καὶ ἢν διακλύσῃς
χυλῷ φακῶν ἢ ἐρεβίνθων, καὶ οὕτω πεπαύσεται.

Τῇ δὲ χολέρῃ συμφέρει, ἢν μὲν ὀδύνη ἔχῃ, διδόναι ἃ γέγραπται ἐν τοῖς Φαρμάκοισι παύοντα
τῆς ὀδύνης, τήν τε κοιλίην θεραπεύειν τήν τε
ἄνω καὶ τὴν κάτω, ὑγραίνοντα πώμασι, καὶ μα-
λάσσοντα τὸ σῶμα λουτροῖσι θερμοῖσι, πλὴν τῆς
κεφαλῆς. καὶ ὅ τε ἔμετος οὕτως εὐπετέστερος
γίνεται, ἢν ἐσίῃ τι ὑγρόν· καὶ τὰ προσεστηκότα
ἄνω ἀπεμεῖται, καὶ ἡ³ κάτω ὑποχώρησις μᾶλλον
διαχωρέει· ἢν δὲ κενὸς ᾖ, ἐμέεται βιαίως, καὶ
ὑποχωρέει βιαιότερον. ἐς ἑσπέρην δὲ διδόναι καὶ
τούτῳ, ὅσαπερ φαρμακοπό|τῃ.

Γίνεται δὲ ταῦτα τὰ ἀλγήματα, ὅσα ἐκ ποσίων
γίνεται ἢ ἐξ εὐωχίης, ὅταν τὰ σιτία καὶ τὰ ποτὰ
πλείω τοῦ εἰωθότος ἐς τὴν κοιλίην ἐσέλθῃ, καὶ τὰ
ἔξωθεν εἰωθότα ὑπερθερμαίνοντα τὸ σῶμα κινέῃ
χολὴν καὶ φλέγμα.

240

¹ Θ: ἢ Μ.   ² Θ: -ων Μ².   ³ Μ: ἢν Θ.

48

27. When, after wine or feasting, a person is attacked by cholera or diarrhoea, in the case of diarrhoea it helps to have him fast, and, if he is thirsty, to give him sweet wine and sweet pressed grapes; towards evening give the same things as to patients that have been cleaned with a medication. If the diarrhoea does not go away, but you want it to, induce vomiting by means of foods or a decoction of lentils; the downward movement will at once be drawn upward. It will also stop if you employ as enema the juice of lentils or chick-peas.

In the case of cholera, if pain is present it helps to give the things recorded in the *Medications* as stopping pain, and to treat both the upper and the lower cavities by moistening them with drinks and softening the body except for the head with hot baths. Vomiting, too, becomes easier, if fluid enters: the offending substances are vomited upwards, and downward motions pass off more readily; but if a person is empty, he vomits violently and evacuates below even more violently. Towards evening, give this patient, too, the same things as to one that has drunk a medication.

These pains that follow drinking or feasting arise when more food and drink than usual enter the cavity, and these substances from outside, prone to overheating the body, set bile and phlegm in motion.

28. Στραγγουρίης τρόποι μὲν πολλοὶ[1] παντοῖοι. συμφέρει δ' ἔξωθεν μὲν τὸ σῶμα μαλάσσειν
λουτροῖσι θερμοῖσιν, ἔσωθεν δὲ διυγραίνειν τὴν
μὲν κοιλίην σιτίοισιν ὑφ' ὧν εὔροος ἔσται, τὴν δὲ
κύστιν ποτοῖσιν ὑφ' ὧν τὸ οὖρον ὡς πλεῖστον
διείσι. διδόναι δὲ καὶ τῶν διουρητικῶν φαρμάκων, ἃ γέγραπται ἐν τῇ Φαρμακίτιδι παύοντα
τῆς ὀδύνης.

Ἡ δὲ νοῦσος ὑπὸ φλέγματος γίνεται. καὶ
ὅταν μὲν ἡ κύστις ξηρανθῇ, ἢ ψυχθῇ, ἢ κενωθῇ,
ὀδύνην παρέχει· ὅταν δὲ ὑγρή τε καὶ πλήρης ᾖ
καὶ κεχυμένη, ἧσσον. ἡ δὲ νοῦσος τοῖσι μὲν πα-
λαιοτέροισι μακροτέρη γίνεται, τοῖς δὲ νεωτέροισι
βραχυτέρη, θανατώδης δὲ οὐδετέροις.

29. Ἰσχιὰς δὲ ὅταν γένηται, ὀδύνη λαμβάνει
ἐς τὴν πρόσφυσιν τοῦ ἰσχίου καὶ ἐς ἄκρον τὸ πυ-
γαῖον καὶ ἐς τὸν γλουτόν· τέλος δὲ καὶ διὰ
παντὸς τοῦ σκέλεος πλανᾶται ἡ ὀδύνη. τούτῳ
ξυμφέρει, ὅταν ἡ ὀδύνη ἔχῃ, μαλάσσειν καθ'
ὁποῖον ἂν τυγχάνῃ τοῦ σκέλεος στηρίζουσα ἡ ὀδύνη[2] λουτροῖσι καὶ χλιάσμασι καὶ πυρίῃ, καὶ τὴν
κοιλίην ὑπάγειν. ὅταν δὲ λωφήσῃ ἡ ὀδύνη, φάρμακον δοῦναι κάτω καὶ μετὰ ταῦτα πιεῖν ὄνου
γάλα ἐφθόν. διδόναι δὲ τῆς ὀδύνης ἃ γέγραπται
ἐν[3] τοῖς Φαρμάκοις.

---

[1] M adds καί.   [2] τούτῳ ... ὀδύνη M: τοῦτον θεραπεύειν Θ.
[3] Θ: παρὰ M.

AFFECTIONS

28. Of strangury, there are many different forms. It is beneficial to soften the body from the outside with hot baths, and inside thoroughly to moisten the cavity with foods that will make it fluent, and the bladder with drinks that will provoke as much urine as possible. Also give the diuretic medications recorded in the *Medication Book* as stopping pain.

Strangury arises from phlegm: when the bladder is dry, cold or empty, it produces pain; but when it is moist, full, and urine has flowed into it, less so. The disease is longer in older patients, shorter in younger ones, mortal in neither.

29. In sciatica pain occupies the attachment of the hip, the coccyx, and the buttock; finally, it also moves through the whole leg. When the pain is present, it helps to soften this patient with baths, fomentations, and a vapour-bath to whichever part of his leg the pain happens to settle in, and to evacuate his cavity downwards. When the pain goes away, give a medication to act downwards, and afterwards have the patient drink boiled ass's milk. For the pain, give what is recorded in the *Medications*.

ΠΕΡΙ ΠΑΘΩΝ

Ἡ δὲ νοῦσος γίνεται, ἐπειδὰν χολὴ καὶ
φλέγμα ἐς τὴν αἱμόρροον φλέβα καταστηρίξῃ, ἢ
ἐξ ἑτέρης νούσου, ἢ καὶ ἄλλως, ὁπόσον ἂν τοῦ
αἵματος ὑπὸ φλέγματος καὶ τῆς χολῆς νοσήσῃ
συνεστηκός· τοῦτο γὰρ πλανᾶται ἀνὰ τὸ σκέλος
242 διὰ τῆς | φλεβὸς τῆς αἱμορρόου καὶ ὅπου ἂν στῇ,
ἡ ὀδύνη κατὰ τοῦτο ἔνδηλος μάλιστα γίνεται. ἡ
δὲ νοῦσος μακρὴ γίνεται καὶ ἐπίπονος, θανατώδης δὲ οὔ. ἢν δὲ ἐς ἔν τι χωρίον καταστηρίξῃ ἡ
ὀδύνη καὶ στῇ, καὶ τοῖσι φαρμάκοισι μὴ ἐξελαύ-
νηται, καῦσαι καθ᾽ ὁποῖον ἂν τόπον τυγχάνῃ
οὖσα ἡ ὀδύνη· καίειν δὲ τῷ ὠμολίνῳ.

30. Ἀρθρῖτις νοῦσος ὅταν ἔχῃ, λαμβάνει πῦρ,
καὶ ὀδύνη τὰ ἄρθρα τοῦ σώματος, λαμβάνει δὲ
ὀξείη. καὶ ἐς ἄλλοτε ἄλλο τῶν ἄρθρων ὀξύτεραί
τε καὶ μαλακώτεραι καταστηρίζουσιν αἱ ὀδύναι.

Τούτῳ ξυμφέρει προσφέρειν ὅταν ἡ ὀδύνη ἔχῃ
ψύγματα, καὶ ἐκ τῆς κοιλίης ὑπάγειν τὰ ἐνεόντα
κλυσμῷ[1] ἢ βαλάνῳ, καὶ ῥυφεῖν διδόναι καὶ πιεῖν ὅ
τι ἂν δοκῇ σοι. ὅταν δὲ ἡ ὀδύνη ἀνῇ,[2] φάρμακον
πῖσαι κάτω, καὶ μετὰ τοῦτο πιεῖν ὀρὸν ἐφθὸν ἢ
ὄνου γάλα.

Ἡ δὲ νοῦσος γίνεται ὑπὸ χολῆς καὶ φλέγματος, ὅταν κινηθέντα ἐς τὰ ἄρθρα καταστηρίξῃ.
καὶ ὀλιγοχρονίη μὲν γίνεται καὶ ὀξέη, θανατώδης

---

[1] Θ: -μοῖσιν Μ. [2] Θ: ἐνῇ Μ.

AFFECTIONS

Sciatica arises when bile and phlegm are deposited in the blood vessels, either in consequence of another disease or in some other way, and some of the blood, being congealed by the phlegm and bile, ails; for this moves through the leg in its blood vessel, and wherever it stops the pain becomes most manifest. The disease is long and painful, but not mortal. If the pain settles in some spot, remains there, and cannot be driven out by medications, cauterize wherever the pain happens to be; burn with raw flax.

30. In the arthritic disease fever sets in, and sharp pains in the joints of the body. Sometimes these pains settle in one joint, sometimes in another, sometimes they are more violent, at other times milder.

When the pain is present, it benefits this patient to apply cooling agents, to evacuate the contents from his cavity downwards with an enema or suppository, and to give him as gruel and drink whatever you think suitable. When the pain goes away, have him drink a medication to act downwards, and afterwards boiled whey or ass's milk.

Arthritis arises from bile and phlegm, when they are set in motion and settle in the joints. It is of short duration and acute, but not mortal. It tends

δὲ οὔ. νεωτέροισι δὲ εἴωθε μᾶλλον ἢ γέρουσι[1]
γίγνεσθαι.

31. Ποδάγρη δὲ βιαιότατον μὲν τῶν τοιούτων
ἁπάντων ὅσα περὶ[2] τὰ ἄρθρα, καὶ πολυχρονιώτα-
τον καὶ δυσαπαλλακτότατον.

Καὶ ἔστι μὲν ἡ νοῦσος αὕτη τοῦ αἵματος
ἐφθαρμένου τοῦ ἐν τοῖς φλεβίοισιν ὑπὸ χολῆς καὶ
φλέγματος· ὅσῳ δ᾽ ἐν λεπτοτάτοις τε φλεβίοις
καὶ ἐν ἀνάγκῃ πεφυκόσι πλείστῃ τοῦ σώματος,
καὶ ἐν νεύροισι καὶ ὀστέοις πολλοῖσί τε καὶ
πυκνοῖσι, τοσούτῳ παραμονιμώτατόν τέ ἐστι τὸ
νόσημα καὶ δυσαπαλλακτότατον.

Συμφέρει δὲ καὶ ταύτῃ τὰ αὐτά, ἃ καὶ τῇ
ἀρθρίτιδι. καὶ μακρὴ μὲν καὶ αὕτη ἡ νοῦσος καὶ |
244 ἐπίπονος, θανατώδης δὲ οὔ. ἢν δὲ[3] τοῖσι δακτύλοισι τοῖς μεγάλοις ἡ ὀδύνη ἐγκαταλείπηται,
καῦσαι τὰς φλέβας τοῦ δακτύλου ὑπὲρ τοῦ κονδύ-
λου ὀλίγον· καίειν δὲ ὠμολίνῳ.

32. Ἴκτερον δὲ ὧδε χρὴ θεραπεύειν· ἔξωθεν
μὲν τὸ σῶμα μαλθάσσειν λουτροῖσι θερμοῖσι· τὴν
δὲ κοιλίην διυγραίνειν καὶ τὴν κύστιν, καὶ τῶν
διουρητικῶν διδόναι ἃ γέγραπται. ἢν δὲ ἰσχυρὸς
ᾖ, καθήρας τὴν κεφαλήν, φάρμακον πῖσαι κάτω,
ὅ τι χολὴν καθαίρει· ἔπειτα τοῖσι διουρητικοῖσι
χρῆσθαι.

---

[1] Θ: γεραιτέροισι Μ.   [2] Μ: ὅσαπερ Θ.   [3] Θ: δ᾽ ἐν Μ.

to occur more in younger persons than in old ones.

31. Gout is the most violent of all such conditions of the joints, as well as the most chronic and intractable.

In it blood is corrupted by bile and phlegm in the small vessels; inasmuch as this takes place in vessels that are the finest and by nature most critical in the body, as well as in cords and bones that are both many and dense, the condition is most persistent and intractable.

The same things are of benefit in this disease as in arthritis. It too is long and painful, but not mortal. If pain remains as a sequola in the large toes, cauterize the vessels of the toe a little above the knuckle; burn with raw flax.

32. You must treat jaundice as follows: soften the body from the outside with hot baths; moisten the cavity and the bladder thoroughly, and give the diuretics described above. If the condition is severe, after you have cleaned out the patient's head have him drink a medication that cleans bile downwards; then give diuretics.

Ἡ δὲ νοῦσος γίνεται ὅταν χολὴ κινηθεῖσα ὑπὸ τὸ δέρμα τράπηται.

33. Ταῦτα[1] ἐπιστάμενος, ἀνὴρ ἰδιώτης οὐκ ἂν ὁμοίως ἐμπίπτοι εἰς ἀνήκεστα νοσήματα, ἃ[2] εἴωθεν ἀπὸ σμικρῶν προφασίων μεγάλα καὶ πολυχρόνια γίνεσθαι.

Καὶ ὅσα μὲν σίτων ἢ ποτῶν ἐχόμενά ἐστιν, ἢ ῥυφημάτων ἢ φαρμάκων, ὅσα ὀδύνης εἴνεκα δίδοται, ἀκίνδυνά ἐστιν ἅπαντα ἃ δεῖ[3] προσφέρειν, ἐὰν κατὰ τὰ γεγραμμένα προσφέρῃς. ὅσα δὲ καθαίρει τῶν φαρμάκων χολὴν ἢ φλέγμα, ἐν τούτοισιν οἱ κίνδυνοι γίνονται καὶ αἱ αἰτίαι τοῖσι θεραπεύουσι· φυλάσσεσθαι οὖν χρὴ ταῦτα μάλιστα.

Ταῦτα μὲν ὅσα κατὰ κοιλίην γίνεται νοσήματα, πλὴν περὶ[4] ἐμπύων καὶ φθινόντων καὶ τῶν γυναικείων· ταῦτα δὲ χωρὶς γεγράψεται.

34. Φύματα πάντα ὅσα φύεται, ὑπὸ φλέγματος ἢ αἵματος φύεται. εἰ γὰρ ὑπὸ φλέγματος[5] . . . ὅταν ὑπὸ τραύματος ἢ πτώματος ἀθροισθῇ.

Συμφέρει δὲ τούτων, τὰ μὲν καταπλάσσοντα καὶ φάρμακα πιπίσκοντα διαχεῖν, τὰ δὲ καταπλάσσοντα πεπαίνειν. καὶ διαχεῖ[6] μὲν τῶν καταπλασμάτων ὅσα θερμὰ ἐόντα ὑγραίνει, καὶ μὴ σπᾷ ἐς ἑωυτά· πεπαίνει[7] δὲ ὅσα | θερμαίνοντα συνάγει. ὅταν δὲ τμηθῇ, ἢ αὐτόματον ῥαγῇ,[8]

[1] M adds δέ.   [2] Θ: δ' M.   [3] Θ: ἀεὶ M.   [4] περὶ Θ M[2];

56

AFFECTIONS

Jaundice arises when bile that has been set in motion invades beneath the skin.

33. Through understanding these things, a layman will be less likely to fall into incurable diseases that tend, from minor provocations, to become serious and chronic.

Of the foods, drinks, gruels or medications given against pain, all that you have to administer are safe, if you administer them as prescribed. But medications that clean bile or phlegm are a source of danger, and of blame for the person treating; thus, with these especially, care must be taken.

These are the diseases that arise in the cavity, except for patients that suppurate internally, consumptives, and diseases of women, which will be described separately.[1]

34. All tubercles that form do so because of phlegm or blood. For if, because of phlegm ... when they gather because of a wound or a fall.

Some tubercles it benefits to disperse by applying plasters and having the patient drink medications, others it helps to bring to maturity with plasters. Those plasters disperse which, being hot, moisten and do not attract; those mature which collect by heating. When tubercles are incised or

[1] The author may be referring to any number of Hippocratic works. Ermerins (II. LVII ff.) identifies this *Internal Suppuration* (Περὶ ἐμπύων) with *Diseases I* 11–22.

I would prefer τῶν. here.   [6] M: -χεῖν Θ.   [5] There seems to be a lacuna   [7] Θ: -νειν M.   [8] Θ adds ἤ.

φαρμάκῳ ἀνακαθαίρειν τὸ πύον·[1] ὅταν δὲ πυορ-
ροοῦντα παύσηται, ὡς ἕλκος ἰᾶσθαι.

35. Λέπρη καὶ κνισμὸς καὶ ψώρη καὶ λειχῆνες
καὶ ἀλφὸς καὶ ἀλώπεκες ὑπὸ φλέγματος γίνονται. ἔστι δὲ ταῦτα ἀεικέα[2] μᾶλλον ἢ νοσήματα.

Κηρίον καὶ χοιράδες καὶ φύγετρα[3] καὶ δοθιῆνες
καὶ ἄνθρακες[4] ὑπὸ φλέγματος φύεται.

36. Τουτοῖσι τοῖσι φαρμάκοισιν ἀποκαθαίρον-
τα ὧδε χρῆσθαι· ὅσοι μὲν χολώδεές εἰσι, διδόναι
ὑφ' ὧν χολὴ καθαίρεται· ὅσοι δὲ φλεγματώδεες,
ὑφ' ὧν φλέγμα. [ὅσοι δὲ μελαγχολῶσιν, ὑφ' ὧν
μέλαινα χολή· τοῖς δὲ ὑδρωπιῶσιν, ὑφ' ὧν
ὕδωρ.][5] ὅσα δὲ δίδοται φάρμακα ποτὰ καὶ μὴ
καθαίρει[6] μήτε χολὴν μήτε φλέγμα, ὅταν ἐς τὸ
σῶμα ἐσέλθῃ τὴν δύναμιν αὐτὰ παρέχεσθαι δεῖ ἢ
ψύχοντα ἢ θερμαίνοντα ἢ ξηραίνοντα ἢ ὑγραίνοντα ἢ συνάγοντα ἢ διαχέοντα. ὅσα δὲ[7] ὕπνον ποιεῖ
ἀτρεμίην δεῖ τῷ σώματι[8] παρέχειν τὸ φάρμακον.

37. Ὅταν δὲ ἐπὶ νοσέοντα ἀφίκῃ, ἐπανέρε-
σθαι[9] χρὴ ἃ πάσχει, καὶ ἐξ ὅτου, καὶ ποσταῖος,
καὶ τὴν κοιλίην εἰ διαχωρέει, καὶ δίαιταν ἥντινα
διαιτᾶται. καὶ ἐνθυμέεσθαι πρῶτον μὲν τὸ νόσημα

---

[1] Θ: ὑγρόν Μ.   [2] Θ: αἶσχος Μ.   [3] Θ: φύγεθλα Μ.
[4] Θ: ἄνθραξ Μ.   [5] Del. Artelt.   [6] Θ: -ειν Μ.
[7] Μ: δεῖ Θ.   [8] Θ: αἵματι Μ.   [9] Θ: ἐπανερωτᾶν Μ.

---

[1] W. Artelt (*Studien zur Geschichte der Begriffe Heil-
mittel und Gift*, Leipzig, 1937, 87) deletes these two

58

break open spontaneously, clean the pus out completely with a medication; when they stop producing pus, treat them in the same way as an ulcer.

35. Lepra, prurigo, psora, lichen, alphos, and alopecia arise because of phlegm. These are disfigurements rather than diseases.

Favus, scrofula, panus, boils, and anthraces are formed as the result of phlegm.

36. In cleaning, employ medications according to the following principle: when patients are bilious, give medications that clean out bile; when they are phlegmatic, give medications that clean out phlegm. [When they are melancholic, give medications that clean out dark bile; when they suffer from dropsy, give medications that clean out water.][1] Medicinal drinks that are not given to clean out bile or phlegm must, when they enter the body, exercise their faculty by cooling, warming, drying, moistening, collecting or dispersing. A medication that brings about sleep must provide the body with calm.

37. When you come to a patient, you must question him thoroughly about what he is suffering, in consequence of what, for how many days, whether his cavity has passed anything, and what regimen he is following. Consider first whether the disease

clauses because they contain the sole reference in the treatise to the humours "dark bile" and "water", in contradiction to the two-humour theory expounded in chs. 1 and 37, and otherwise followed.

πότερον ἀπὸ χολῆς ἢ φλέγματος γεγένηται ἢ
ἀμφότερα, καὶ τοῦτο εὖ εἰδέναι ὅτι ἀνάγκην ἔχει
ὥστε ὑπὸ τούτων τοῦ ἑτέρου ἢ ἀμφοτέρων
γίνεσθαι. ἔπειτα πότερον ὑγρασίης ἢ ξηρασίης
χρήζει, ἢ τὰ μὲν τοῦ σώματος ξηρασίης, τὰ δὲ
ὑγρασίης. ἔπειτα τὴν νοῦσον, εἴτε ἄνω δεῖ
θεραπεύειν, εἴτε κάτω, εἴτε διὰ τῆς κύστιος, καὶ
εἴτε αὔξεται ἡ νοῦσος εἴτε μαραίνεται, εἴτε
τελευτᾷ, εἴτε μεταπίπτει εἰς ἑτέρην νοῦσον.

38. Τοὺς τρωματίας λιμοκτονέειν, καὶ ἐκ τῆς
κοιλίης ὑπάγειν τὰ ἐνεόντα, ἢ ὑποκλύζοντα ἢ
248 φάρμακον κάτω διδόντα, καὶ πίνειν | ὕδωρ καὶ
ὄξος καὶ ῥυφεῖν ὕδωρ. τὰ φλεγμαίνοντα ψύχειν
καταπλάσμασι· τὰ δὲ καταπλάσματα εἶναι ἢ
σεῦτλα ἑφθὰ ἐν ὕδατι, ἢ σέλινον, ἢ ἐλαίης φύλλα,
ἢ συκῆς φύλλα, ἢ ἀκτῆς φύλλα ἢ βάτου, ἢ ῥοιῆς
γλυκείης· ἑφθοῖς μὲν τούτοισι χρῆσθαι. ὠμοῖσι δὲ
ῥάμνου φύλλοισιν, ἢ ἄγνου, ἢ ἐλελισφάκου, ἢ τι-
θυμάλλου, ἢ γληχὼ χλωρήν, ἢ πράσα, ἢ σέλινα,
ἢ κόριον, ἢ ἰσάτιος φύλλα. ἢν δὲ μηδὲν τούτων
ἔχῃς μήτε ἄλλο μηδὲν κατάπλασμα, ἄλφιτον φυ-
ρήσας ὕδατι ἢ οἴνῳ κατάπλασαι. τοσοῦτον δὲ
χρόνον τὰ καταπλάσματα ὠφελέει, ἐφ᾽ ὅσον[1] ἂν
χρόνον ψυχρότερα ᾖ ἢ τὸ ἕλκος· ὅταν δὲ ᾖ θερμότερα ἢ ὁμοίως θερμά, βλάπτει.

Τὰ λιπαρὰ πρὸς τὰ φλεγμαίνοντα οὐ συμ-

[1] ἐφ᾽ ὅσον Θ· ὁπόσον Μ.

has arisen from bile or from phlegm or from both, and have full confidence that it must be because of these, either one or both of them. Then, see whether the patient has need of moisture or dryness, or whether one part of the body needs dryness and another part moisture. Finally, determine whether you must treat the disease upwards or downwards or via the bladder, and whether the disease is increasing, diminishing, ending or changing into some other disease.

38. Treat persons suffering from wounds by having them abstain from food, by administering an enema or giving a medication to evacuate the contents downwards from their cavity, and by having them drink water and vinegar, and take watery gruel. If the wound is inflamed, cool it with plasters; let the plasters be made from beets boiled in water, or celery, or olive leaves, or fig leaves, or leaves of elder or bramble, or sweet pomegranate; apply these boiled. Raw, use buckthorn, chaste-tree, salvia or spurge leaves, green pennyroyal, leeks, celery, coriander, or woad leaves. If you do not have any of these, nor any other plaster, mix meal in water or wine and apply this as a plaster. Such plasters are of benefit only as long as they are colder than the wound; when they are warmer or equally warm, they do harm.

Fat substances are of no benefit to wounds that

φέρει, οὐδὲ πρὸς τὰ ἀκάθαρτα, οὐδὲ πρὸς τὰ σηπόμενα· ἀλλὰ πρὸς μὲν τὰ φλεγμαίνοντα, ψυχρά, πρὸς δὲ τὰ ἀκάθαρτα καὶ σηπόμενα, δριμέα καὶ ὅσα δῆξιν παρεχόμενα καθαίρει. ὅταν δὲ σαρκοφυῆσαι βούλῃ, τὰ λιπαρὰ καὶ τὰ θερμὰ μᾶλλον ξυμφέρει, πρὸς ταῦτα γὰρ ἡ σὰρξ θάλλει.

39. † Ὅσοις[1] ἄνθρωποι σίτοισιν ἢ ποτοῖσιν ὑγιαίνοντες ἐς δίαιταν χρῶται, ἐκ τούτων χρὴ τῶν παρεόντων χρῆσθαι πρὸς τοὺς νοσέοντας σκευάζοντα[2] καὶ ψυχρὰ καὶ θερμὰ καὶ ὑγρὰ καὶ ξηρά· ἐκ μὲν ψυχρῶν θερμά, καὶ θερμὰ[3] ἐκ μὴ θερμῶν, καὶ ξηρὰ ἐκ μὴ ξηρῶν, καί τὰ λοιπὰ κατὰ τὸν αὐτὸν τρόπον. ἀπορέειν δὲ οὐ[4] χρή, 250 οὐδὲ τοῖσι | παρεοῦσι μὲν μὴ δύνασθαι, τὰ ἀπόντα δὲ ζητοῦντα μηδὲν ὠφελέειν τὸν κάμνοντα οἷόν τε εἶναι.[5] εὑρήσεις δέ, ἢν ὀρθῶς σκοπήσῃς,[6] ἔξω τούτων οἷσι[7] πρὸς τὸν νοσέοντα χρῶνται, ὀλίγα.†

40. Ῥυφήματα τάδε[8] ἐν τῇσι νούσοισι πάσῃσι διδόναι· πτισάνην, κέγχρον, ἄλητον, χόνδρον. τούτων ὅσα μὲν πρὸς[9] διαχώρησιν, λεπτὰ διδόναι καὶ διεφθότερα, καὶ γλυκύτερα ἢ ἁλυκώτερα ἢ θερμότερα· ὅσα δὲ ἐς ἰσχὺν ἢ ἀνακομιδήν, παχύτερα καὶ λιπαρώτερα καὶ μετρίως ἑφθά.

Ποτοῖσι δὲ χρῆσθαι, ἢν μὲν ὑπάγειν ἐθέλῃς τὴν κοιλίην καὶ τὴν κύστιν, γλυκὺν οἶνον ἢ μελί-

---

[1] Potter: ὅσοι Θ: ὁπόσοις Μ. [2] Μ: -ζονται Θ. [3] Θ adds μή. [4] οὐ eras. Μ. [5] Θ adds χρή. [6] Θ: σκοπέῃς Μ. [7] Μ: τοῖσι Θ. [8] Θ: δ' Μ. [9] Θ: ἐς Μ.

are inflamed, or to those that are unclean and suppurating; rather, it is cold substances that clean
inflammations, and sharp irritating substances
that clean wounds that are unclean and suppurating. When you wish to promote the growth of
tissue, fat and warm substances are more useful,
since tissue thrives on these.

39. Of the foods or drinks that people employ in
their regimen when they are in a state of health,
for the ill you must use whichever are available,
preparing cold, hot, moist and dry; instead of cold,
hot, and hot instead of not hot; dry instead of not
dry, and the rest according to the same fashion.
You must not be at a loss, or incapable with those
available to help the patient, instead demanding
ones not available. You will gain few, if you investigate carefully, outside the ones people use for the
ill person.[1]

40. Give the following gruels in all diseases:
barley, millet, flour, or spelt. Gruels that are laxative give thin and more thoroughly boiled, also
more sweet, salty or hot; those that strengthen or
restore give thicker, richer, and moderately boiled.

As for drinks, if you wish to evacuate the cavity
and bladder, use sweet wine or melicrat; if you wish

[1] The text of this whole chapter is very doubtful.
Where the Greek seems not to offer any clear sense, I
have translated word for word.

κρητον· ἢ δὲ στύφειν, αὐστηρόν, λεπτόν, λευκόν,
ὑδαρέα· ἢν δὲ ἐς ἰσχύν, αὐστηρὸν μέλανα. ὅσοι
τὸν οἶνον πίνουσιν ἀνηλεῶς, τούτοις διδόναι ἅ γέγραπται ἐν τῇ Φαρμακίτιδι ποτὰ σκευαζόμενα.

41. Τοῖσι φαρμακοποτέουσι διδόναι μετὰ τὴν
κάθαρσιν, τοῖσι μὲν πυρέσσουσιν, ἢ φακὸν ἢ
κέγχρον λεπτὸν ἢ πτισάνης χυλόν· διδόναι δὲ πτισάνην μὲν καὶ κέγχρον ὡς κοῦφα ὄντα, χόνδρον
δὲ ὡς ἰσχυρότερον τούτων,[1] καὶ ἄλητον ὡς ἰσχυρότατον πάντων·[2] φακὸν δὲ ὡς[3] εὐώδεα σκευάσαι,
καὶ ὀλίγον δεύτερον διδόναι ὡς καὶ κοῦφον ῥόφημα
καὶ εὐκάρδιον ἄνω· παραμίσγειν δὲ ἢ ἅλας ἢ μέλι
καὶ κύμινον καὶ ἔλαιον τῷ[4] φακῷ, ἢ χλόης
γλήχωνος καὶ ὄξους[5] ὀλίγον. τοῖσι δὲ ἀπυρέτοις
ἄρτου καθαροῦ τὸ ἔσωθεν ἐνθρύψας ἐν ζωμῷ, ἢ
μᾶζαν καὶ τέμαχος ἑφθόν, ἢ κρέας οἰὸς ὡς νεωτάτης, ἢ ὄρνιθος, ἢ σκύλακος ἑφθά, ἢ τεῦτλα ἢ
κολοκύντην ἢ βλίτον· καὶ μετὰ τὸ σιτίον πίνειν
οἶνον εὐώδεα,[6] παλαιόν, λευκόν, ὑδαρέα.

252   42. Ὅσοις λούεσθαι μὴ συμφέρει, ἀλείφειν
οἴνῳ καὶ ἐλαίῳ θερμῷ, καὶ ἐκμάσσειν διὰ τρίτης.

43. Ὅταν κοιλίην ὑγραίνειν ἀπὸ σιτίων ἀσθε-
νέοντος ἐθέλῃς, διδόναι μᾶζαν καὶ ὄψα, θαλασ-

---

[1] Littré: ἰσχυρότατον πάντων ΘΜ.   [2] Littré: ἰσχυρότατον τούτων
Θ: ἰσχυρότερον τούτων Μ.   [3] ὡς om. Μ.   [4] Θ: τῷδε Μ.
[5] Littré: ὄξος ΘΜ.   [6] Θ: οἰνώδεα Μ.

to contract the body, use a light dry white wine diluted; if you wish to give strength, use a dry dark wine. To those who are harmed by drinking wine give the prepared drinks recorded in the *Medication Book*.

41. To persons that have drunk a medication, if after the cleaning they have fever, give lentils, thin millet or barley-water; give the barley-water and millet as light drinks, spelt as stronger than these, and flour as the strongest of all; prepare the lentils so that they are savoury, and give a small amount a second time, since it is a light gruel and pleasant to the upper region about the cardia; add cummin, oil and either salt or honey to the lentils, or a little green pennyroyal and vinegar. If the patients do not have fever give the inner part of a loaf of white bread crumbled into soup, or barley-cake and a slice of boiled fish, or meat of very young lamb, fowl or puppy, these boiled, or beets or gourd or blites; after the meal let them drink dilute fragrant old white wine.

42. Persons whom it does not benefit to wash, anoint every second day with warm wine and oil, and wipe dry.

43. When you wish to moisten a patient's cavity by means of foods, give barley-cake and, as main-

σίων μὲν τεμάχη ἑφθὰ ἐν ὑποτρίμματι, κρέας[1] δὲ
οἰὸς ὡς νεωτάτης, ἢ ἐρίφου, ἢ σκύλακος, ἢ ὄρνιθος ἑφθά, καὶ τεῦτλα ἢ[2] βλίτα ἢ λάπαθα ἢ κολοκύντην, ἢν ὥρη ᾖ. λάχανα δέ σοι εἶναι[3] σέλινα
καὶ ἄνηθα καὶ ὤκιμα· καὶ τὸν οἶνον μελιχρόν, παλαιόν, λευκόν, ὑδαρέα.

Ὅταν δὲ ξηραίνῃς τὸ σῶμα, διδόναι ἄρτον, καὶ
ὄψα ὀπτὰ καὶ ξηρά, καὶ θερμότερα ταῦτα πάντα·
κρεῶν τὰ μέλεα ὅσα σαρκώδη, ἰχθύων τοὺς
πετραίους, λάχανα πήγανον ἢ θύμον ἢ ὀρίγανον,
τὸν δὲ οἶνον μέλανα αὐστηρόν. σκευάζειν δὲ τὰ
ὄψα ἁλσὶ καὶ κυμίνῳ, καὶ τοῖς ἄλλοις ἀρτύμασιν
ὡς[4] ἐλαχίστοις χρῆσθαι.

Ὅταν δ' ἀνακομίσαι ἐκ νούσου θέλῃς, διδόναι
τὰ μὲν ἄλλα τὰ αὐτά, ἃ καὶ ὅταν διυγραίνῃς τὴν
κοιλίην, τὰ δὲ κρέα ἀντὶ τῶν γαλαθηνῶν ἰσχυρότερα, καὶ ἀντὶ τῶν κυνείων ὀρνίθεια καὶ λάγεια,
καὶ τούτων ἔνια ὀπτὰ καὶ τῶν κρεῶν καὶ τῶν
ἰχθύων καὶ ἐσκευασμένα ὡς ἄριστα.

Ὁπόσοισι τῶν νοσημάτων ξηρασίη συμφέρει,
μονοσιτίη συμφέρει, καὶ τὰ σιτία καὶ τὰ ποτὰ
ἐλάσσω τελέειν ἢ ὥστε πλήρη εἶναι, καὶ ταῦτα
ἐκπονεῖν, καὶ περιπατεῖν, καὶ κοιμᾶσθαι ὡς ἐλάχιστα. ὅσοις δὲ ὑγρασίη, μὴ ἀσιτέειν, καὶ τοῦ
σίτου καὶ τοῦ ποτοῦ μὴ ἐνδεᾶ εἶναι, μηδὲ πονεῖν,
καὶ κοιμᾶσθαι ὅσ' ἂν ἐθέλῃ.

[1] Θ: κρέα Μ.　　[2] Μ: καὶ Θ.　　[3] σοι εἶναι om. Μ.　　[4] Μ: οἷς Θ.

dishes, slices of seafoods in a mixed dish, meat of very young lamb, kid, puppy or fowl, these boiled, and beets, blites, docks or gourd if they are in season. Let your vegetables be celery, dill and basil, and your wine a dilute old honeyed white.

To dry the body give bread and roasted dry main-dishes, all quite hot; of meats give the fleshiest parts, of fish those of the rocks, of vegetables rue, thyme or marjoram, and as wine a dry dark one. Prepare the main-dishes with salt and cummin, but employ other seasonings as little as possible.

When you wish to build up a patient after a disease, give the same things as to moisten the cavity, but strong meats rather than those of young animals, and fowl and hare rather than dog; let these, both the meat and the fish, be roasted and very well prepared.

In diseases where dryness benefits, it is best for the person to eat only once a day, to consume less foods and drinks than would fill, to work these off by taking walks, and to sleep as little as possible. Where moistness benefits, it is best not to fast or lack for food or drink, not to be subject to exertions, and for the person to sleep as much as he wants.

254   44.   Ὅσων ἐπιθυμέουσιν οἱ κάμνοντες ἢ σιτίων
ἢ ὄψων ἢ ποτῶν, ὑπαρχέτω ταῦτα, ἢν μή τι
μέλλῃ τῷ σώματι βλάβος ἔσεσθαι.

Ὅταν ἢ σιτίων ἢ ποτῶν προστιθέναι ἄρξῃ ἢ
ἀφαιρέειν, κατ᾽ ὀλίγον χρὴ καὶ τὰς προσθέσιας[1]
ποιέεσθαι καὶ τὰς ἀφαιρέσιας.

Ὅσοι σιτία ἱκανὰ οἷοί τέ εἰσι τελεῖν, ῥυφήματα
μὴ διδόναι· ἀποκλείει γὰρ τοῦ σίτου· ὅσοι δὲ μὴ
οἷοί τε, τούτοισι διδόναι.

Ἢν δέ τι διδόναι ἐθέλῃς κομιδῆς ἕνεκα, διδόναι χόνδρον ἢ πτισάνην πυρίνην· ταῦτα γὰρ τῶν
ῥυφημάτων ἰσχυρότερα·[2] καὶ διδόναι μετὰ τὸ
δεῖπνον.

45.   Τὰ φάρμακα, ὅσα ποτὰ καὶ ὅσα πρὸς τὰ
τραύματα προσφέρεται, μανθάνειν[3] ἄξιον παρὰ
παντός· οὐ γὰρ ἀπὸ γνώμης ταῦτα εὑρίσκουσιν οἱ
ἄνθρωποι, ἀλλὰ μᾶλλον ἀπὸ τύχης, οὐδέ τι[4] οἱ
χειροτέχναι μᾶλλον ἢ οἱ ἰδιῶται. ὅσα δὲ ἐν τῇ
τέχνῃ τῇ ἰητρικῇ γνώμῃ εὑρίσκεται, ἥπερ[5] σίτων
ἢ φαρμάκων, παρὰ τῶν οἵων τε διαγινώσκειν
τὰ ἐν τῇ τέχνῃ μανθάνειν χρή, ἤ τι θέλῃς
μανθάνειν.

46.   Μετὰ τὰ ῥυφήματα διδόναι τὸν σῖτον
τοῖσιν ἀσθενέουσιν· ἐπιπίνειν δὲ οἶνον εὐώδεα.[6]

---

[1] M adds καὶ.   [2] Θ: -ότατα Μ.   [3] Μ adds χρὴ.
[4] Μ: οὐδ᾽ ὅτι Θ.   [5] Θ: ἢ περι Μ.   [6] Θ: οἰνώδεα Μ.

AFFECTIONS

44. Let the cereals, main-dishes or drinks that patients set their heart on be granted unless they are likely to be injurious to the body.

When you are beginning to add foods or drinks or to withdraw them, you must make the addition or the withdrawal a little at a time.

To patients able to take an adequate amount of food do not give gruel, for it turns them away from food; but to those that are not able to take food, give gruel.

If you wish to give something restorative, give spelt or wheat gruel, as these are among the stronger gruels; give them after dinner.

45. About medications that are drunk or applied to wounds it is worth learning from everyone; for people do not discover these by reasoning but by chance, and experts not more than laymen. But whatever is discovered in medicine by reasoning, whether about foods or about medications, you must learn from those that have discernment in the art, if you wish to learn anything.

46. After gruels, give food to patients; then have them drink fragrant wine. Before the foods and

πρὸ δὲ τῶν σιτίων καὶ ποτῶν ἢ ῥυφημάτων[1] . . .
καὶ μετὰ ταῦτα[2] ὅ τι ἄν σοι δοκέῃ. τοῖς ἀσθενέουσιν ἐσορῶν τὸ σῶμα καὶ τὴν ψυχὴν προσφέρειν
καὶ τὸ σιτίον καὶ τὸ ποτόν· μάλιστα γὰρ οὕτως
ὠφελέονται.

47. Καὶ[3] τῶν σιτίων, ἃ δύναμιν ἕκαστα ἔχει,
τεκμαίρεσθαι χρὴ ἀπὸ τῶν φανερὴν τὴν δύναμιν
ἐχόντων,[4] ὅσα ἢ φῦσαν ἢ δῆξιν ἢ πλη|σμονὴν ἢ
ἐρευγμὸν παρέχει ἢ στρόφον, ἢ διαχωρέει ἢ μὴ
διαχωρέει, καὶ φανερά ἐστιν[5] ὅτι ταῦτα ἐργάζεται. ἀπὸ τούτων χρὴ τὰ ἄλλα σκοπεῖν· ἔχει γὰρ
τὰ[6] ἕκαστα τῶν ἐδεσμάτων, διότι[7] ὠφελέει καὶ
βλάπτει· ἀλλὰ τὰ μὲν φανερώτερά[8] ἐστιν ἐργα-
ζόμενα ἃ ἐργάζεται, τὰ δὲ ἀμυδρότερα.

Τὰ σῖτα καὶ τὰ ὄψα σκευάζειν καὶ διδόναι τοῖς
ἀσθενοῦσιν, ὑφ' ὧν μήτε φῦσα ἔσται, μήτε ὀξυ-
ρεγμίη, μήτε στρόφος, μήτε λίην διαχωρέει, μήτε
λίην ξηραίνεται. ταῦτα δὲ γίνεται ὧδε· ὅσα μὲν
ἡ κοιλίη κρατέει, καὶ τὸ σῶμα αὐτὰ ἀναδέχεται,
ταῦτα μὲν οὔτε φῦσαν παρέχεται οὔτε στρόφον.
ἢν δὲ μὴ ἡ κοιλίη κρατῇ, ἀπὸ τούτων καὶ φῦσα
καὶ στρόφος καὶ τἆλλα τὰ τοιαῦτα γίνεται.

---

[1] The text of this whole chapter is doubtful. My punctua-
tion makes the first and third sentences understandable,
but requires the assumption of a lacuna in the
second.   [2] καὶ μετὰ ταῦτα om. Θ.   [3] Καὶ om. M.
[4] Θ: παρεχομένων M.   [5] ἐστιν om. M.   [6] τὰ om. M.
[7] M udds καὶ.   [8] Θ: -τατα M.

drinks, or gruels ... and after that whatever you think suitable. Administer both food and drink to patients in accordance with their body and their spirit; for in this way they are helped most.

47. About foods, all of which have some faculty, you must judge on the basis of those that have an evident faculty, for example, those that produce flatulence, irritation, fullness, belching, or colic, or those that pass off below or do not pass off, being clearly seen to do these things. Beginning with these, you must go on to examine the others; for each food has some faculty by which it helps or harms, although some are more evident in doing what they do, while others are more obscure.

Prepare and give to patients cereals and main-dishes from which there will be neither flatulence, oxyrygmia nor colic, and that will neither pass off too much nor dry too much. This occurs as follows: whichever foods the cavity masters and the body accepts produce neither flatulence nor colic; those, however, that the cavity does not master produce flatulence, colic and the rest of those sorts of things.

ΠΕΡΙ ΠΑΘΩΝ

Κοῦφα[1] τῶν σιτίων καὶ τῶν ὄψων καὶ τῶν ποτῶν, ὅσα μέτρια ἐσιόντα ἐς τὸ σῶμα ἢ ὀλίγῳ πλείω τῶν μετρίων, μήτε πλήρωσιν παρέχει, μήτε στρόφον, μήτε φῦσαν, μήτε ἄλλο τῶν τοιούτων μηδέν· καὶ πέσσεται τάχιστα, καὶ πεσσόμενα διαχωρεῖ· καὶ ἀνὰ πᾶσάν τε ἡμέρην ἐσιόντα ἐς τὴν κοιλίην ἀλυπότατά ἐστι,[2] καὶ ὅταν διὰ παλαιοῦ ἐσέλθῃ. βαρέα δέ, ὅσα μέτρια τελεύμενα ἢ ἐλάσσω τῶν μετρίων, πλήρωσιν καὶ πόνον παρέχει. καθ᾿ ἡμέρην δὲ[3] μηδὲ οἷόν τε ἐσθίειν αὐτὰ μηδὲ πίνειν, ἀλλὰ πόνον παρέχει. διὰ χρόνου δὲ ἤν[4] τις αὐτὰ ἢ πίνῃ ἢ ἐσθίῃ, καὶ οὕτω πόνον παρέχει, καὶ οὐ διαχωρέει ἀνὰ λόγον.

Ἐς ὑγιείην δὲ ἄριστα, ὅσα ὀλίγα[5] ἐσιόντα αὐτάρκη ἐστὶ καὶ λιμοῦ καὶ δίψους ἄκος εἶναι· καὶ πλεῖστον χρόνον τὸ σῶμα αὐτὰ δέχεται, | καὶ διαχωρέει κατὰ λόγον. ἐς ἰσχὺν δὲ ἄριστα, ὅσα σάρκα φύει[6] πλείστην καὶ πυκνοτάτην, καὶ τὸ αἷμα παχύνει, καὶ διαχωρέει κατὰ λόγον τῶν ἐσιόντων, καὶ τὸ σῶμα πλεῖστον χρόνον ἀναδέχεται.

Τὰ λιπαρὰ καὶ πίονα καὶ τυρώδη καὶ μελιτώδεα[7] καὶ τὰ σησαμοῦντα ὀξυρεγμίην μάλιστα

258

---

[1] Θ: -ότατα Μ. [2] Μ: ἔσται Θ. [3] καθ᾿ ἡμέρην δὲ Littré: καὶ μὴ τελέει(ν) ΘΜ. [4] Potter: ἄν Θ: αὖ- Μ. [5] Θ: ὀλίγιστα Μ. [6] φύει om. Θ. [7] καὶ μελιτώδεα om. Θ.

AFFECTIONS

Light cereals, main-dishes and drinks are ones that, on entering the body in moderate amounts or slightly more than moderate, produce neither fullness, colic, flatulence, nor any other of those kinds of things; that are digested very rapidly and, on being digested, pass off below; that entering the cavity daily are most harmless, and also when they enter at greater intervals. Heavy are ones that, when consumed in moderate amounts or less than moderate, produce fullness and distress; these it is not possible to eat or drink daily because they produce pain: if a person drinks or eats them over a period of time, even in that case they produce pain and do not pass off as they should.

The best foods for health are ones that, entering in a small quantity, are sufficient in themselves to satisfy both hunger and thirst; the body accepts them at most times, and they pass off as they should. For strength the best are those that produce the greatest amount of and the densest flesh, that thicken the blood, that pass off in the stools in proper measure of their ingestion, and that the body accepts at most times.

In particular, it is foods that are rich and fat and contain cheese, honey and sesame that produce

παρέχει καὶ χολέρην καὶ στρόφον καὶ φῦσαν καὶ
πλησμονήν. ποιεῖ δὲ τοῦτο τὸ αὐτὸ καὶ ὅταν τις
πλείω καταφάγῃ ἢ πίῃ ἢ ὅσα οἵη τε πέψαι ἡ
κοιλίη. τοῖς ἀσθενέουσιν ἢν μὲν κατὰ λόγον
τῆς νούσου καὶ τοῦ σώματος διδῷς ἃ ἂν διδῷς,
ὑπαναλίσκει τὸ σῶμα, καὶ οὔτε ἐνδεές ἐστιν οὔτε
πλῆρες· ἢν δὲ ἁμαρτάνῃς τοῦ καιροῦ ἢ ἐπὶ τὰ ἢ
ἐπὶ τά,[1] βλάβος ἐπ᾽ ἀμφότερα. ὅσα τῶν ὄψων ἢ
τῶν σίτων ἢ τῶν ποτῶν τὸ σῶμα ἀναδέχεται
μάλιστα, ἀπὸ τούτων οὔτε στρόφος γίνεται οὔτε
φῦσα οὔτε ὀξυρεγμίη· ὅταν γὰρ ἐς τὴν κοιλίην
ἐσέλθῃ,[2] ἀπ᾽ αὐτοῦ σπᾷ τὸ σῶμα τὸ ἐπιτήδειον
ὄν,[3] καὶ ἀσθενέστερον ἤδη τὸ λοιπὸν ἀνάγκη
εἶναι, ὥστε στρόφον ἢ φῦσαν ἢ ἄλλο τι τῶν
τοιούτων ἐν τῇ κοιλίῃ μὴ ποιῆσαι.

48. Τῶν οἴνων οἱ γλυκεῖς καὶ αὐστηροὶ καὶ
μελιχροὶ καὶ παλαιοὶ τὴν κοιλίην ὑπάγουσι μάλιστα καὶ[4] διουρέονται καὶ τρέφουσι, καὶ οὔτε
φῦσαν παρέχουσιν οὔτε στρόφον οὔτε πλησμονήν.

49. Κρεῶν τὰ δίεφθα καὶ ἔξοπτα ἀσθενέστερα[5] μὲν πρὸς τὴν ἰσχὺν ἀμφότερα, ἐς δὲ τὴν διαχώρησιν τὰ μὲν δίεφθα ἐπιτήδεια, τὰ δὲ ὀπτὰ
στασιμώτερα· τὰ δὲ μετρίως ἔχοντα καὶ ἐψήσιος
καὶ ὀπτήσιος μετρίως καὶ ἐς τὴν ἰσχὺν ἔχει καὶ
τὴν διαχώρησιν· τὰ δὲ ἐνωμότερα πρὸς μὲν τὴν

---

[1] ἐπὶ τὰ ἢ ἐπὶ τά Coray in Littré: ἔπειτα ΘΜ. [2] Θ adds τὰ.
[3] τὸ ἐπιτήδειον ὄν Μ: ἐπιτήδειον ***** Θ. [4] Θ adds οὐ.
[5] Θ: ἀσθενέα Μ.

oxyrygmia, cholera, colic, flatulence and fullness. The same also result when a person eats or drinks a greater quantity than his cavity is able to digest. If you make your administrations to patients in accordance with their disease and their body, the body will consume the foods in due course and be neither in want nor overfull; if, however, you miss the right measure either in the one direction or in the other, in both cases harm will be done. From the main-dishes, cereals and drinks that the body accepts best there arise neither colic, flatulence, nor oxyrygmia; for when one of these comes into the cavity, the body draws out of it what is suitable to itself, and what remains must by this very fact be weaker and so not produce colic, flatulence or anything else of that sort in the cavity.

48. Of wines the sweet dry honeyed and aged are the most laxative, diuretic, and nourishing, and these do not produce flatulence, colic or fullness.

49. Of meats both the well-boiled and the well-roasted are too weak to give strength, the well-boiled being suitable as laxatives, the well-roasted tending more towards constipation; meats boiled or roasted moderately are of moderate strength and

ἰσχὺν ἐπιτήδεια, πρὸς δὲ τὴν διαχώρησιν οὐκ ἐπιτήδεια.

260   50. Τῶν σιτίων καὶ τῶν ποτῶν ἃ προσφορώ-
τατα ἐν[1] τῷ σώματι καὶ μάλιστα αὐτάρκη καὶ ἐς
τροφὴν καὶ ἐς ὑγιείην, ἀπὸ τούτων τῶν αὐτῶν,
ὅταν τις αὐτοῖς μὴ ἐν τῷ καιρῷ χρῆται ἢ πλείωσι
τοῦ καιροῦ, αἵ τε νοῦσοι καὶ ἐκ τῶν νούσων οἱ θά-
νατοι γίνονται. τὰ δὲ ἄλλα σιτία καὶ ποτὰ ὅσα
μὴ τοιαύτην δύναμιν ἔχει, σμικρὸν μέν τι ὠφελέει, ἤν τις καὶ πάντα αὐτοῖς ἐν καιρῷ χρέηται,
σμικρὰ δὲ καὶ βλάπτει· ἐπ' ἀμφότερα δέ ἐστιν
ἀσθενέα, ὥστε καὶ ἀγαθόν τι ποιῆσαι καὶ ὥστε
καὶ κακόν. ἔστι δὲ τῶν σιτίων καὶ ποτῶν ἃ ταύτην τὴν δύναμιν ἔχει τάδε, ἄρτος, μᾶζα, κρέα,
ἰχθύες, οἶνος, τούτων μέντοι τὰ μὲν μᾶλλον, τὰ
δὲ ἧσσον.

51. Ὅσοι ξηρὴν δίαιταν διαιτῶνται, τούτοις
μὴ σὺν τῷ σίτῳ τὸ ποτὸν διδόναι, ἀλλὰ μετὰ τὸ
σιτίον, διαλιπὼν πολὺν χρόνον. καὶ οὕτω μὲν ξηρὴ ἡ ἰκμὰς ἀπὸ ξηρῶν τῶν σιτίων γινομένη τὸ
σῶμα[2] ξηραίνει· ἢν δὲ ἅμα τῷ σίτῳ πίνῃ, νοτιω-
τέρη[3] ἡ τροφὴ ἐοῦσα ὑγρότερον τὸ σῶμα ποιέει.
ἄρτος ὁ θερμὸς καὶ τὰ κρέα τὰ θερμὰ αὐτὰ ἐφ'
ἑαυτῶν ἐσθιόμενα ξηραίνει· ἢν δὲ ἢ[4] ξὺν ὑγρῷ

───────────

[1] ἐν om. M.   [2] Foes (n. 46 corpus Cornarius): τῷ σίτῳ Θ:
τῷ σιτίῳ M.   [3] Θ: νοτερωτέρη M.   [4] ἢ om. M.

76

moderately laxative; rarer meats are suitable for strengthening but not as laxatives.

50. If the foods and drinks that are most nourishing to the body and most sufficient for nourishment and health are employed at an inopportune moment or in an excessive amount, diseases result and, from the diseases, deaths. Those foods and drinks, however, that lack such strength provide little benefit, even if consumed together and at the right moment, but also do little harm; in both directions they are weak: to do good, and to do harm. The foods and drinks that possess the strength mentioned are bread, barley-cake, meats, fish and wine, some of them, of course, more, and others less.

51. To persons employing a drying regimen do not give any drink with their meals, but only later, after waiting a good long time. In this way the dry emanations arising from the dry foods will dry the body; if, however, a person takes drink with his meal, the food becomes moister and so makes the body moister. Hot bread and hot meats, if eaten alone, dry; but, if you give them together with a

διδῷς ἢ ἐπιπίνῃ[1] παραχρῆμα ἐπὶ τῷ σίτῳ, οὐ
ξηραίνει.

52. Ὁ ἄρτος καθαρῶν τῶν ἀλεύρων ἐς ἰσχὺν
καὶ κομιδὴν συμφορώτερος ἢ ἀνέρικτος, καὶ
πρόσφατος ἢ ἕωλος, καὶ τῶν ἀλεύρων[2] προσφάτων ἢ παλαιοτέρων. τὰ ἄλφιτα ἀβρέκτων[3]
τῶν κριθῶν περίχυδα ἐπτισμένων ἰσχυρότερα ἢ
βεβρεγμένων, καὶ πρόσφατα παλαιοτέρων.[4] καὶ
ἡ μᾶζα προπεφυρημένη ἰσχυροτέρη ἢ μὴ προ-
262 πεφυ|ρημένη.

Οἶνος διαχεόμενος καὶ[5] ἀποψυχόμενος καὶ δι-
ηθεόμενος λεπτότερος καὶ ἀσθενέστερος γίνεται.

Τὰ κρέα τὰ μὲν ἑφθά, ἢν μὲν δίεφθα ποιήσῃς,
ἀσθενέστερα καὶ κουφότερα, τὰ δ' ὀπτά, ἢν
ἔξοπτα· καὶ τὰ παλαιὰ[6] ἐξ ὄξους ἢ ἁλῶν, ἀσθενέστερα καὶ κουφότερα τῶν προσφάτων.

Τὰ δὲ[7] ἀσθενῆ τῶν σιτίων καὶ τὰ κοῦφα τὴν
μὲν κοιλίην οὐ λυπέει οὐδὲ τὸ σῶμα, διότι οὐκ
ἀνοιδέει θερμαινόμενα οὐδὲ πληροῖ, ἀλλὰ πέσσεται ταχὺ καὶ πεσσόμενα διαχωρεῖ. ἡ δὲ ἰκμὰς
ἀπ' αὐτῶν τῷ σώματι ἀσθενὴς γίνεται, καὶ οὔτε
αὔξην οὔτε ἰσχὺν ἀξίην λόγου παρέχει. τὰ δὲ
ἰσχυρὰ τῶν σιτίων ἀνοιδέει τε ὅταν ἐς τὴν κοιλίην ἐσέλθῃ, καὶ πλήρωσιν παρέχει· καὶ πέσσεται

---

[1] Potter: -ειν ΘΜ.   [2] Θ adds ἢ.   [3] Κ[2] (= Par. Graec.
2145, 15th cent.), Littré: ἀποβρέχων ΘΜ.   [4] Θ: ἢ παλαιότερα Μ.
[5] Μ: ἢ Θ.   [6] Μ: λ**πὰ Θ.   [7] δὲ om. Μ.

liquid, or the patient drinks immediately after eating them, they do not dry.

52. Bread made from fine wheat-meal strengthens and restores better than that from unground meal, fresh bread more than day-old, and bread made from freshly ground meal more than that from older meal. Barley meal that has not been soaked but peeled by being sprinkled is stronger than barley meal that has been soaked, and fresh it is stronger than when it is too old. Barley-cakes mixed a time before baking are stronger than those that are not.

Wine poured from one vessel into another, chilled and strained becomes thinner and weaker.

Boiled meats, if you boil them well, are weaker and lighter, as are roasted meats if you roast them well; meats aged in vinegar or salt are weaker and lighter than fresh ones.

Foods that are weak and light neither distress the cavity or the body, since they do not swell up on being heated, nor do they fill, but rather they are quickly digested and, being digested, pass off in the stools. The emanation from them in the body is weak and does not produce any growth or strength to speak of. Strong foods, on the other hand, swell up when they enter the cavity and produce fullness; they are digested more slowly, and pass off below

μὲν σχολαίτερον καὶ διαχωρέει· ἡ δὲ ἰκμὰς ἀπ'
αὐτῶν ἰσχυρὴ καὶ ἀκήρατος προσγινομένη ἰσχύν
τε παρέχει τῷ σώματι πολλὴν καὶ αὔξην.

Κρεῶν κουφότατα ἐς τὸ σῶμα κύνεια καὶ ὀρνίθεια καὶ λαγῷα τὰ[1] δίεφθα· βαρέα δὲ τὰ βόεια
καὶ τὰ χοίρεια· μετριώτατα δὲ πρὸς τὴν φύσιν
καὶ ἑφθὰ καὶ ὀπτὰ καὶ ὑγιαίνουσι καὶ ἀσθενέουσι
τὰ μήλεια. τὰ δὲ ὕεια ἐς εὐεξίην μὲν καὶ ἰσχὺν
πονοῦσι καὶ γυμναζομένοις ἀγαθά, ἀσθενέουσι δὲ
καὶ ἰδιώτῃσιν ἰσχυρότερα. καὶ τὰ θηρία τῶν ἡμέρων κουφότερά ἐστι, διότι καρπὸν οὐχ ὅμοιον
ἐσθίει. διαφέρει δὲ τὰ κρέα τῶν κτηνέων καὶ ὅσα
καρπὸν ἐσθίει καὶ ὅσα μὴ ἐσθίει. καὶ ὁ καρπὸς οὐ
τὸ αὐτὸ ἅπασι ποιέει, ἀλλ' ὁ μὲν πυκνήν τε τὴν
σάρκα τοῦ ἱερείου παρέχει καὶ ἰσχυρήν, ὁ δὲ
ἀραιήν τε καὶ ὑγρὴν καὶ ἀσθενέα.

264 Ὡς | μὲν τὸ σύμπαν κοῦφον ἰχθύες εἰρῆσθαι
ἔδεσμα καὶ ἑφθοὶ καὶ ὀπτοί, καὶ αὐτοὶ ἐφ' ἑαυτῶν
καὶ μεθ' ἑτέρων σιτίων. αὐτοὶ δὲ ἑαυτῶν διαφέ-
ρουσι· καὶ οἱ μὲν λιμναῖοι καὶ πίονες καὶ ποτάμιοι
βαρύτεροι, οἱ δὲ ἀκταῖοι κουφότεροι, καὶ ἑφθοὶ[2]
ὀπτῶν κουφότεροι.

Τούτων τὰ μὲν ἰσχυρὰ διδόναι, ὅταν ἀνακομίσαι τινὰ βούλῃ, τὰ δὲ κοῦφα, ὅταν ἰσχνὸν δέῃ ἢ
λεπτὸν ποιῆσαι.

[1] τὰ om. M.   [2] Θ: δίεφθοι M.

later; the emanation accrued from them, being strong and pure, gives the body great strength and growth.

The lightest meats for the body are well-boiled dog, fowl and hare; heavy are beef and pork; most moderate in their nature, for both the healthy and the sick, are boiled and roasted mutton. Pork is good for creating top condition and strength in labourers and athletes, but too strong for the sick or even normal person. Game has lighter flesh than domestic animals because the two do not eat the same fruit: the meat of animals varies according to whether or not they eat fruit; and fruit does not have the same effect in them all, but in one case makes the animal's flesh dense and strong, in another case rarified, moist and weak.

On the whole, fish are agreed to be a light food, both boiled and baked, and both alone and together with other foods. They vary one from another: those of the marsh and river and the fat ones are heavier, while those of the shore are lighter; fish boiled are lighter than when baked.

Give strong foods when you wish to restore a patient, light ones when you must make him lean and thin.

53. Τὸ λουτρὸν τὸ θερμόν, τὸ μὲν μέτριον
μαλάσσει τὸ σῶμα καὶ αὔξει· τὸ δὲ πλεῖον τοῦ
καιροῦ τὰ μὲν ξηρὰ τοῦ σώματος διυγραίνει, τὰ δὲ
ὑγρὰ ἀποξηραίνει· καὶ τὰ μὲν ξηρὰ ὑγραινόμενα
ἀσθενείην καὶ λιποθυμίην παρέχει, τὰ δὲ ὑγρὰ
ξηραινόμενα ξηρασίην καὶ δίψος.

54. Λαχάνων σκόροδα καὶ ἑφθὰ καὶ ὀπτὰ καὶ
διουρητικὰ καὶ ὑποχωρητικὰ καὶ πρὸς τὰ γυναι-
κεῖα σύμφορα. κρόμμυα ἐς τὰ οὖρα ἐπιτήδεια· ὁ
γὰρ ὀπὸς δριμύτητά τινα παρέχει ὥστε διαχωρέειν. τούτοις ὧδε χρῆσθαι, ἀλλὰ τοῖς ἀσθενέουσι
μὴ προσφέρειν. σέλινα ἑφθὰ καὶ ὠμὰ διουρητικά·
καὶ τῶν σελίνων τὰ ἕλεια τῶν ἡμέρων[1] πλείω
ἔχει δύναμιν. κορίαννον καὶ εὐκάρδιον καὶ διαχωρητικόν, καὶ ἑφθὸν καὶ ὠμόν. ὤκιμον καὶ ὑγρὸν
καὶ ψυχρὸν καὶ εὐκάρδιον.[2] πράσα τὰ μὲν ἑφθὰ
καὶ διουρητικὰ καὶ διαχωρητικά, ὠμὰ δὲ καυμα-
τώδη καὶ φλεγματώδη. ῥοιὴ κομιστικὸν καὶ
φλεγματῶδες, καὶ σὺν μὲν τῷ πυρῆνι στάσιμον,
ἄνευ δὲ τοῦ πυρῆνος διαχωρητικόν.

55. Τὰ θερμὰ τῶν σιτίων ξηρὰ μὲν ἵστησι, τὸ
γὰρ ὑγρὸν τὸ ἐν τῇ κοιλίῃ ἀναξηραίνει, ὑγρὰ δὲ
ὄντα, διυγραίνοντα τῇ θερμότητι ὑπάγει. τὰ δὲ
στρυφνὰ ξηραίνει καὶ ξυνάγει[3] τὸ σῶμα καὶ
266 στά|σιμα· τὰ ὀξέα λεπτύνει, δῆξιν ἐμποιέοντα·
τὰ ἁλμυρὰ διαχωρέει καὶ διουρέεται· τὰ λιπαρὰ

[1] M adds ᾶ.   [2] M: εὔκαρπον Θ.   [3] M: ὑπ- Θ.

AFFECTIONS

53. The hot bath, when employed in moderation, softens the body and increases it; when used to excess, it moistens the dry parts of the body, and dries out the moist ones; when the dry parts are moistened, it brings on weakness and fainting; when the moist parts are dried, they produce dryness and thirst.

54. Of vegetables the garlic, both boiled and baked, is diuretic and laxative and promotes menstruation. Onions are diuretic because their juice possesses a certain acridness that makes urine flow; use garlic and onions for this purpose, but do not administer them to the ill. Celery, both boiled and raw, is diuretic; meadow-celery has more strength than the cultivated variety. Coriander, both boiled and raw, is both pleasant to the cardia and laxative. Basil is moist, cool and pleasant to the cardia. Leeks, when boiled, are diuretic and laxative; raw, they heat and promote phlegm. The pomegranate strengthens and promotes phlegm; administered with its stones it is constipating, but without the stones it is laxative.

55. Hot foods, if dry, are constipating, since they dry out the moisture in the cavity; but if moist, they moisten with their heat, and promote evacuation. Sour foods dry, contract the body, and constipate; sharp ones thin the body by causing irritation; salty ones are laxative and diuretic; those that

καὶ τὰ πίονα καὶ τὰ γλυκέα ὑγρασίην μὲν καὶ φλέγμα παρέχει, κομιστικὰ δέ.

Κολοκύντη καὶ τεῦτλα καὶ βλίτα καὶ λάπαθα τῇ ὑγρότητι διαχωρητικά ἐστι· κράμβη δὲ ἔχει τινὰ[1] δριμύτητα ἐς τὸ διαχωρέειν, καὶ ἅμα ἔγχυλος·[2] τυρὸς καὶ σήσαμα καὶ ἀσταφίς, κομιστικὰ καὶ φλεγματώδεα.

Γλυκεῖς οἶνοι καὶ μελιηδεῖς καὶ κομιστικοὶ καὶ οὐρητικοὶ[3] καὶ φλεγματώδεις· οἱ δὲ αὐστηροὶ ἐς ἰσχὺν καὶ ξηρασίην ἐπιτήδειοι· οὐρητικοὶ δὲ καὶ τῶν αὐστηρῶν ὅσοι λεπτοί τε καὶ παλαιοὶ καὶ λευκοί.

Ἔλαιον καὶ ὅσα τοιαῦτα, κομιστικὰ καὶ φλεγματώδεα.

56. Λαχάνων τῶν ἐφθῶν διαχωρεῖ, ὅσα φύσει ὑγρότατά ἐστιν ἢ δριμύτητα ἢ θερμότητα ἔχει· διδόναι δὲ ταῦτα <ἃ>[4] σύμφορά ἐστιν ἐς ἄφοδον καὶ[5] χλιερώτερα καὶ τακερώτερα.

57. Σίκυος πέπων καὶ διουρητικὸς καὶ διαχωρητικὸς καὶ κοῦφος. ὁ δ' ἕτερος ψύξιν τε παρέχει καὶ δίψος παύει. τροφὴ δὲ ἀπὸ οὐδετέρου αὐτῶν γίνεται εἰ μὴ λεπτή τις, οὐδὲ φλαῦρον ἀπ' οὐδετέρου οὐδὲν[6] ὅ τι καὶ ἄξιον λόγου.

58. Τὸ μέλι σὺν μὲν ἑτέροις ἐσθιόμενον καὶ τρέφει καὶ εὔχροιαν παρέχει, αὐτὸ δὲ ἐφ' ἑωυτοῦ

---

[1] Θ adds καί.   [2] Potter: -λον Θ: εὔχυμος M.   [3] Θ: δι- M.
[4] Potter.   [5] καί om. M.   [6] οὐδὲν om. M.

are fat, rich, and sweet promote moistness and phlegm, and strengthen.

The gourd, beets, blites and docks are laxative on account of their moistness; cabbage has a certain sharpness that promotes evacuation, and at the same time it is succulent; cheese, sesame and raisins strengthen and promote phlegm.

Sweet and honeyed wines strengthen and promote urine and phlegm; dry wines are suited to promote strength and dryness; among dry wines the thin aged whites are also diuretic.

Olive oil and the like strengthen and promote phlegm.

56. Laxative among boiled vegetables are the naturally very moist or those possessing some acridness or heat; give laxative vegetables quite warm and soft.

57. The common melon is diuretic, laxative and light; the cucumber cools and stops thirst; no nourishment comes from either of them, or at most very little, nor any harm worth speaking of.

58. Honey eaten together with other foods nourishes and promotes good colour, but alone it thins

λεπτύνει μᾶλλον ἢ κομίζει, καὶ γὰρ διουρέεται
καὶ διακαθαίρεται μᾶλλον τοῦ μετρίου.

268   59. Τὰ διαχωρητικὰ θερμαινόμενα ἐν τῇ κοιλίῃ θερμαίνεταί τε ταχὺ καὶ θερμαινόμενα μαραίνεται καὶ τήκεται· καὶ τὴν διαχώρησιν διὰ τοῦτο
ταχείαν παρέχει. ὅσα δὲ στάσιμα τῶν σιτίων,
καὶ θερμαίνεταί τε βραδέως καὶ[1] θερμαινόμενα
ξηραίνεται καὶ συνίσταται, καὶ διὰ τὸ τοιοῦτο
περίσκληρα γινόμενα οὐ διαχωρέει. τὰ διαχωρητικὰ ἔγχυλά τέ ἐστι καὶ φύσει θερμά, τὰ δὲ οὐρη-
τικὰ ψυχρὰ καὶ ξηρά.

60. Ὁ σῖτος καὶ ὁ οἶνος διαφέρουσι μὲν καὶ
αὐτοὶ ἑωυτῶν φύσει ἐς ἰσχὺν καὶ ἀσθενείην καὶ
κουφότητα καὶ βαρύτητα. διαφέρει δὲ καὶ χώρη
χώρης ἐξ ὁποίης ἂν ᾖ, καὶ εὔυδρος ἐοῦσα καὶ ἄνυ-
δρος, καὶ εὐήλιος καὶ παλίνσκιος, καὶ ἀγαθὴ καὶ
φλαύρη, ὥστε ἅπαντα ταῦτα συμβάλλεται ἐς τὸ
ἰσχυρότατα[2] τῶν σιτίων εἶναι ἕκαστα καὶ ἀσθενέστερα.

61. Ὅσοι ὑγιαίνοντες ἀρτοφαγέειν εἰώθασι,
ταὐτὰ διδόναι τούτοισι καὶ ἐν ταῖς νούσοις.

Ὅταν ἢ σῖτα ἢ ποτὰ πλείω τοῦ εἰωθότος ἢν
μὴ τὰ εἰωθότα τελέσῃ, ἀπεμέσαι[3] παραχρῆμα
ἄριστον.

Ὀπώρη καὶ ἀκρόδρυα διὰ τόδε μετὰ τὸ σιτίον

[1] τε and καὶ om. M.   [2] Θ: -τερα Μ.   [3] Μ: -ση Θ.

more than it strengthens, since it passes off excessively in the urine and the stools.

59. Laxative foods, on being heated in the cavity, heat up quickly and, once heated, waste and melt away; for this reason, they bring about a rapid movement. Foods that constipate are heated slowly and, on being heated, dry up and congeal; this is why they become very hard and do not pass off. Laxatives are succulent and warm by nature; diuretics are cold and dry.

60. Cereals and wines differ even among themselves in their nature with regard to strength and weakness, and to lightness and heaviness; also, the places where they grow differ, one place being well-watered and another unwatered, one sunny and another thickly shaded, one favourable and another indifferent; thus it follows that all these factors contribute to each kind of food's being stronger or weaker.

61. To persons accustomed to eat breads when they are healthy give the same in diseases.

When a person consumes foods or drinks in greater amounts than is his habit, or of a different kind, it is best for him to vomit them up immediately.

Fruits and nuts are less injurious after the meal

ἀλυπότερά[1] ἐστι καὶ ὑγιαίνοντι καὶ ἀσθενέοντι·
ὅτι βεβρωκότος μὲν ὀλίγην[2] ἀπ' αὐτῶν ἰκμάδα
σπᾷ τὸ σῶμα, ἢν δὲ νῆστις ἐσθίῃ, πλείω.

Ὅσα τῶν σιτίων ἢ φῦσαν ἢ καῦμα ἢ δῆξιν ἢ
πλησμονὴν ἢ στρόφον παρέχει· ὁ οἶνος ὁ ἐπιπινό-
μενος ὁ ἄκρητος ἀπαλλάσσει τῶν τοιούτων· τὸ
γὰρ σῶμα διαθερμαινόμενον ὑπὸ τοῦ οἴνου ἀπαλλάσσεται τὰ ἐνεόντα θερμότητι.

Ἀπὸ τῶν σιτίων τε καὶ τῶν ποτῶν[3] τῶν
270 ὁμοίων ἐνίοτε | μὲν διαταράσσεται ἡ κοιλίη, ἐνί-
οτε δὲ ἵσταται, ἐνίοτε δὲ κατὰ λόγον διαχωρέει.
διότι δὲ ταῦτα οὕτως ἔχει· πρῶτον μὲν ἡ κοιλίη
ὅταν ὑγροτέρη ἐοῦσα καὶ ὅταν ξηροτέρη ὑποδέξη-
ται τὸ σιτίον διαφθείρει· ἔπειτα ὅταν μεταβολὴ
γένηται εἴτ' ἐκ ψύξεως[4] εἰς θάλπος, εἴτ' ἐκ θάλ-
ψιος[5] ἐς ψύχος, διαφθείρει. ὥστε ἀνάγκη τὴν
κοιλίην ἀπὸ τῶν αὐτῶν σιτίων καὶ ποτῶν δι'
αὐτὰ ταῦτα καὶ μαλακωτέρην γίνεσθαι καὶ
σκληροτέρην.

Τῶν σίτων καὶ τῶν ποτῶν καὶ τῶν ὄψων,
πλὴν ἄρτου καὶ μάζης καὶ κρεῶν καὶ ἰχθύων καὶ
οἴνου καὶ ὕδατος, τἆλλα πάντα λεπτὰς μὲν καὶ
ἀσθενέας [καὶ][6] τὰς ὠφελείας παρέχει ἐς τὴν
αὔξησιν καὶ τὴν ἰσχὺν καὶ ἐς τὴν ὑγιείαν· λεπτὰ
δὲ καὶ ἀσθενέα καὶ τὰ κακὰ ἀπ' αὐτῶν γίνεται.

---

[1] Θ: λυπηρότερά M. [2] ὀλίγην om. M. [3] M adds καί.
[4] Θ: -χεος M. [5] Θ: -πεος M. [6] Del. Littré.

in both the healthy and the sick, for the following reason: because, in a person that has eaten, the body attracts little emanation[1] from them, whereas if someone eats them in a fasting state, it attracts more.

Foods that produce flatulence, heat, irritation, fullness or colic: unmixed wine drunk afterwards provides relief; for the body, being heated through by the wine, is relieved of its contents as a result of this heat.

From the same foods and drinks the cavity is sometimes set in motion, sometimes stands still, and sometimes functions as it should. This is so for the following reasons: first, if the cavity is either too moist or too dry when it receives the meal, it spoils it; or then again, when a change occurs from cold to hot or from hot to cold, the same thing happens. Thus it follows that the cavity may become both softer and harder from the same foods and drinks.

All cereals, drinks, and main-dishes except bread, barley-cake, meat, fish, wine and water provide little and weak support for growth, strength and health; but little and weak, too, are the ills that come from them.

[1] The process of digestion is here imagined as occurring through the active absorption by the body of certain particularly potent or characteristic components of the food passing off as "emanations". Cf. chs. 51 and 52.

ΠΕΡΙ ΠΑΘΩΝ

Ὅσους τῶν νοσεόντων μὴ συνεχῶς οἱ πυρετοὶ
ἔχουσιν, ἀλλὰ διαλείποντες λαμβάνουσι, τούτοισι
τὰ σιτία διδόναι μετὰ τὴν λῆψιν, τεκμαιρόμενος
ὡς μὴ ἔτι[1] νεοβρῶτι ὁ πυρετὸς ἐπιπεσεῖται, ἀλλὰ
ἤδη πεπεμμένων τῶν σιτίων.

Οἶνος καὶ μέλι κάλλιστα κέκριται[2] ἀνθρώποις,
πρὸς τὴν φύσιν, καὶ ὑγιαίνουσι καὶ ἀσθενοῦσι, σὺν
καιρῷ καὶ μετριότητι προσφερόμενα· καὶ ἀγαθὰ
μὲν αὐτὰ ἐφ' ἑωυτῶν, ἀγαθὰ δὲ συμμισγόμενα,
τὰ δ' ἄλλα ὅσα τε καὶ ἀξίην λόγου ὠφελείην ἔχει.

Ὅσα ὑγιαίνουσι σύμφορα, ταῦτα ἐν ταῖς νούσοις
προσφερόμενα ἰσχυρότερά ἐστι, καὶ δεῖ[3] αὐτῶν
ἀφαιροῦντα τὴν ἀκμὴν διδόναι· ἢ οὐ φέρει αὐτὰ
τὸ σῶμα, ἀλλὰ βλάπτει μᾶλλον ἢ ὠφελέει.

---

[1] Θ: ἐπὶ Μ.   [2] Later mss, Ermerins: κέκρηται ΘΜ,
Littré.   [3] Θ: δι' Μ.

AFFECTIONS

To patients with fevers that are not continuous but intermittent, give meals after the attack, watching carefully to make sure that fever does not attack a person that has just eaten, but only when his meal has been digested.

Wine and honey are held to be the best things for human beings, so long as they are administered appropriately and with moderation to both the well and the sick in accordance with their constitution; they are beneficial both alone and mixed, as indeed is anything else that has a value worth mentioning.

Things beneficial to persons in health are too strong if administered in diseases, and you must give them only with their major strength removed; otherwise the body cannot stand them, and they will harm rather than help

DISEASES I

INTRODUCTION

Many terms from *Diseases I* are to be found in the Hippocratic glossaries of Erotian and Galen, and there is a possibility that the work was already known to the medical glossator Bacchius of Tanagra in the second century B.C.[1] One of the Galenic glosses includes a short quotation from the text:

> ἀμαλῶς: weakly, which can be the same as moderately, as in *Internal Suppuration*: "people praise him ἀμαλῶς". . . .[2]

From this passage and the following two out of his Hippocratic commentaries, it would appear that Galen preferred the title *Internal Suppuration* (Περὶ ἐμπύων) to *Diseases I* for the treatise, although the latter was obviously in general use:

> Thus, in the preface of the work <not> rightly entitled *Diseases I*, it is written that fever inevitably follows a chill. . . .[3]

[1] See Wittern pp. LX–LXIII.

[2] Kühn XIX. 76. Galen's text is quite different from ours (*Diseases I* 8), but close enough to allow confident identification; cf. Wittern p. 117 n. 1.

[3] Kühn XVII(1). 276 = CMG V 10,1 p. 138. The reference is to *Diseases I* 4.

DISEASES I

It is also stated in *Diseases I*, not properly so entitled, which begins: "Anyone who wishes to ask correctly about healing, and, on being asked, to reply". . . .[1]

In support of the title *Diseases I* are the following: first, if my argument elsewhere[2] is correct, this treatise was *Diseases I* for Erotian; second, a papyrus of the second century A.D. uses the title *Diseases I*:

κ αι εν τω πρω[τω]
[π]ερ[ι] νουσω[ν] οτα[ν] λε[γη αι μεν ουν]
[νο]υσοι γιγνον[ι ωι η]μι[ν απα]
[σα]ι των μεν [εν τω σωματι εν][3] . . .

finally, Caelius Aurelianus ascribes the opinion that venesection is beneficial in cases of bleeding to "Hippocrates, writing in *On Diseases*",[4] a reference to *Diseases I* 14.

Diseases I can be divided into two parts. The first (1–10) consists of general remarks on the medical art meant, according to the first chapter, to

---

[1] Kühn XVIII(1). 513.

[2] Potter (op. cit. vol. VI p. 5) 55 ff.

[3] Pap. gr. 26 Strasb. 16–19. See J. Jouanna, "Un nouveau témoinage sur la collection hippocratique: P. gr. inv. 26, col. III, de Strasbourg" *Zeitschr. f. Papyrologie u. Epigraphik* 8, 1971, 147–60.

[4] *Chronic Diseases II* 184, ed. I. E. Drabkin, Chicago, 1950, 686.

prepare the reader to be able to state and defend his views. Despite the use of the terms "rebut" and "in your rebuttal", which would suggest a rhetorical disputation, the author's concentration on the actual material discussed, rather than on techniques of argumentation, and the candidness with which he handles many difficult aspects of medical practice indicate that the discussion intended is a professional one. Of the topics announced in the first chapter, most, but not all, appear in chapters 2–10. The second part of *Diseases I*, which has as its subjects internal suppurations (11–22) and the acute diseases pleurisy, pneumonia, ardent fever and phrenitis (23–34), concentrates exclusively on the aetiology and pathogenesis of these conditions.

The relationship between the two parts of *Diseases I* has been the subject of much scholarly debate.[1] The following points seem clear:

The transition from chapter 10 to chapter 11 is abrupt.

The subject matter and the purpose of the two parts is fundamentally different.

The basic theory of disease in the two parts is compatible, if not identical.

---

[1] E.g. Littré (VI. 138) and Wittern (pp. LXXI ff.) hold them to be the complementary general and special parts of a textbook of pathology. Ermerins (II. LVI f.) and Fuchs (II. 377), on the other hand, claim that they have nothing to do with one another.

DISEASES I

The style and vocabulary of the two parts have much in common.[1]

Diseases I is included in the two renaissance works devoted to the Hippocratic books on Diseases:

> *Hippocratis Coi de morbis libri quatuor* Georgio Pylandro interprete. ... Paris, 1540.
>
> *Commentaria in Hippocratis libros quatuor de morbis luculentissima* ... Petri Salii Diversi. Frankfurt, 1602.

Much more recently, R. Wittern has subjected the treatise to a very thorough study, and it is upon her text and commentary that the present edition for the most part depends;

> Renate Wittern, *Die hippokratische Schrift De morbis I, Ausgabe, Übersetzung und Erläuterungen*, Hildesheim/New York, 1974. (= Wittern)

[1] See Ermerins II. LVIII.

ΠΕΡΙ ΝΟΥΣΩΝ Α

VI 140
Littré

1. Ὃς ἂν περὶ ἰήσιος ἐθέλῃ ἐρωτᾶν τε ὀρθῶς
καὶ ἐρωτώμενος ἀποκρίνεσθαι καὶ ἀντιλέγειν ὀρθῶς, ἐνθυμεῖσθαι χρὴ τάδε. πρῶτον μέν, ἀφ᾽ ὧν
αἱ νοῦσοι γίνονται τοῖσιν ἀνθρώποισι πᾶσαι· ἔπειτα δέ, ὅσα ἀνάγκας ἔχει τῶν νοσημάτων ὥστε
ὅταν γένηται εἶναι ἢ μακρὰ ἢ βραχέα ἢ θανάσιμα
ἢ μὴ θανάσιμα ἢ ἔμπηρόν τι τοῦ σώματος γενέσθαι ἢ μὴ ἔμπηρον· καὶ ὅσα, ἐπὴν γένηται, ἐν-
δοιαστά, εἰ[1] κακὰ ἀπ᾽ αὐτῶν ἀποβαίνει ἢ ἀγαθά·
καὶ ἀφ᾽ ὁποίων νοσημάτων ἐς ὁποῖα μεταπίπτει·
καὶ ὅσα ἐπιτυχίῃ ποιέουσιν οἱ ἰητροὶ θεραπεύοντες
τοὺς ἀσθενέοντας καὶ ὅσα ἀγαθὰ ἢ κακὰ οἱ νο-
σέοντες ἐν τῇσι νούσοισι πάσχουσι· καὶ ὅσα εἰκα-
σίῃ ἢ λέγεται ἢ ποιεῖται ὑπὸ τοῦ ἰητροῦ πρὸς τὸν
νοσέοντα, ἢ ὑπὸ τοῦ νοσέοντος πρὸς τὸν ἰητρόν·
καὶ ὅσα ἀκριβῶς ποιέεται ἐν τῇ τέχνῃ καὶ λέγεται, καὶ ἅ τε ὀρθὰ ἐν αὐτῇ καὶ ἃ μὴ ὀρθά· καὶ ὅ
τι αὐτῆς ἢ ἀρχὴ ἢ τελευτὴ ἢ μέσον ἢ ἄλλο τι
ἀποδεδειγμένον[2] τῶν τοιούτων· ὅ τι καὶ ὀρθῶς
ἐστιν ἐν αὐτῇ εἶναι ἢ μὴ εἶναι· καὶ τὰ σμικρὰ καὶ

---

[1] Littré: ἢ ΘΜ.   [2] Littré places ἀποδ. after ὀρθῶς; cf. ch. 9.

DISEASES I

1. Anyone who wishes to ask correctly about healing, and, on being asked, to reply and rebut correctly, must consider the following: first, whence all diseases in men arise. Then, which diseases, when they occur, are necessarily long or short, mortal or not mortal, or permanently disabling to some part of the body or not, and which other diseases, when they occur, are uncertain as to whether their outcome will be bad or good. From which diseases there are changes into which others. What physicians treating patients achieve by luck. What good or bad things patients suffer in diseases. What is said or done on conjecture by the physician to the patient, or by the patient to the physician. What is said and done with precision in medicine, which things are correct in it, and which not correct. What starting point of medicine, or end, or middle, or any other feature of this kind has been demonstrated; what truly does or does not exist in medicine[1]: the small and the large, the many and

---

[1] The clause ὅ τι καὶ ὀρθῶς ἐστιν ἐν αὐτῇ εἶναι ἢ μὴ εἶναι has traditionally been taken as referring backward; I find that this interpretation (e.g. Wittern: *dessen Existenz in ihr dann auch gesichert ist oder nicht*) makes little sense, and thus prefer to understand the clause in connection with the words that follow it.

τὰ μεγάλα, καὶ τὰ πολλὰ καὶ τὰ ὀλίγα· καὶ ὅ τι
ἅπαν ἐστὶν ἐν αὐτῇ [ἕν καὶ πάντα],[1] καὶ ὅ τι ἕν·
καὶ τὰ ἀνυστὰ νοῆσαί τε καὶ εἰπεῖν καὶ ἰδεῖν καὶ
ποιῆσαι, καὶ τὰ μὴ ἀνυστὰ μήτε νοῆσαι μήτε
εἰπεῖν μήτε ἰδεῖν μήτε ποιῆσαι· καὶ[2] ὅ τι εὐχειρίη
142 ἐν αὐτῇ, καὶ ὅ τι ἀχειρίη· καὶ ὅ τι | καιρός, καὶ ὅ
τι ἀκαιρίη· καὶ τῶν τεχνέων τῶν ἄλλων ᾗσί τε
ἔοικε καὶ ᾗσιν οὐδὲν ἔοικε· καὶ τοῦ σώματος ὅ τι
ἢ ψυχρὸν ἢ θερμὸν ἢ ἰσχυρὸν ἢ ἀσθενὲς ἢ πυκνὸν
ἢ ἀραιὸν ἢ ὑγρὸν ἢ ξηρόν· καὶ ὅσα τῶν πολλῶν
ὀλίγα γίνεται, ἢ ἐπὶ τὸ κάκιον, ἢ ἐπὶ τὸ ἄμεινον·
καὶ ὅ τι καλῶς ἢ αἰσχρῶς ἢ βραδέως ἢ ταχέως ἢ
ὀρθῶς ἢ μὴ ὀρθῶς· καὶ ὅ τι κακὸν ἐπὶ κακῷ γενόμενον ἀγαθὸν ποιέει, καὶ ὅ τι κακὸν ἐπὶ κακῷ[3]
ἀνάγκη γενέσθαι.

Ταῦτ' ἐνθυμηθέντα διαφυλάσσειν δεῖ ἐν τοῖσι
λόγοισιν· ὅ τι ἂν δέ τις τούτων ἁμαρτάνῃ ἢ λέγων ἢ ἐρωτῶν ἢ ὑποκρινόμενος, καὶ ἢν πολλὰ
ὄντα ὀλίγα φῇ εἶναι, ἢ μεγάλα ἐόντα[4] σμικρά,
καὶ ἢν ἀδύνατα ἐόντα δυνατὰ φῇ εἶναι, ἢ[5] ὅ τι ἂν
ἄλλο ἁμαρτάνῃ λέγων, ταύτῃ φυλάσσοντα δεῖ
ἐπιτίθεσθαι ἐν τῇ ἀντιλογίῃ.

2. Αἱ μὲν οὖν νοῦσοι γίνονται ἡμῖν ἅπασαι,
τῶν μὲν ἐν τῷ σώματι ἐνεόντων, ἀπό τε χολῆς
καὶ φλέγματος, τῶν δ' ἔξωθεν, ἀπὸ πόνων καὶ

---

[1] Del. Wittern.   [2] καὶ om. Θ.   [3] γενόμενον . . . κακῷ om. M.
[4] ὀλίγα . . . ἐόντα om. M.   [5] ἢ om. Θ.

many and the few; what is all in it and what is one. What it is possible to perceive, to say, to see, and to do, and what it is not possible to perceive, to say, to see, or to do. What is dexterity in medicine, and what is awkwardness. What the opportune moment is, and what inopportunity. To which of the other arts medicine has similarities, and to which it has none. What in the body is cold or hot, strong or weak, dense or rarified, or moist or dry; which of the many become few, either for worse or for better. What is noble or base, slow or fast, correct or incorrect. Which evil, on following another evil, brings something good, and which evil follows inevitably upon some other evil.

When you have considered these questions, you must pay careful attention in discussions, and when someone makes an error in one of these points in his assertions, questions, or answers—for example, if he asserts that something that is many is few, or something large small, or claims that something impossible is possible, or errs in any other way in his statements—then you must catch him there and attack him in your rebuttal.

2. Now all our diseases arise either from things inside the body, bile and phlegm, or from things outside it: from exertions and wounds, and from

τρωμάτων, καὶ ὑπὸ τοῦ θερμοῦ ὑπερθερμαίνοντος
καὶ τοῦ ψυχροῦ ὑπερψύχοντος.[1]

Καὶ ἡ μὲν χολὴ καὶ τὸ φλέγμα γινομένοισί τε
συγγίνεται καὶ ἔνι αἰεὶ ἐν τῷ σώματι ἢ πλέον ἢ
ἔλασσον· τὰς δὲ νούσους παρέχει, τὰς μὲν ἀπὸ σιτίων καὶ ποτῶν, τὰς δὲ ἀπὸ τοῦ θερμοῦ ὑπερθερμαίνοντος καὶ ἀπὸ τοῦ ψυχροῦ ὑπερψύχοντος.

3. Ἀνάγκη δὲ τὰ τοιάδε ἔχει ὥστε γίνεσθαι, ὅ
τι ἂν[2] γίνηται· ἐν μὲν τοῖσι τρώμασι νεῦρα τὰ
παχέα τιτρωσκομένους χωλοῦσθαι καὶ τῶν μυῶν
τὰς κεφαλάς, καὶ μάλιστα τῶν ἐν τοῖσι μηροῖσιν·
144 ἀπο|θνήσκειν δέ, ἤν τις ἐγκέφαλον τρωθῇ ἢ
ῥαχίτην μυελὸν ἢ κοιλίην[3] ἢ ἧπαρ ἢ φρένας ἢ
κύστιν ἢ φλέβα αἱμόρροον ἢ καρδίην· μὴ ἀπο-
θνήσκειν δὲ τιτρωσκόμενον ἐν οἷσι ταῦτα τῶν
μελέων μὴ ἔνι ἢ τούτων προσωτάτω ἐστίν.

Τῶν δὲ νοσημάτων τὰ τοιάδε ἔχει ἀνάγκας
ὥστε ὑπ᾽ αὐτῶν ἀπόλλυσθαι, ὅταν ἐπιγένηται·
φθίσις, ὕδρωψ ὑποσαρκίδιος, καὶ γυναῖκα ὅταν
ἔμβρυον ἔχουσαν περιπλευμονίη ἢ καῦσος λάβῃ ἢ
πλευρῖτις ἢ φρενῖτις, ἢ ἐρυσίπελας ἐν τῆσιν
ὑστέρῃσι γένηται.

Ἐνδοιαστὰ δὲ τὰ τοιάδε ἀπολλύναι τε καὶ μή·
περιπλευμονίη, καῦσος, πλευρῖτις, φρενῖτις, κυ-

---

[1] M adds καὶ τοῦ ξηροῦ ὑπερξηραίνοντος.   [2] ὅ τι ἂν Θ: ὁκόταν M.
[3] ἢ κοιλίην om. M.

heat that makes it too hot, and cold that makes it too cold.

Bile and phlegm come into being together with man's coming into being, and are always present in the body in greater or lesser amounts. They produce diseases, however, partly because of the effects of foods and drinks, and partly as the result of heat that makes them too hot, or cold that makes them too cold.

3. It is inevitable, in the following conditions, for that to occur which does occur: for patients injured by wounds to the thick cords and the insertions of the muscles, especially the ones in the thighs, to become lame; if a person is wounded in the brain, spinal marrow, cavity, liver, diaphragm, bladder, blood vessel, or the heart, for him to die, but if he is wounded in areas in which these organs are not present or that are farthest from them, not to die.

The following diseases are such that, when they occur, the patient inevitably perishes from them: consumption, dropsy beneath the tissue, and when pneumonia, ardent fever, pleurisy or phrenitis befalls a pregnant woman, or if erysipelas arises in the uterus.

The following diseases are uncertain with regard to mortality: pneumonia, ardent fever, pleurisy,

ΠΕΡΙ ΝΟΥΣΩΝ Α

νάγχη, σταφυλή, σπληνῖτις,[1] νεφρῖτις, ἡπατῖτις,
δυσεντερίη, γυναικὶ ῥόος αἱματώδης.

Τὰ δὲ τοιάδε οὐ θανάσιμα, ἢν μή τι αὐτοῖς
προσγένηται· κέδματα, μελαγχολίη, ποδάγρη,
ἰσχίας, τεινεσμός, τεταρταῖος, τριταῖος, στραγ-
γουρίη, ὀφθαλμίη, ἀρθρῖτις, λέπρη, λειχήν.[2]

Ἔμπηροι δὲ ἀπὸ τῶνδε γίνονται· ἀπόπληκτοι
μὲν καὶ χεῖρας καὶ πόδας καὶ φωνῆς ἀκρατέες
καὶ παραπλῆγες ὑπὸ μελαίνης χολῆς, χωλοὶ δὲ
ὑπὸ ἰσχιάδων, ὄμματα δὲ καὶ ἀκοὴν <κατά-
πηροι>[3] ὑπὸ φλέγματος καταστηρίξαντος.

Μακρὰ δὲ τάδε ἀνάγκη εἶναι· φθόην, δυσεντερίην, ποδάγρην, κέδματα, φλέγμα λευκόν, ἰσχιάδα,
στραγγουρίην, γεραιτέροισι δὲ νεφρῖτιν, γυναικὶ[4]
δὲ ῥόον αἱματώδη, αἱμορροΐδας, σύριγγας. καῦσος
δέ, φρενῖτις, περιπλευμονίη, κυνάγχη, σταφυλή,
πλευρῖτις ταχέως κρίνει.

146 Μεταπίπτει δὲ τάδε· ἐκ πλευρί|τιδος ἐς καῦ-
σον, καὶ ἐκ φρενίτιδος[5] ἐς περιπλευμονίην· ἐκ δὲ
περιπλευμονίης καῦσος οὐκ ἂν γένοιτο· τεινεσμὸς
ἐς δυσεντερίην, ἐκ δὲ δυσεντερίης λειεντερίη,[6] ἐκ
δὲ λειεντερίης ἐς ὕδρωπα, καὶ ἐκ λευκοῦ φλέγματος ἐς ὕδρωπα, καὶ σπληνὸς οἴδημα ἐς ὕδρωπα·
ἐκ περιπλευμονίης καὶ πλευρίτιδος ἐς ἔμπυον.

[1] σπληνῖτις om. Θ.   [2] λειχήν om. Θ.   [3] Wittern
[4] Θ: -ξὶ Μ.   [5] Μ: νεφρίτιδος Θ.   [6] τεινεσμὸς ... λειεντερίη Μ
τεινεσμὸς ἐκ λειεντερίης Θ.

104

phrenitis, angina, staphylitis, splenitis, nephritis, hepatitis, dysentery, a haemorrhage in a woman.

The following diseases are not fatal unless complications develop: swellings at the joints (*kedmata*), melancholy, gout, sciatica, tenesmus, quartan fever, tertian fever, strangury, ophthalmia, arthritis, lepra, lichen.

Patients become permanently disabled because of the following: they have strokes that affect the movement of their arms and legs, they lose command over their voice, and they become paralysed as the result of dark bile; they become lame from sciaticas; and they lose their sight and hearing from phlegm being deposited.

The following diseases are inevitably long: consumption, dysentery, gout, swellings at the joints (*kedmata*), white phlegm, sciatica, strangury, nephritis in older patients, a haemorrhage in a woman, haemorrhoids, and fistulas. But ardent fever, phrenitis, pneumonia, angina, staphylitis, and pleurisy reach their crises quickly.

These changes occur: from pleurisy to ardent fever and from phrenitis to pneumonia; but ardent fever does not arise from pneumonia; from tenesmus to dysentery, from dysentery to lientery, from lientery to dropsy, from white phlegm to dropsy, and from swelling of the spleen to dropsy; from pneumonia and pleurisy to internal suppuration.

4. Τάδε ἐπὶ κακοῖσιν ἀνάγκη κακὰ γίνεσθαι·
ῥῖγος ἢν λάβῃ, πῦρ ἐπιλαβεῖν·[1] καὶ νεῦρον ἢν
διακοπῇ μὴ ξυμφῦναι[2] ἐπιφλεγμῆναί τε ἰσχυρῶς·
καὶ ἢν ὁ ἐγκέφαλος σεισθῇ τε καὶ πονήσῃ πλη-
γέντος, ἄφωνον παραχρῆμα ἀνάγκη γενέσθαι,
καὶ μήτε ὁρῆν μήτε ἀκούειν· ἢν δὲ τρωθῇ, πυρετόν τ᾽ ἐπιγενέσθαι καὶ χολῆς ἔμετον, καὶ ἀπό-
πληκτόν τι τοῦ σώματος γενέσθαι, καὶ ἀπολέ-
σθαι· ἐπιπλοῖον ἢν ἐκπέσῃ, ἀνάγκη τοῦτο ἀπο-
σαπῆναι· κἢν αἷμα ἐκ τρώματος ἢ φλεβὸς ῥυῇ ἐς
τὴν ἄνω κοιλίην, ἀνάγκη τοῦτο πύον γενέσθαι.

5. Καιροὶ δέ, τὸ μὲν καθάπαξ εἰπεῖν, πολλοί
τ᾽ εἰσὶν ἐν τῇ τέχνῃ καὶ παντοῖοι, ὥσπερ καὶ τὰ
νοσήματα καὶ τὰ παθήματα καὶ τούτων αἱ θερα-
πεῖαι.

Εἰσὶ δὲ ὀξύτατοι μέν, ὅσοισιν ἢ ἐκψύχουσι δεῖ
τι ὠφελῆσαι ἢ οὐρῆσαι ἢ ἀποπατῆσαι μὴ δυναμέ-
νοισιν ἢ πνιγομένοισιν ἢ γυναῖκα τίκτουσαν ἢ
τρωσκομένην ἀπαλλάξαι ἢ ὅσα τοιαῦτά ἐστιν.
καὶ οὗτοι μὲν οἱ καιροὶ ὀξέες, καὶ οὐκ ἀρκέει[3] ὀλίγῳ ὕστερον· ἀπόλλυνται γὰρ οἱ πολλοὶ ὀλίγῳ
148   ὕστερον. ὁ μέντοι | καιρός ἐστιν, ἐπὴν πάθῃ τι
τούτων ὤνθρωπος· ὅ τι ἄν τις πρὸ τοῦ τὴν ψυχὴν
μεθεῖναι ὠφελήσῃ, τοῦθ᾽ ἅπαν ἐν καιρῷ ὠφέλησεν. ἔστι μὲν οὖν σχεδόν τι οὗτος ὁ καιρὸς καὶ ἐν

[1] Θ: -λαμβάνει Μ.   [2] μὴ ξυμφῦναι Θ: σπασμόν· καὶ μήτε συμφῦναι
διακοπὲν Μ.   [3] Θ: ἀρκέσει Μ.

4. The following evils follow inevitably upon one another: if there is a chill, fever follows. If a cord is severed, it does not reunite, and a violent swelling supervenes. If the brain is shaken and suffers damage as the result of a blow, the patient immediately loses his speech, sight, and hearing; if the brain is wounded, fever and the vomiting of bile ensue, the patient becomes paralysed in some part of his body, and he dies. If a fold of peritoneum becomes exposed, it must putrefy. If blood flows from a wound or a vessel into the upper cavity, it must turn to pus.

5. Opportune moments in medicine, generally speaking, are many and varied, just as are the diseases and affections and their treatments.

The most acute ones are when you must help patients that are losing consciousness, that are unable to pass urine or stools, that are choking, or when you must deliver a woman that is giving birth or aborting, or in other cases like these. These opportune moments are acute, and a little later does not suffice, for a little later most patients die. The opportune moment is when a person is suffering one of the above: whatever aid anyone gives before the patient's spirit departs he gives at the opportune moment. Generally speaking, such opportune moments exist in other diseases as well,

τοῖσιν ἄλλοισι νοσήμασιν· αἰεὶ γάρ, ὅταν τις ὠφελήσῃ, ἐν καιρῷ ὠφέλησεν.

Ὅσα δὲ τῶν νοσημάτων ἢ τρωμάτων μὴ ἐς
θάνατον φέρει, ἀλλὰ καίριά ἐστιν, ὀδύναι δ' ἐγγί-
νονται ἐν αὐτοῖσι, καὶ οἷά τέ ἐστιν, ἤν τις ὀρθῶς
θεραπεύῃ, παύεσθαι, τούτοισι δὲ οὐκ ἀρκέουσι γι-
νόμεναι αἱ ὠφέλειαι ἀπὸ τοῦ ἰητροῦ, ὅταν γένων-
ται· καὶ γὰρ καὶ μὴ παρεόντος τοῦ ἰητροῦ
ἐπαύσαντο ἄν.

Ἕτερα δ' ἔστι νοσήματα, οἷσι καιρός ἐστι θεραπεύεσθαι τὸ πρωῒ τῆς ἡμέρης, διαφέρει δ' οὐδὲν
ἢ πάνυ πρωῒ ἢ ὀλίγῳ ὕστερον· ἕτερα δὲ νοσή-
ματά ἐστιν, οἷσι καιρὸς θεραπευθῆναι ἅπαξ τῆς
ἡμέρης, ὁπηνίκα δ' οὐδὲν διαφέρει· ἕτερα δὲ διὰ
τρίτης ἡμέρης ἢ τετάρτης· καὶ ἕτερά γε ἅπαξ
τοῦ μηνός· καὶ ἕτερα διὰ τριῶν μηνῶν, καὶ τοῦ
τρίτου ἢ ἱσταμένου ἢ φθίνοντος, οὐδὲν διαφέρει.
τοιοῦτοι οἱ καιροί εἰσιν ἐνίοισι, καὶ ἀκριβείην οὐκ
ἔχουσιν ἄλλην ἢ ταύτην.

Ἀκαιρίη δ' ἐστὶ τὰ τοιάδε· ὅσα μὲν πρωῒ
δεῖ θεραπεύεσθαι ἢν μεσαμβρίῃ θεραπεύηται,
ἀκαίρως θεραπεύεται·[1] ἀκαίρως δὲ ταύτῃ, ἐπεὶ
ῥοπὴν[2] ἴσχει εἰς τὸ κάκιον διὰ τὴν οὐκ ἐν καιρῷ
θεραπείην· ὅσα δὲ τάχα, ἤν τε μεσαμβρίης ἢν
τ' ὀψὲ ἤν τε τῆς νυκτὸς | θεραπεύηται, ἀκαίρως
θεραπεύεται· καὶ ἢν τοῦ ἦρος δέῃ θεραπεύεσθαι,

150

[1] ἀκαίρως θεραπεύεται om. M.   [2] Cornarius· ῥώμην ΘΜ.

for whenever a person provides help, he is helping at an opportune moment.

There are also non-mortal diseases and wounds that have opportune moments; these are diseases that involve suffering and that, if treated properly, can be made to go away; however, in this case the help the physician gives is not truly saving these patients, since the diseases would also have gone away even if no physician had been in attendance.

There are other diseases which have their opportune time for treatment early in the day, it making no difference whether very early or a little later. Other diseases have their opportune time for treatment once a day, although the particular time is unimportant, others every second or every third day, others once a month, and still others once every three months, it not mattering whether at the beginning or at the end of the third month. These are the opportune times of some diseases, and opportunity has no other kind of precision than this.

Inopportunity is as follows: if diseases that should be treated early in the day are handled at midday, they are treated inopportunely; inopportunely, since they have a turn for the worse because their treatment was not opportune. Those that should be treated immediately are treated inopportunely, if they are treated at noon, in the evening, or at night, those that should be treated in

θεραπεύηται δὲ χειμῶνος, ἢ τοῦ μὲν χειμῶνος
δέῃ, τοῦ δὲ θέρεος θεραπεύηται· ἢ ὅ τι ἤδη δεῖ
θεραπεύεσθαι, τοῦτο ἀναβάλληται, ἢ ὅ τι ἀνα-
βάλλεσθαι δεῖ, τοῦτ᾽ ἤδη θεραπεύηται, τὰ τοιαῦτα ἀκαίρως θεραπεύεται.

6. Ὀρθῶς δ᾽ ἐν αὐτῇ καὶ οὐκ ὀρθῶς τὰ τοιάδε·
οὐκ ὀρθῶς μέν, τήν τε νοῦσον ἑτέρην ἐοῦσαν ἑτέρην φάναι εἶναι, καὶ μεγάλην οὖσαν σμικρὴν φάναι εἶναι, καὶ σμικρὴν ἐοῦσαν μεγάλην, καὶ περι-
εσόμενον μὴ φάναι περιέσεσθαι, καὶ μέλλοντα
ἀπολεῖσθαι μὴ φάναι ἀπολεῖσθαι, καὶ ἔμπυον ἐόντα μὴ γινώσκειν, μηδὲ νούσου μεγάλης τρεφομέ-
νης ἐν τῷ σώματι γινώσκειν, καὶ φαρμάκου δεό-
μενον, ὁποίου του δεῖ, μὴ γινώσκειν, καὶ τὰ δυνατὰ μὴ ἐξιᾶσθαι, καὶ τὰ ἀδύνατα φάναι ἐξιήσε-
σθαι.

Ταῦτα μὲν οὖν ἐστι κατὰ γνώμην οὐκ ὀρθῶς,
κατὰ δὲ χειρουργίην τάδε· πύον ἐν ἕλκει ἐνεὸν ἢ
ἐν φύματι μὴ γινώσκειν, καὶ τὰ κατήγματα καὶ
τὰ ἐκπεπτωκότα[1] μὴ γινώσκειν, καὶ μηλῶντα
κατὰ κεφαλὴν μὴ γινώσκειν, εἰ τὸ ὀστέον κατέη-
γε, μηδ᾽ ἐς κύστιν αὐλίσκον καθιέντα δύνασθαι
καθιέναι, μηδὲ λίθου ἐνεόντος ἐν κύστι γινώσκειν,
μηδὲ πύον[2] διασείοντα γινώσκειν, καὶ τάμνοντα

---

[1] Θ: ἐκπτώματα Μ.   [2] Θ: μηδ᾽ ἔμπυον ἐόντα Μ.

the spring, if they are treated in winter, those that should be treated in winter, if they are treated in summer; if what should be treated at once is put off, or if what should be put off is treated at once: things of this sort constitute treating inopportunely.

6. Correctness and incorrectness in medicine are as follows: it is incorrect to say that a disease is different from what it really is, to say that a major disease is minor, or to say that a minor disease is major; not to tell a patient that is going to survive that he will survive, not to tell a patient about to die that he will die; not to recognize a patient that has internal suppuration; not to recognize a serious disease developing in the body; not to recognize which medication is required by a patient that needs one; not to cure what can be cured; to say that what cannot be cured will be cured.

These, then, are incorrect with regard to understanding, whereas surgically incorrect are the following: not to recognize that there is pus in an ulcer or tubercle; not to recognize fractures or dislocations; not to recognize when probing the skull whether the bone is fractured; not to be able to succeed in inserting a tube into the bladder; not to recognize that there is a stone in the bladder; not to recognize pus by succussion; when incising or

ἢ καίοντα ἐλλείπειν ἢ τοῦ βάθεος ἢ τοῦ μήκεος,
ἢ καίειν τε καὶ τάμνειν ἃ οὐ χρή.

Καὶ ταῦτα μὲν οὐκ ὀρθῶς· ὀρθῶς δέ, τά τε
νοσήματα γινώσκειν ἅ τέ ἐστι καὶ ἀφ' ὧν ἐστιν,
καὶ τὰ μακρὰ αὐτῶν καὶ τὰ βραχέα, καὶ τὰ θανάσιμα καὶ τὰ μὴ θανάσιμα, καὶ τὰ μεταπίπτοντα
καὶ τὰ αὐξανόμενα καὶ τὰ μαραινόμενα, καὶ τὰ
μεγάλα καὶ τὰ σμικρά, καὶ θεραπεύοντα τὰ μὲν
152 ἀνυστὰ ἐκθερα|πεύειν, τὰ δὲ μὴ ἀνυστὰ εἰδέναι
διότι οὐκ ἀνυστά, καὶ θεραπεύοντα τοὺς τὰ τοιαῦτα ἔχοντας ὠφελέειν ἀπὸ τῆς θεραπείης ἐς τὸ
ἀνυστόν.

Τὰ δὲ προσφερόμενα τοῖσι νοσέουσιν ὧδε χρὴ
φυλάσσειν τά τε ὀρθῶς καὶ τὰ μὴ ὀρθῶς· ἤν τις ἃ
δεῖ ξηραίνειν ὑγραίνῃ, ἢ ἃ δεῖ ὑγραίνειν ξηραίνῃ,[1]
ἢ[2] παχῦναι δέον[3] μὴ προσφέρῃ ἀφ' ὧν δεῖ παχύνειν, ἢ ἃ δεῖ λεπτύνειν μὴ λεπτύνῃ, ἢ ψύχων μὴ
ψύχῃ, ἢ θερμαίνων μὴ θερμαίνῃ, ἢ σήπων μὴ σή-
πῃ, καὶ τὰ λοιπὰ κατὰ τὸν αὐτὸν λόγον τούτοις.

7. Τὰ δὲ τοιάδε ἀνθρώποισιν ἀπὸ τοῦ αὐτομάτου ἐν τῇσι νούσοισι γίνεται καὶ ἀγαθὰ καὶ κακά.

Πυρέσσοντι μὲν καὶ χολῶντι σκεδασθεῖσα ἔξω
ἡ χολή, ἀγαθόν, ὑπὸ τὸ δέρμα κεχυμένη καὶ ἐσκεδασμένη καὶ εὐπετεστέρη ἔχειν τε τῷ ἔχοντι καὶ
τῷ ἰωμένῳ ἰᾶσθαι· κεχυμένη δὲ καὶ ἐσκεδασμένη
πρὸς ἕν τι τοῦ σώματος προσπεσοῦσα, κακόν.

[1] ἢ ... ξηραίνῃ om. M.   [2] Θ adds ἃ.   [3] Potter: δέῃ ΘΜ.
112

cauterizing to lack depth or width, or to incise or cauterize where you should not.

These are all incorrect. Correct is to recognize what diseases are and whence they come; which are long and which are short; which are mortal and which are not; which are in the process of changing into others; which are increasing and which are diminishing; which are major and which are minor; to treat the diseases that can be treated, but to recognize the ones that cannot be, and to know why they cannot be; by treating patients with the former, to give them the benefit of treatment as far as it is possible.

In the administration of treatment to patients, observe what is correct or incorrect as follows: if a person moistens what he should dry, or dries what he should moisten; if he should fatten, but does not administer the treatment by which he should fatten; if he does not attenuate what he should attenuate, or cooling does not cool, or heating does not heat, or promoting maturation of pus does not do so, and so on in like manner.[1]

7. The following things, both good and bad, happen to patients spontaneously during their illnesses.

In a patient suffering from fever or from bile, it is good if the bile is dispersed externally, for when it is exuded and dispersed beneath the skin, this is less troublesome both for the patient to bear and for the physician to heal. But if, after it has been exuded and dispersed, it falls upon any particular part of the body, this is bad.

[1] Understand: "these are all incorrect."

Κοιλίη ταραχθεῖσα ὑπὸ πλευρίτιδος ἐχομένῳ ἢ περιπλευμονίης ἢ ἐμπύῳ ἐόντι, κακόν· πυρέσσοντι δὲ ἢ τρῶμα τετρωμένῳ ἀποξηρανθεῖσα, κακόν· ὑφύδρῳ καὶ σπληνώδει καὶ ὑπὸ λευκοῦ φλέγματος ἐχομένῳ ταραχθεῖσα ἡ κοιλίη ἰσχυρῶς, ἀγαθόν.

Ἐρυσίπελας ἢν ἔξω κατακεχυμένον ἔσω τράπηται, κακόν· ἔσω δὲ κατακεχυμένον ἔξω τραπῆναι, ἀγαθόν.

Διαρροίη δ᾽ ἐχομένῳ ἰσχυρῇ ἄνω ἔμετος γενόμενος, ἀγαθόν.

Γυναικὶ αἷμα ἐμεούσῃ τὰ καταμήνια ῥαγῆσαι, ἀγαθόν· ὑπὸ ῥόου δὲ πιεζομένη ἐς τὰς ῥῖνας ἢ εἰς τὸ στόμα μεταπεσεῖν τὸν ῥόον, ἀγαθόν· γυναικὶ ὑπὸ σπασμοῦ ἐχομένη ἐκ τόκου πυρετὸν ἐπιγενέ-
154 σθαι, ἀγαθόν· καὶ τετάνου | ἔχοντος καὶ σπασμοῦ πῦρ ἐπιγενέσθαι, ἀγαθόν.

Τὰ τοιαῦτα δι᾽ οὐδεμίαν οὔτε ἀμαθίην οὔτε σοφίην ἱ ων γίνεταί τε καὶ οὐ γίνεται, ἀλλ᾽ ἀπὸ τοῦ αὐτομάτου καὶ ἀπὸ ἐπιτυχίης, καὶ γενόμενά τε ὠφελέει ἢ βλάπτει καὶ οὐ γενόμενα ὠφελέει ἢ βλάπτει[1] κατὰ τὸν αὐτὸν λόγον.

8. Ἐπιτυχίῃ δὲ τὰ τοιάδε οἱ ἰητροὶ ποιέουσιν ἐν τῇ θεραπείῃ ἀγαθά· ἄνω φάρμακον δόντες καθαίρουσι καὶ ἄνω καὶ κάτω καλῶς· καὶ γυναικὶ φάρμακον δόντες κάτω χολῆς ἢ φλέγματος, ἐπι-

---

[1] καὶ οὐ γενόμενα . . . βλάπτει om. M.

The cavity being set in motion, in a patient suffering from pleurisy, pneumonia, or internal suppuration, is bad; in a patient with a fever or a wound, if the cavity is dried out, that is bad. But in patients with dropsy, a disease of the spleen, or white phlegm, the cavity being set in violent motion is good.

If erysipelas turns inward after having been dispersed externally, this is bad; if it turns outward after being spread internally, good.

Vomiting in a patient suffering from severe diarrhoea is good.

When the menses break forth in a woman that is vomiting blood, this is good; if, in a woman suffering from haemorrhage, the haemorrhage is transferred to her nose or mouth, also good. It is good when fever occurs in a woman that has had a convulsion after giving birth, and when fever follows upon tetanus or a convulsion.

Such things occur or do not occur, not through any ignorance or knowledge of physicians, but spontaneously and by chance; and, when they do occur, it may help or harm; likewise, when they do not occur, it may help or harm.

8. Physicians achieve the following good results in their therapy by luck: by giving a medication to clean upwards, they clean both upwards and downwards to good effect. By giving to a woman a medication meant to clean downwards of bile or phlegm,

μήνια οὐ γινόμενα κατέρρηξαν· καὶ σπλῆνα ἔμπυον ἔχοντι κάτω φάρμακον δόντες ὥστε χολὴν
καὶ φλέγμα καθῆραι, πύον κάτω ἐκάθηραν ἐκ τοῦ
σπληνὸς καὶ ἀπήλλαξαν τῆς νούσου· καὶ λιθιῶντι
φάρμακον δόντες, τὴν[1] λίθον ἐς τὸν οὐρητῆρα
προέωσαν ὑπὸ βίης τοῦ φαρμάκου, ὥστε ἐξουρη-
θῆναι· καὶ πύον ἔχοντι ἐν τῇ ἄνω κοιλίῃ ἐν φύματι, οὐκ εἰδότες ὅτι ἔχει, δόντες ἄνω φάρμακον
ὅ τι φλέγμα καθαίρει, ἐξ οὖν ἤμεσε τὸ πύον καὶ
ἐγένετο ὑγιής· καὶ ἐκ φαρμάκου ὑπερκαθαιρό-
μενον ἄνω θεραπεύοντες, καταρραγείσης τῆς
κοιλίης ἀπὸ τοῦ αὐτομάτου, ὑγιέα ἐποίησαν τοῦ
ἐμέτου.

Κακὰ δὲ τάδε ἀπεργάζονται ἀπὸ ἀτυχίης·
φάρμακον δόντες ἄνω χολῆς ἢ φλέγματος, φλέβα
ἐν τοῖσι στήθεσιν ἔρρηξαν ὑπὸ τοῦ ἐμέτου, οὐδὲν
ἔχοντος πρόσθεν ἄλγημα ἐν τῷ στήθει φανερόν,
καὶ ἐγένετο νοῦσος· καὶ γυναικὶ ἐν γαστρὶ ἐχούσῃ
ἄνω φάρμακον δόντες, κάτω ῥαγεῖσα ἡ κοιλίη ἐξ-
έτρωσε τὸ[2] ἔμβρυον· καὶ ἔμπυον θεραπεύοντι κοιλίη ῥυεῖσα διαφθείρει· καὶ ὀφθαλμοὺς θεραπεύοντι
καὶ ὑπαλείψαντι ὀδύναι ἐνέπεσον ὀξύτεραι, καὶ ἢν
οὕτω τύχῃ, ῥήγνυταί τε ὁ ὀφθαλμὸς καὶ ἀμαυροῦ-
ται,[3] καὶ αἰτιῶνται τὸν ἰητρόν, ὅτι ὑπήλειψεν·

---

[1] Θ: τὸν Μ.   [2] Μ: ἐξέτρωται Θ.   [3] Θ: ῥήγνυνται οἱ ὀφθαλμοὶ
καὶ ἀμαυροῦνται Μ.

they have caused the absent menses to break forth. By giving a medication intended to clean bile and phlegm downwards to a patient with a suppurating spleen, they have cleaned pus out of the spleen in the same direction, and cured the disease. By giving a medication to a patient with a stone, they have propelled the stone into the urethra by the force of the medication, so that it has passed with the urine. After their having given a medication to clean phlegm upwards to a patient that, unknown to them, had pus in a tubercle in his upper cavity, the patient has vomited up the pus, and recovered. In attending a patient that had been cleaned upwards to excess by the use of medications, they have cured him of his vomiting when a spontaneous evacuation of the cavity took place.

They bring about the following bad results through misfortune: by giving a medication meant to clean upwards of bile or phlegm to a patient that previously had no obvious pain in his chest, they have caused a vessel in his chest to rupture from the vomiting, and a new disease to arise. When they had given a medication of the kind that acts upward to a pregnant woman, the lower cavity, being evacuated, has made the fetus miscarry. If, when a person is attending a patient with internal suppuration, the cavity has a flux, it is fatal. It has happened to a physician treating the eyes by anointing them that the pains have become sharper; if this happens, the eye can rupture and become blind, and people hold the physician to blame, because he was anointing. If a physician

117

ΠΕΡΙ ΝΟΥΣΩΝ Α

156 καὶ λεχοῖ γαστρὸς ὀδύνης ἦν | δῷ τι ὁ ἰητρὸς καὶ κακῶς ἔχῃ ἢ καὶ ἀπόληται, ὁ ἰητρὸς αἴτιος.

Σχεδὸν δέ, ὅσα ἀνάγκας ἔχει ὥστε γίνεσθαι ἐν τοῖσι νοσήμασι καὶ τρώμασι κακὰ ἐπὶ κακοῖσι, τὸν ἰητρὸν αἰτιῶνται τούτων γινομένων, καὶ τὴν ἀνάγκην τὴν τὰ τοιαῦτα ἀναγκάζουσαν γίνεσθαι οὐ γινώσκουσιν. καὶ ἢν ἐπὶ πυρέσσοντα ἢ τρῶμα ἔχοντα ἐσελθὼν καὶ προσενέγκας τὸ πρῶτον μὴ ὠφελήσῃ, ἀλλὰ τῇ ὑστεραίῃ κάκιον ἔχῃ, τὸν ἰητρὸν αἰτιῶνται· ἢν δ' ὠφελήσῃ, τοῦτο δὲ οὐχ ὁμαλῶς ἐπαινέουσιν· χρεὼν γὰρ πεπονθέναι αὐτὸν δοκέουσιν. τὰ δ' ἕλκεα φλεγμαίνειν καὶ ἐν τῇσι νούσοισιν ἔστιν ἧσιν ὀδύνας γίνεσθαι, ταῦτα δὲ οὐ δοκέουσι χρεὼν εἶναι γίνεσθαι[1] αὐτοῖσιν, οὐδὲ τὰ τοιάδε ὥστε γίνεσθαι· νεῦρον ἢν διακοπῇ μὴ ξυμφῦναι μηδὲ κύστιν μηδ' ἔντερον, ἢν ᾖ τῶν λεπτῶν, μηδὲ φλέβα αἱμόρροον μηδὲ γνάθου τὸ λεπτὸν μηδὲ τὸ ἐπὶ τοῦ αἰδοίου δέρμα.

9. Ἀρχὴ δὲ ἰήσιος ἀποδεδειγμένη μὲν οὐκ ἔστιν, ἥτις ὀρθῶς ἀρχή ἐστι πάσης τῆς τέχνης, οὐδὲ δεύτερον οὐδὲν οὐδὲ μέσον οὐδὲ τελευτή· ἀλλὰ ἀρχόμεθά τε αὐτῆς ἄλλοτε λέγοντες ἄλλοτε ἐργαζόμενοι, καὶ τελευτῶμεν ὡσαύτως· καὶ οὔτε λέγοντες ἀρχόμεθα ἐκ τῶν αὐτῶν λόγων, οὐδ' ἢν περὶ τῶν αὐτῶν λέγωμεν, οὐδ' ἐς τοὺς αὐτοὺς τελευτῶμεν· καὶ ἐργαζόμενοι κατὰ τὸν αὐτὸν

---

[1] ταῦτα . . . γίνεσθαι om. M.

118

gives anything to a woman in childbed for the pain in her belly, and she becomes worse or even dies, the physician is blamed.

Generally speaking, people blame the physician, in diseases and wounds, even for the evils that follow of necessity from other evils, when these occur, not recognizing the constraint that makes such things happen. If he attends a patient with a fever or a wound, and fails at first to help him by his administration, but on the next day the patient is worse, people blame the physician; but if he does help the patient, people do not praise him in due proportion, for they hold the patient's improvement to have been a matter of course. That ulcers become swollen, and that in certain diseases pains occur, such things patients refuse to accept as necessary events in their own cases, nor that such things as the following occur: if a cord is severed, it does not reunite, nor does the bladder, the intestine—if it is part of the thin one—a blood vessel, the narrow part of the jaw, or the skin of the genital organs.

9. There is no demonstrated starting point of healing, which truly is the starting point of the whole art, nor any second point, nor any middle, or end. Instead, we start out in medicine sometimes by speaking, at other times by acting, and we end in like manner; nor, when we begin by speaking, do we begin with the same words, not even if we are speaking about the same thing, nor do we end with the same words. In the same way, when we begin

λόγον οὔτε ἀρχόμεθα ἐκ τῶν αὐτῶν ἔργων οὔτε
τελευτῶμεν ἐς ταὐτά.

158   10. Εὐχειρίη δ' ἐστὶ τὰ τοιάδε· ὅταν τις τάμνῃ
ἢ καίῃ,[1] μήτε νεῦρον ταμεῖν[2] μήτε φλέβα· καὶ ἢν
ἔμπυον καίῃ, τυγχάνειν τοῦ πύου, καὶ τάμνοντα
κατὰ τὸν αὐτὸν λόγον· καὶ τὰ κατήγματα συν-
τιθέναι ὀρθῶς· καὶ ὅ τι ἂν τοῦ σώματος ἐκπέσῃ
ἐκ τῆς φύσιος, ὀρθῶς ἐς τὴν φύσιν τοῦτ' ἀπῶσαι·
ἐμβάλλειν[3] δὲ ἃ δεῖ ἰσχυρῶς, καὶ λαμβάνοντα
πιέζειν, ἃ καὶ ὅσα ἀτρέμα λαβεῖν τε δεῖ, καὶ
λαβόντα μὴ πιέζειν.[4] καὶ ἐπιδέοντα στρεβλὰ μὴ
ποιεῖν ἐξ εὐθέων, μηδὲ πιέζειν[5] ἃ μὴ δεῖ· καὶ
ψαύοντα, ὅτου ἂν ψαύῃ, μὴ ὀδύνην παρέχειν ἐκ
περισσοῦ.

Ταῦτα μέν ἐστιν εὐχειρίη· τὸ δὲ τοῖσι δακτύλοισιν εὐσχημόνως λαμβάνειν ἢ καλῶς ἢ μὴ
καλῶς ἢ μακροῖς ἢ βραχέσιν, ἢ καλῶς ἐπιδεῖν καὶ
ἐπιδέσιας παντοίας, οὐ πρὸς τῆς τέχνης κρίνεται
εὐχειρίης πέρι, ἀλλὰ χωρίς.

11.   Ὅσοι ἔμπυοι γίνονται τὸν πλεύμονα ἢ
τὴν ἄνω ἢ τὴν κάτω κοιλίην, ἢ φύματα ἴσχουσιν
εἴτ' ἐν τῇ ἄνω κοιλίῃ εἴτ' ἐν τῇ κάτω ἢ ἐν τῷ
πλεύμονι, ἢ ἕλκεα εἴσω,[6] ἢ αἷμα ἐμέουσιν ἢ
πτύουσιν, ἢ ἄλγημά τι ἔχουσιν εἴτ' ἐν τοῖσι

---

[1] Θ: -ων ἢ -ων Μ.   [2] Potter: τάμῃ Θ: τάμῃ ἢ καύσῃ Μ.   [3] Θ:
λαβεῖν Μ.   [4] ἃ καὶ ... πιέζειν om. Θ.   [5] Μ: πιεζέοντα Θ.
[6] Θ: ἔνδοθεν Μ.

by acting, we do not begin with the same actions, nor do we end with the same ones.

10. Dexterity is as follows: when a person is incising or cauterizing, that he does not cut a cord or vessel; if he is cauterizing a patient with internal suppuration, that he hits the pus, and when cutting, the same; to reduce fractures correctly; to return any part of the body that has fallen out of its normal position to that position correctly; what you must reduce forcefully, to take hold of and to press tight, what you must take hold of gently, to take hold of and not to press tight; when bandaging, not to make uneven twists or to apply pressure where you should not; when palpating, wherever you do, not to cause unnecessary pain.

These things are dexterity; but taking hold with the fingers gracefully poised, elegantly or inelegantly, with them outstretched or folded, or bandaging elegantly and all the possible sorts of bandages, these things are not judged in the eyes of the art as dexterity, but separately.

11. Patients that suppurate in the lung or in the upper or lower cavity, or have tubercles in the upper or lower cavity or the lung, or have ulcers internally, or vomit blood or expectorate it, or have

στήθεσιν εἶτ᾽ ἐν τοῖσιν ὄπισθεν [ἐν τῷ νώτῳ],[1]
πάντα ταῦτα ἴσχουσι, τῶν μὲν ἐν τῷ σώματι
ἐνεόντων, ἀπὸ χολῆς καὶ φλέγματος, τῶν δ᾽
ἔξωθεν, ἀπὸ τοῦ ἠέρος ἐπιμιγνυμένου τῷ
συμφύτῳ θερμῷ, ἀτὰρ καὶ ἀπὸ[2] πόνων καὶ
τρωμάτων.

12. Καὶ ὅσοι μὲν τὸν πλεύμονα ἔμπυοι γίνονται, ἀπὸ τῶνδε γίνονται· ἢν περιπλευμονίη ληφθεὶς μὴ καθαρθῇ ἐν τῇσι κυρίῃσιν ἡμέρῃσιν,
ἀλλ᾽ ὑπολειφθῇ ἐν τῷ πλεύμονι πύον τε καὶ
160   φλέγμα, | ἔμπυος γίνεται· καὶ ἢν μὲν αὐτίκα θε-
ραπευθῇ, διαφεύγει ὡς τὰ πολλά· ἢν δ᾽ ἀμεληθῇ,
διαφθείρεται, διαφθείρεται δὲ ὧδε· τοῦ φλέγματος ἐν τῷ πλεύμονι ἐνισταμένου τε καὶ σηπομέ-
νου ἑλκοῦταί τε ὁ πλεύμων καὶ διάπυος γίνεται,
καὶ οὔτ᾽ ἔτι ἔσω ἕλκει ἐς ἑωυτὸν ὅ τι καὶ ἄξιον
λόγου τῆς τροφῆς, οὔτ᾽ ἔτι ἀποκαθαίρεται ἀπ᾽
αὐτοῦ ἄνω οὐδέν. ἀλλὰ πνίγεταί τε καὶ δυσπνοεῖ
αἰεὶ ἐπὶ μᾶλλον, καὶ ῥέγκει ἀναπνέων, καὶ ἀναπνεῖ αὐτόθεν ἄνωθεν ἐκ τῶν στηθέων· τέλος
δὲ ἀποφράσσεται ὑπὸ τοῦ πτύσματος καὶ ἀποθνήσκει.

13. Γίνεται δ᾽ ἔμπυος καὶ ἢν ἀπὸ τῆς κεφαλῆς φλέγμα οἱ καταρρυῇ ἐς τὸν πλεύμονα. καὶ
τὸ μὲν πρῶτον ὡς τὰ πολλὰ λανθάνει καταρρέον,
καὶ βηχά τε παρέχει λεπτὴν καὶ τὸ σίελον πικρό-
τερον ὀλίγῳ τοῦ ἐωθότος καὶ ἄλλοτε θέρμην

[1] Del. Wittern.   [2] τοῦ ἠέρος . . . ἀπὸ om. Θ.

pains in the chest or the back suffer all these either from things inside the body, bile and phlegm, or from things outside it: from air being mixed with the natural heat, or also from exertions and wounds.

12. Patients that suppurate in the lung do so as a result of the following: if a person with pneumonia is not cleaned out on the critical days, but pus and phlegm are taken up into the lung, he suppurates internally. If he is treated at once, he usually escapes, but if he is neglected, he perishes in the following way: when the phlegm becomes fixed in his lung and putrefies, the lung ulcerates, becomes purulent, and no longer draws into itself any nourishment worth mentioning; nor can anything further be cleaned upwards from it, but the patient chokes, and has more and more difficulty breathing; his breathing is stertorous, and he exhales only from the upper part of his chest. In the end, he becomes completely blocked up by the sputum, and dies.

13. Internal suppuration also occurs if phlegm streams down from the patient's head into his lung. At first, though, this flux usually goes unnoticed, producing only a slight cough, sputum that is slightly more bitter than normal, and sometimes a

λεπτήν· ὅταν δὲ ὁ χρόνος προΐῃ, τρηχύνεταί τε
ὁ πλεύμων καὶ ἑλκοῦται ἔσωθεν ὑπὸ τοῦ φλέγματος ἐνισταμένου καὶ ἐνσηπομένου, καὶ βάρος τε
παρέχει ἐν τοῖσι στήθεσι καὶ ὀδύνην ὀξέην καὶ
ἔμπροσθεν καὶ ὄπισθεν, θέρμαι τε ὀξύτεραι ἐμπί-
πτουσιν ἐς τὸ σῶμα· καὶ ὁ πλεύμων ὑπὸ τῆς θερμασίης ἄγει ἐς ἑωυτὸν ἐκ παντὸς τοῦ σώματος
φλέγμα, καὶ μάλιστα ἐκ τῆς κεφαλῆς· ἡ δὲ κεφαλὴ θερμαινομένη ἐκ τοῦ σώματος· καὶ τοῦτο
σηπόμενον πτύει ὑπόπαχυ· ὅσῳ δ' ἂν ὁ χρόνος
προΐῃ, εἰλικρινὲς πύον πτύει, καὶ οἱ πυρετοὶ ὀξύτεροι γίνονται, καὶ ἡ βὴξ πυκνή τε καὶ ἰσχυρή,
καὶ ἡ ἀσιτίη διακναίει· καὶ[1] ἡ κοιλίη ἡ κάτω ταράσσεται, ταράσσεται δὲ ὑπὸ τοῦ φλέγματος· τὸ
δὲ φλέγμα ἐκ τῆς κεφαλῆς καταβαίνει. οὗτος,
ὅταν ἐς τοῦτο ἀφίκηται, ἀπόλλυται· ἀπόλλυται
δέ, καθάπερ ἐν τοῖσι πρόσθεν εἴρηται, διαπύου
τοῦ πλεύμονος[2] γενομένου ἢ τῆς γαστρὸς ῥυείσης
κάτω.

162   14. Γίνεται δὲ καὶ ἀπὸ τῶνδε ἔμπυος ὁ πλεύ-
μων· ὅταν τι τῶν ἐν αὐτῷ φλεβίων ῥαγῇ, ῥήγνυται δὲ ὑπὸ πόνων, καὶ ὅταν ῥαγῇ, αἱμορροεῖ τὸ
φλέβιον· καὶ ἢν μὲν παχύτερον ᾖ, μᾶλλον, ἢν δὲ
λεπτότερον, ἧσσον· καὶ τὸ μὲν παραυτίκα τοῦ
αἵματος πτύει· τὸ δέ, ἢν μὴ στεγνωθῇ ἡ φλέψ,
χεῖταί τε εἰς τὸν πλεύμονα καὶ σήπεται ἐν αὐτῷ,

---

[1] M adds τέλος.   [2] M adds καὶ σαπροῦ.

mild feverish heat. However, as time goes on, the lung becomes rough and ulcerates internally because of the phlegm standing and putrefying in it, and this produces heaviness in the chest and sharp pain both anteriorly and posteriorly; also, very high fevers attack the body. The lung, because of its heat, attracts phlegm from the whole body, most especially from the head; the head, in turn, is heated from the body. The patient expectorates this material mature and somewhat thickened. As more time passes, he expectorates pus proper, his fevers become higher, his cough is frequent and violent, and fasting wears him down; in the end, his lower cavity is set in motion by the phlegm that descends out of the head. This patient, when he has reached such a state, succumbs, and for the reasons indicated above: either because his lung becomes purulent, or because of a downward flux from his belly.

14. The lung also suppurates when one of the small vessels in it ruptures. Such a vessel ruptures because of exertions, and when it does, if it happens to be a wider one, it bleeds more, if a narrower one, less. Some of the blood the patient expectorates immediately, but, unless the vessel closes, other blood is poured into the lung and putrefies there;

καὶ ὅταν σαπῇ, πύον πτύει,[1] προϊόντος δὲ τοῦ χρόνου ἄλλοτε πύον εἰλικρινές, ἄλλοτε πύον ὕφαιμον, ἄλλοτε αἷμα. καὶ ἢν μᾶλλον πληρωθῇ τὸ φλέβιον, ἀπεμεῖ τὸ πλήρωμα ἀφ' ἑωυτοῦ ἁλὲς τοῦ αἵματος, τὸ δὲ πύον πτύεται παχὺ ὑπὸ τοῦ προσγινομένου καὶ ἐνσηπομένου φλέγματος.

Οὗτος ἢν καταληφθῇ ἀρχομένου τοῦ νοσήματος πρὶν ἢ τὴν φλέβα αἱμορροεῖν ἢ χαλᾶν ἰσχυρῶς, καὶ πρὶν ἢ λεπτυνθῆναί τε καὶ κλινοπετῆ γενέσθαι καὶ τὴν κεφαλὴν ἄρξασθαι φθίνειν καὶ τὸ ἄλλο σῶμα τήκεσθαι, ἐξάντης γίνεται. ἢν δ' ἀμεληθῇ καὶ ταῦτα καταλάβῃ, ὥστε παθεῖν ἢ πάντα ἢ τὰ πλεῖστα, ἀπόλλυται· ἀπόλλυται δὲ οὗτος ἢ ὑπὸ τῶν αὐτῶν ἃ εἴρηκα ἐν τῇ πρόσθεν, ἢ ὑπὸ ἐμέτου αἵματος πολλοῦ καὶ πολλάκις ἐμευμένου.

Ἢν δὲ τὸ φλέβιον παντάπασι μὲν μὴ διαρραγῇ, σπαδὼν δ' ἐν αὐτῷ ἐγγένηται, γίνεται δὲ μάλιστα οἷον κιρσός, ὃ παραυτίκα μὲν ὅταν γένηται ὀδύνην τε παρέχει λεπτὴν καὶ βῆχα ξηρήν. ἢν δὲ χρονίσῃ τε καὶ ἀμεληθῇ, διαδιδοῖ αἷμα, τὸ μὲν πρῶτον ὀλίγον καὶ ὑπομέλαν, ἔπειτα δὲ ἐπὶ πλέον τε καὶ εἰλικρινέστερον, καὶ πάσχει τε |
164   ὅσαπερ ἐν τῇ πρόσθεν εἴρηται.

Ξυμφέρει δὲ τοῖσι τοιούτοισιν, ἢν κατ' ἀρχὰς λάβῃς ὥστε θεραπεύειν, φλέβες ἐξιέμεναι ἐκ τῶν

---

[1] Θ: ποιεῖ Μ.

when this has putrefied, the patient expectorates pus: with the passage of time, sometimes pus proper, sometimes pus charged with blood, and sometimes blood. If the vessel is greatly filled, the quantity of blood filling it is vomited up in a mass. The pus expectorated is thick because phlegm has been added to it and putrefied in it.

This patient, if caught at the beginning of the disease, before the vessel either bleeds or grows very slack, before he becomes lean and bed-ridden, and before his head begins to be consumed and the rest of his body to melt away, recovers; if, however, he is neglected, so that he suffers many or all of these things, he dies. Death results either from the things mentioned in the preceding disease, or due to the frequent vomiting of much blood.

If the small vessel is not completely ruptured, but a tear arises in it, this develops very much like a varix, which, immediately on its formation, produces a mild pain and a dry cough. If the tear persists for a time and is neglected, it exudes blood—at first little and darkish, but then more and of a purer kind—and the patient suffers the things mentioned in the preceding disease.

Of benefit to such patients, if you take them for treatment at the beginning, is to let blood from the

χειρῶν καὶ δίαιτα, ὑφ' ἧς ἔσται ὡς ξηρότατός τε
καὶ ἀναιμότατος.

Τὸν αὐτὸν δὲ τρόπον τοῦτον καὶ τὰ ἐν τῷ
πλευρῷ φλέβια πάσχει, ὅσα ἔσω ἀκρόπλοά ἐστιν·
ὅταν οὖν πονήσῃ, κιρσοειδέα τε γίνεται καὶ με-
τέωρα ἔνδον· καὶ ἢν μὲν ἀμεληθῇ, τάδε πάσχει·
ἐκρήγνυται, καὶ πτύουσί τε ἀπὸ σφῶν αἷμα, καὶ
ἐνίοτε καὶ ἐμέουσι, καὶ ἔμπυοι γίνονται καὶ ὡς τὰ
πολλὰ δι' οὖν ἐφθάρησαν· ἢν δὲ θεραπευθῶσιν ἀρχομένου τοῦ νοσήματος, πάλιν κατὰ χώρην προσπίπτει τε[1] πρὸς τὸ πλευρὸν τὰ φλέβια καὶ γίνεται ταπεινά.

Καὶ ὁ μὲν πλεύμων ἀπὸ τούτων ἔμπυος γίνεται, καὶ τὰ ἀπ' αὐτοῦ πάσχουσί τε τοιαῦτα καὶ
τελευτῶσιν οὕτως.

15. Τὴν δ' ἄνω κοιλίην ἔμπυοι γίνονται
πολλαχῶς· καὶ γὰρ ὅταν φλέγμα ρυῇ ἐκ τῆς
κεφαλῆς ἁλὲς ἐς τὴν ἄνω κοιλίην, σήπεταί τε
καὶ γίνεται πύον· σήπεται δ' ἐπὶ τῶν φρενῶν
κεχυμένον·[2] σήπεται δ' ἐν ἡμέρῃσι μάλιστα δυοῖν
καὶ εἴκοσι.[3] τοῦτ' οὖν διασείεται καὶ ἐγκλυδάζεται
τὸ πύον πρὸς τὰ πλευρὰ προσπῖπτον· οὗτος ἢν
καυθῇ ἢ τμηθῇ, πρὶν ἢ χρονίσαι τὸ πύον, ὑγιὴς
γίνεται ὡς τὰ πολλά.

Γίνονται δὲ τὴν ἄνω κοιλίην ἔμπυοι καὶ ἐκ

---

[1] προσπίπτει τε Θ: ἰζάνουσι Μ.   [2] σήπεται δ' ἐπὶ ... κεχυμένον
om. Μ.   [3] δυοῖν καὶ εἴκοσι Θ: μιῇ καὶ εἰκοστῇ Μ.

vessels of the arms, and to employ a regimen that will make them as dry and bloodless as possible.

The vessels in the side, too, suffer in the same way, inasmuch as they are on the surface[1]; for, whenever they are strained, they become varicose and prominent within, and if they are neglected, they suffer the following: they rupture; patients expectorate blood from such ruptured vessels, and sometimes even vomit it, they suppurate internally, and in many cases they have actually perished. If they are attended at the onset of the disease, though, the small vessels fall back into place against the side and become flat.

The lung, too, suppurates from these things, and in that case patients suffer the same kinds of things, and die in the same way.

15. Suppuration in the upper cavity arises in many ways. For example, when a large amount of phlegm flows down out of the head into the upper cavity, it putrefies and turns to pus as it collects on the diaphragm, and this process usually occurs in twenty-two days. Succussion is employed here, and the pus makes a splashing sound as it strikes the sides. If this patient is cauterized or incised before the pus has become old, he usually recovers.

Suppuration in the upper cavity also develops

[1] I.e. on the interior surface of the chest wall.

πλευρίτιδος, ὅταν ἰσχυρὴ γένηται καὶ ἐν τῇσι
κυρίῃσιν ἡμέρῃσι μήτε σαπῇ μήτε πτυσθῇ, ἀλλ᾽
ἑλκωθῇ τὸ πλευρὸν ὑπὸ τοῦ προσπεπτωκότος[1]
166 φλέγματός τε καὶ χολῆς. | καὶ ὅταν ἕλκος γένη-
ται, ἀναδίδοται[2] ἀπό τε αὐτοῦ[3] ἑωυτοῦ πύον, καὶ
ἐκ τῶν πλησίον χωρίων ὑπὸ θερμασίης ἄγει ἐφ᾽
ἑωυτὸ φλέγμα· καὶ τοῦτο ὅταν σαπῇ, πτύεται
πύον. ἐνίοτε δὲ καὶ ἐκ τῶν φλεβίων διαδιδοῖ ἐς
τὸ ἕλκος αἷμα, καὶ γίνεται σηπόμενον πύον· οὗτος ἢν μὲν παραχρῆμα ὑποληφθῇ, ὑγιὴς γίνεται
ὡς τὰ πολλά· ἢν δ᾽ ἀμεληθῇ, διαφθείρεται.

Γίνονται δ᾽ ἔμπυοι καὶ ἢν φλέγμα ἐκ τῆς
κεφαλῆς ῥυὲν πρὸς τὸ πλευρὸν προσπαγῇ καὶ
σαπῇ· τό τε[4] πλευρὸν ὡς τὰ πολλὰ καίεται,
καὶ πάσχει ὅσαπερ ἐκ πλευρίτιδος, ὅταν ἔμπυος
γένηται.

Γίνονται δὲ καὶ ὅταν ὑπὸ ταλαιπωρίης ἢ ἐκ
γυμνασίης ἢ ἄλλως πως ῥαγῇ ἢ ἔμπροσθεν ἢ
ὄπισθεν, ῥαγῇ δὲ ὥστε μὴ παραυτίκα πτύσαι
αἷμα, ἀλλ᾽ ἐν τῇ σαρκὶ σπαδὼν γένηται, καὶ ἡ
σὰρξ σπασθεῖσα εἰρύσῃ ἰκμάδα ὀλίγην καὶ γένηται ὑποπέλιος. καὶ παραυτίκα μὲν μὴ αἰσθάνη-
ται παθὼν ὑπὸ ῥώμης καὶ εὐεξίης, ἢν δὲ καὶ
αἴσθηται, μηδὲν πρᾶγμα ἡγήσηται· οὗτος ὅταν
καταλάβῃ ὥστ᾽ αὐτὸν ὑπὸ πυρετῶν ληφθέντα

[1] Θ: -πεπηγότος Μ.   [2] Θ: -διδοῖ Μ.   [3] F. Kudlien: τε αὐτὸ Θ:
om. Μ.   [4] τό τε Μ: πρὸς τὸ Θ.

out of pleurisy, when the pleurisy is severe, and on the critical days the pus fails to become mature and to be coughed up, but the side ulcerates because of the phlegm and bile that invade it. When such an ulcer arises, it gives off pus from itself and, because of its heat, attracts phlegm from the areas near it; when this putrefies, it is coughed up as pus; sometimes blood, too, is exuded from small vessels into the ulcer, and putrefies to become pus. If this patient is taken in hand immediately, he usually recovers; if neglected, he perishes.

Internal suppuration also occurs if phlegm that has flowed out of the head and towards the side becomes fixed, and putrefies. In most cases the side becomes warm, and the patient suffers the same things as when a person suppurates internally after pleurisy.

Internal suppuration also arises when, as the result of exertions, either in athletics or otherwise, a rupture occurs anteriorly or posteriorly, a rupture such that the patient does not expectorate blood at once, but a tear arises in his tissue. The tissue, being torn, attracts a small amount of moisture, and becomes somewhat livid. At first the patient has no sensation of illness, because of his strength and good condition, and even if he does sense something, he holds it to be unimportant. But when it so happens that this person becomes lean as the result of being seized by fevers, or from drinking, or

λεπτυνθῆναι ἢ ποσίων ἢ λαγνείης ἢ ἄλλου του, ἡ
σὰρξ ἡ τετρωμένη ὑποξηραίνεταί τε καὶ ὑποθερ-
μαίνεται, καὶ ἕλκει ἰκμάδα ἐς ἑωυτὴν ἀπὸ τῶν
πλησίον καὶ φλεβῶν καὶ σαρκῶν· ὅταν δ᾽ εἰρύσῃ,
οἰδίσκεταί τε καὶ φλεγμαίνει, καὶ ὀδύνην παρέχει
λεπτὴν καὶ βῆχα ἀραιήν τε καὶ ξηρὴν τὸ πρῶτον,
ἔπειτα ἐπὶ μᾶλλον ἕλκει τε ἐς ἑωυτήν, καὶ ὀδύνην παρέχει ἰσχυροτέρην[1] καὶ βῆχα πυκνοτέρην·
καὶ πτύει τὸ μὲν πρῶτον ὑπόπυον, ἐνίοτε δὲ καὶ
ὑποπέλιον καὶ ὕφαιμον· ὅσῳ δ᾽ ἂν ὁ χρόνος προΐῃ,
ἕλκει τε μᾶλλον ἐς ἑωυτὴν καὶ σήπει· καὶ αὐτῆς
τῆς σαρκός, ὅσον πελιδνὸν ἐγένετο τὴν ἀρχήν,
τοῦτο πᾶν ἕλκος γίνεται, καὶ ὀδύνην παρ-
168 έχει ἰσχυρὴν[2] καὶ πυρετὸν | καὶ βῆχα πολλήν τε
καὶ πυκνήν, καὶ τὸ πτύσμα εἰλικρινὲς πτύει
πύον. ἢν δὲ χρονίσῃ τὸ πύον ἐν τῇ κοιλίῃ, διαθερμαίνεται[3] αὐτοῦ τὸ σῶμα πᾶν, μάλιστα δὲ τὰ
ἐγγυτάτω· θερμαινομένου δὲ τοῦ σώματος ἐκτήκεται τὸ ὑγρόν, καὶ τὸ μὲν ἀπὸ τῶν ἄνω ἐς τὴν
ἄνω κοιλίην μάλιστα συρρεῖ καὶ γίνεται πύον
πρὸς τῷ ἐνεόντι, τὸ δὲ καὶ ἐς τὴν κάτω κοιλίην
ῥεῖ, καὶ ἐνίοτε ταράσσεται ἡ κοιλίη ὑπ᾽ αὐτοῦ, καὶ
δι᾽ οὖν ἔφθειρε τὸν ἄνθρωπον. τὰ γὰρ εἰσιόντα
τῶν σιτίων διαχωρέει ἄσηπτα,[4] καὶ τροφὴ ἀπ᾽
αὐτῶν οὐ γίνεται τῷ σώματι· καὶ ἡ τοῦ πτύσμα-

[1] M: ἰσχυρὴν Θ.   [2] Θ: ὀξείην M.   [3] M adds ὑπ᾽.
[4] Θ: ἄπεπτα M.

132

from venery or anything else, the injured tissue becomes slightly dry and warm, and attracts moisture from the surrounding vessels and tissues. As it does, it enlarges, swells, and at first produces light pain and a cough that is infrequent and dry; then, it attracts more moisture, and produces severer pain and more frequent coughing; at first the sputum is somewhat purulent, and sometimes livid and charged with blood. The more time passes, the more moisture the tear draws to itself and turns to pus. The part of the tissue itself that became livid at the start all ulcerates and gives rise to severe pain, fever, and violent frequent coughing; in this case the sputum is pus proper. If the pus remains in the cavity for long, the whole body becomes heated, especially the parts that are nearest to it. As the body is heated, its moist part is melted: part of this flows from the upper regions mainly into the upper cavity, and becomes pus additional to what was already there; the other part flows into the lower cavity, and the cavity is sometimes set in motion by it, and so kills the person; for the food taken in passes through undigested, and there is no nourishment from it for the body. Also, upward

τος ἄνω κάθαρσις οὐχ ὁμαλῶς γίνεται, ἅτε δια-
τεθερμασμένης τῆς κοιλίης καὶ ἀγούσης πάντα
κάτω ἐφ' ἑωυτήν· καὶ ὑπὸ μὲν τοῦ πτύσματος
πνίγεταί τε καὶ ῥέγκει οὐ καθαιρόμενος, ὑπὸ δὲ
τῆς γαστρὸς ῥεούσης ἐξασθενέει, καὶ ὡς τὰ πολλὰ διαφθείρεται.

Μάλιστα δ' ἐν τῇσι τοιαύτῃσι τῶν νούσων τὸ
ῥεῦμα τοῦτο ἡ κεφαλὴ παρέχει, ἅτε γὰρ κοίλη
ἐοῦσα καὶ ἄνω ἐπικειμένη· ὅταν διαθερμανθῇ ὑπὸ
τῆς κοιλίης, ἕλκει ἐς ἑωυτὴν ἐκ τοῦ σώματος τὸ
λεπτότατον τοῦ φλέγματος· ὅταν δ' ἁλισθῇ ἐν
αὐτῇ, ἀποδιδοῖ πάλιν ἁλές τε καὶ παχύ, καὶ
ὥσπερ εἴρηται, τὸ μὲν αὐτοῦ ἐς τὴν ἄνω κοιλίην
καταρρεῖ, τὸ δ' ἐς τὴν κάτω· ὅταν οὖν ἄρξηται ἥ
τε κεφαλὴ ῥεῖν καὶ τὸ ἄλλο σῶμα τήκεσθαι, οὐκέτι ὁμαλῶς οὐδὲ καυθέντες περιγίνονται· κρατέει
γὰρ πρὸς μὲν τὸ πύον τὰ ἐπιρρέοντα κακὰ ἢ τὰ
ἀπορρέοντα, αἱ δὲ σάρκες τηκόμεναι μᾶλλον ὑπὸ
τῶν κακῶν ἢ τρεφόμεναι ὑπὸ τῶν ἐσιόντων.

16. Οὗτοι ὅσοι τοιουτότροπα νοσήματα ἴσχουσι
καὶ ἀπὸ τούτων, ἔνιοι μὲν δι' ὀλίγου ἀπόλλυνται,
ἔνιοι δὲ πολὺν χρόνον ἕλκουσιν. διαφέρει γὰρ καὶ
170 σῶμα σώματος[1] καὶ πάθημα | παθήματος καὶ
ὥρη ὥρης, ἐν ᾗ ἂν νοσέωσι· καὶ οἱ μὲν ταλαιπω-
ρότεροί εἰσιν ἐν τῇσι νούσοισιν, οἱ δὲ παντάπασι
ταλαιπωρέειν ἀδύνατοι.

---

[1] M adds καὶ ἡλικίη ἡλικίης.

cleaning through expectoration does not proceed adequately, inasmuch as the cavity is heated and draws everything down into itself. Thus, on the one hand, the patient is choked by his expectoration, and his breathing is stertorous because he is not being cleaned out, on the other hand, he is weakened by the flux from his belly, and so he usually perishes.

Generally, in these kinds of diseases it is the head that produces the flux, inasmuch as it is hollow and situated in the superior position. When the head is heated by the cavity, it attracts the finest part of the phlegm from the body; when this has been collected in it, the head returns it in a thick mass, and, as has been said, part flows down into the upper cavity, and part into the lower cavity. Thus, when the head has begun with its flux, and the rest of the body to melt, patients no longer have a decent chance of survival, even if they are cauterized. For the harmful afflux to the pus surpasses what flows off, and the tissues are more wasted by the disease than nourished by the food taken in.

16. Some patients, that have diseases of this kind and from these factors, succumb within a short time, others drag on much longer. For one body differs from another, one affection from another, and one season in which to be ill from another; some patients are more able to endure the stress of diseases, while others are totally incapable of enduring.

ΠΕΡΙ ΝΟΥΣΩΝ Α

Οὔκουν ἐστὶ τὸ ἀκριβὲς εἰδέναι καὶ τυχεῖν εἴπαντα[1] τοῦ χρόνου, ἐν ὁπόσῳ ἀπόλλυται,[2] οὔτ' εἰ πολλόν, οὔτ' εἰ ὀλίγον· οὐδὲ γὰρ οὗτος ὁ χρόνος ἀκριβής, ὃν ἔνιοι λέγουσιν, ὡς τὰ πολλά, οὐδὲ αὐτὸ τοῦτο ἐκποιεῖ· διαφέρει γὰρ καὶ ἔτος ἔτεος καὶ ὥρη ὥρης. ἀλλ' ἤν τις ἐθέλῃ περὶ αὐτῶν ὀρθῶς γινώσκειν καὶ λέγειν, γνώσεται οὕτω πᾶσαν ὥρην καὶ ἀπολλυμένους καὶ περιγινομένους καὶ πάσχοντας ἃ ἂν πάσχωσιν.

17. Τὴν δὲ κάτω κοιλίην ἔμπυοι γίνονται μάλιστα μέν, ὅταν φλέγμα ἢ χολὴ συστῇ ἁλὲς μεσηγὺ τῆς τε σαρκὸς καὶ τοῦ δέρματος· γίνονται δὲ καὶ ἀπὸ σπασμῶν· καὶ ὅταν φλέβιον σπασθὲν ῥαγῇ, τὸ αἷμα ἐκχυθὲν σήπεταί τε καὶ ἐκπυεῖ· ἢν δὲ ἡ σὰρξ σπασθῇ ἢ φλασθῇ, ἕλκει ἐκ τῶν παρ' ἑωυτῇ φλεβίων αἷμα, καὶ τοῦτο σήπεταί τε καὶ ἐκπυεῖ.

Τούτοις ἢν μὲν ἔξω ἀποσημήνῃ καὶ τὸ πύον ἐξέλθῃ, ὑγιέες γίνονται· ἢν δ' ἐκραγῇ ἔσω αὐτόματον, ἀπόλλυνται.

Κεχυμένον δὲ πύον ἐν τῇ κάτω κοιλίῃ, ὥσπερ ἐν τῇ ἄνω εἴρηται ἐγγίνεσθαι, οὐκ ἂν δύναιτο ἐγγενέσθαι, ἀλλ' ὥσπερ[3] εἴρηται, ἐν χιτῶσί τε καὶ ἐν φύμασιν ἐγγίνεται. καὶ ἢν μὲν ἔσω ἀποσημήνῃ, δυσπετὲς γνῶναι· οὐδὲ γὰρ διασείσαντά ἐστι

---

[1] εἰδέναι . . . εἴπαντα om. M.   [2] Θ: -νται Μ.   [3] M adds μοι.

136

It is certainly not possible to know precisely and to state correctly the period within which a patient will die, not even whether it will be long or short. For the period of time that some people give is not precise in most cases, nor does this information, of itself, suffice; for one year differs from another, and one season from another. If anyone wishes to recognize the truth on this subject and to say it, he will recognize that patients both perish and survive, and suffer whatever they suffer, in every season.

17. Suppuration in the lower cavity usually occurs when phlegm or bile congeals in a mass between the tissues and the skin. It can also result from tears: when a small vessel is torn and ruptures, the blood that is poured out putrefies and suppurates; furthermore, if the tissue is torn or contused, it draws blood from the surrounding small vessels, and this too putrefies and suppurates.

In these patients, if the abscess points outward,[1] and the pus comes out, they survive; but if the pus ruptures spontaneously inward, they die.

A collection of pus in the lower cavity, as it was described occurring in the upper cavity, cannot take place, but rather, as I indicated, it occurs within membranes and tubercles. If this points inward, it is difficult to perceive, since it cannot be detected by succussion. In most cases, it is to be

[1] I.e. forms a head on the surface.

γνῶναι. γινώσκεται δὲ μάλιστα τῇ ὀδύνῃ, ἔνθα
ἂν ᾖ· καὶ ἢν καταπλάσῃς γῇ[1] κεραμίτιδι ἢ ἄλλῳ
τῳ τοιούτῳ, ἀποξηραίνει δι᾽ ὀλίγου.

172   18. Ἐρυσίπελας δ᾽ ἐν τῷ πλεύμονι γίνεται,
ὅταν ὑπερξηρανθῇ ὁ πλεύμων· ὑπερξηραίνεται δὲ
καὶ ὑπὸ καύματος καὶ ὑπὸ πυρετῶν καὶ ὑπὸ
ταλαιπωρίης καὶ ἀκρασίης· καὶ ὅταν ὑπερξηρανθῇ,
ἕλκει τὸ αἷμα[2] ἐφ᾽ ἑωυτόν, μάλιστα μὲν καὶ πλεῖστον ἐκ τῶν μεγάλων φλεβῶν· αὗται γὰρ αὐτῷ
ἐγγυτάτω εἰσὶ καὶ ἐπίκεινται ἐπ᾽ αὐτῷ· ἕλκει δὲ
καὶ ἐκ τῶν ἄλλων τῶν πλησίον· ἕλκει δὲ τὸ
λεπτότατον.[3]

῞Οταν δ᾽ εἰρύσῃ, πυρετὸς ἀπ᾽ αὐτοῦ γίνεται
ὀξὺς καὶ βὴξ ξηρὴ καὶ πληθώρη ἐν τοῖσι στήθεσι
καὶ ὀδύνη ὀξέη ἔμπροσθέν τε καὶ ὄπισθεν, μάλιστα δὲ κατὰ τὴν ῥάχιν, ἅτε τῶν φλεβῶν τῶν
μεγάλων διαθερμαινομένων· καὶ ἐμέουσιν ἄλλοτε
ὕφαιμον, ἄλλοτε πελιδνόν· ἐμέουσι δὲ καὶ
φλέγμα καὶ χολήν· καὶ ἐκψύχουσι πυκινά, ἐκψύχουσι δὲ διὰ τοῦ αἵματος τὴν μετάστασιν ἐξαπίνης γινομένην· καὶ μάλιστα διασημαίνει τοῦτο,
ὅταν ἐπὶ τοῦ πλεύμονος ἐπιγένηται ἐρυσίπελας
καὶ τοῦ πυρετοῦ ᾖ συνεχὴς λῆψις.

Τούτῳ ἢν μὲν δύο ἢ τριῶν ἢ τεσσέρων τὸ
πλεῖστον ἡμερέων διαχυθῇ καὶ μεταστῇ τὸ ἔνδον[4]

---

[1] Foes: τῇ ΘΜ.   [2] τὸ αἷμα Θ: τοῦ αἵματος πλεῖστον Μ.
[3] Μ adds καὶ ἀσθενέστατον.   [4] Μ adds ἐς τό.

138

recognized by where the pain happens to be; also, if you plaster the patient over with potter's earth or some other such material, the pus dries it up in a short time.

18. Erysipelas arises in the lung, when the lung becomes too dry; this happens as the result of burning heat, fevers, exertion, and intemperance. When the lung becomes too dry, it attracts blood, most frequently and in the greatest quantity from the large vessels, since these are nearest to it and lie over it, but also from the other vessels around it; it is the blood's finest component that is attracted.

When this attraction occurs, it gives rise to a high fever, as well as a dry cough, fullness in the chest, and sharp pains both anteriorly and posteriorly, especially along the spine, inasmuch as the large vessels become heated. Sometimes patients vomit material charged with blood, sometimes livid material, also phlegm and bile. They lose consciousness frequently, because of some sudden migration of the blood; in most cases, this happens when erysipelas is attacking the lung, and there is a continuing accession of fever.

If within two, three, or at most four days, this patient has a dispersion, and what is within moves

ἔξω, ὑγιὴς γίνεται ὡς τὰ πολλά· ἢν δὲ μὴ διαχυθῇ καὶ μεταστῇ, ἐνσήπεταί τε καὶ ἔμπυος γίνεται
καὶ ἀπόλλυται· ἀπόλλυται δὲ δι' ὀλίγου, ἅτε τοῦ
πλεύμονος διαπύου ἐόντος ὅλου καὶ σαπροῦ· ἢν δ'
ἔξω κατακεχυμένον ἔσω τράπηται καὶ λάβῃ τοῦ
πλεύμονος, τοῦτον οὐδεμία ἐλπὶς περιγενέσθαι.
ὅταν γὰρ προαπεξηρασμένος ὁ πλεύμων εἰρύσῃ ἐς
ἑωυτόν, οὐκ ἂν ἔτι μεταστάιη, ἀλλὰ παραχρῆμα
ὑπὸ τοῦ καύματος καὶ τῆς ξηρασίης οὔτε[1] δέχεται
οὐδὲν οὔτε ἄνω ἀναδιδοῖ οὐδέν, ἀλλὰ δι' οὖν
ἔφθειρεν.

19. Φῦμα δὲ γίνεται ἐν τῷ πλεύμονι ὧδε·
ὅταν φλέγμα ἢ | χολὴ ξυστραφῇ, σήπεται, καὶ
ἕως μὲν ἂν ἔτι ὠμότερον ᾖ, ὀδύνην τε παρέχει
λεπτὴν καὶ βῆχα ξηρήν· ὅταν δὲ πεπαίνηται,
ὀδύνη τε γίνεται καὶ πρόσθεν καὶ ὄπισθεν ὀξέη,
καὶ θέρμαι λαμβάνουσι καὶ βὴξ ἰσχυρή.

Καὶ ἢν μὲν ὅτι τάχιστα πεπανθῇ καὶ ῥαγῇ καὶ
ἄνω τράπηται τὸ πύον καὶ ἀναπτυσθῇ πᾶν καὶ ἡ
κοιλίη, ἐν ᾗ τὸ πύον ἐνῇ,[2] προσπέσῃ τε καὶ ἀναξηρανθῇ, ὑγιὴς γίνεται παντελῶς.

Ἢν δὲ ῥαγῇ μὲν[3] τάχιστα καὶ πεπανθῇ καὶ
ἀνακαθαίρηται, ἀποξηρανθῆναι δὲ παντάπασι μὴ
δύνηται, ἀλλ' αὐτὸ ἀφ' ἑωυτοῦ τὸ φῦμα ἀναδιδῷ
τὸ πύον, καὶ ἀπὸ τῆς κεφαλῆς τε καὶ τοῦ ἄλλου
σώματος φλέγμα καταρρέον ἐς τὸ φῦμα σήπηται

---

[1] Θ: οὐκέτι M.   [2] Θ: ἐνῆν Wittern: om. M.   [3] M adds ὅτι.

outward, he usually recovers; but if no dispersion and movement occurs, there is putrefaction, the patient suppurates internally, and he dies. Death occurs in a short time, since the lung is already totally purulent and putrid. If, after having been dispersed externally, the disease turns inward and seizes the lung, there is no hope for the patient's survival. For when the lung, which has been dried out previously, attracts, there can no longer be any movement, but, owing to its burning heat and dryness, it immediately becomes unable either to accept anything or to give anything off, and so the patient perishes.

19. A tubercle in the lung arises as follows: when phlegm or bile collects there, it putrefies, and, as long as it is still in a raw state, it produces mild pain and a dry cough. When it becomes mature, sharp pains arise, both anteriorly and posteriorly, and feverish heat sets in together with violent coughing.

If the tubercle matures very quickly, and ruptures, if the pus turns upwards and is all coughed up, and if the cavity that the pus occupied collapses and is dried out, the patient recovers completely.

However, if the tubercle ruptures very quickly, matures, and is cleaned upwards, but, because its cavity cannot be dried up completely, the tubercle itself continues to give off pus, and besides phlegm pouring down from the head and the rest of the body into the tubercle putrefies, turns to pus, and is

τε καὶ πύον γίνηται καὶ πτύηται, δι᾽ οὖν ἐφθάρη·
διαφθείρεται δὲ ὑπὸ τῆς γαστρὸς ῥυείσης ἢ ἀφ᾽
ὧνπερ τὸ πρόσθεν εἴρηται· λεσχηνευομένου δὲ
αὐτοῦ καὶ φρονέοντος[1] πάντα χρήματα ὁμαλῶς[2]
καὶ ἐν τῷ πρὶν χρόνῳ, ἀποξηραίνεταί τε καὶ ἀπο-
ψύχεται, καὶ ξυμμύει τὰ φλέβια τὰ ἐν τῷ σώματι
πάντα, ἅτε τοῦ αἵματος ἐξ αὐτῶν[3] ἐκκεκαυμένου
ὑπὸ πυρετῶν, ἐνίοτε δὲ ὑπὸ χρόνου τε πλήθεος
καὶ μεγέθεος τῆς νούσου καὶ τῶν ἐνεόντων κακῶν
καὶ τῶν προσεπιγινομένων.

Ἢν δὲ μὴ δύνηται πολλοῦ χρόνου[4] ῥαγῆναι,
μήτε ἐκ τοῦ αὐτομάτου μήτε ὑπὸ φαρμάκων,
τήκεται ὁ ἀσθενέων ὑπό τε ὀδυνέων ἰσχυρῶν καὶ
ἀσιτίης καὶ βηχὸς καὶ πυρετῶν καὶ ὡς τὰ πολλὰ
διαφθείρεται.

Ἢν δ᾽ ἤδη λελεπτυσμένῳ καὶ κλινοπετεῖ
ἐόντι ῥαγῇ τὸ πύον, οὐδ᾽ οὕτω μάλα ἀναφέρουσιν,
ἀλλὰ διαφθείρονται τρόπῳ τῷ αὐτῷ.

Ἢν δὲ ῥαγῇ μὲν ὅτι τάχιστα καὶ πεπανθῇ,
πεπανθὲν δὲ ἐκχυθῇ ἐπὶ τὰς φρένας τὸ πολλὸν
αὐτοῦ, τὸ παραυτίκα μὲν δοκέει ῥάων εἶναι·
176 προϊόντος | δὲ τοῦ χρόνου, ἢν μὲν ἀναπτύσῃ πᾶν
καὶ ἡ κοιλίη, ἐν ᾗ τὸ πύον ἐνῇ,[5] προσπέσῃ τε καὶ
ἀναξηρανθῇ, ὑγιὴς γίνεται· ἢν δὲ ὅ τε χρόνος
πλείων γένηται, καὶ αὐτὸς ἀσθενέστερος, καὶ

---

[1] M: ἀφρον- Θ.   [2] M: ὁμαλῶσα Θ.   [3] αἵματος ἐξ αὐτῶν M:
σώματος Θ.   [4] τε πλήθεος ... χρόνου om. Θ.   [5] ΘM: ἐνῆν
Wittern.

expectorated, then the patient perishes; death results from a flux of the belly, or from the factors mentioned before. While the patient is chatting and still retains an understanding of every subject just as he had before,[1] he is dried up and breathes out his spirit, and all the small vessels in his body close, inasmuch as the blood from them is burnt out by the fevers, and sometimes also by the extent of time, by the magnitude of the disease, by the evils first present, and by those added.

Now, if the tubercle fails to rupture for a long time, either spontaneously or with the help of a medication, the weakened patient melts away as the result of his violent pains, fasting, cough and fevers, and usually he perishes.

If pus breaks through in a patient that is already emaciated and bed-ridden, they do not recover very often in this case, either, but perish in the same way.

If the tubercle ruptures very quickly and matures, but, when it has matured, most of its pus is poured out on to the diaphragm, for the moment the patient seems better; and if, with time, he coughs everything up, and the cavity in which the pus resided collapses and is dried out, he recovers. However, if the time increases, if the patient

[1] An incomprehensible passage, presumably already so in Erotian's time; see Nachmanson, *Erotianstudien* 401 f. and Wittern, 142 f.

ἀναπτύσαι μὴ δύνηται, ἀλλὰ καυθῇ ἢ τμηθῇ, καὶ
τὸ πύον ἐξέλθῃ, παραυτίκα μὲν καὶ οὕτω δοκέει
δή τι ῥάων εἶναι, προϊόντος δὲ τοῦ χρόνου διαφθείρεται ὑπὸ τῶν αὐτῶν, ὑφ' ὧνπερ καὶ ἐν τῇ
πρώτῃ εἴρηται.

20. Ἐν δὲ τῷ πλευρῷ γίνεται μὲν φύματα
καὶ ἀπὸ φλέγματος καὶ ἀπὸ χολῆς κατὰ τὸν αὐτὸν λόγον τοῖσιν ἐν τῷ πλεύμονι· γίνεται δὲ καὶ
ἀπὸ πόνων, ὅταν τι τῶν φλεβίων σπασθὲν ῥαγῇ,
ἢ σπασθῇ μέν, ῥαγῇ δὲ μὴ παντελῶς, ἀλλὰ σπαδὼν ἐν αὐτῷ γένηται· ἢν μὲν οὖν ῥαγῇ παραυτίκα, τὸ αἷμα ἐκχυθὲν ἐκ τοῦ φλεβίου σήπεταί τε
καὶ ἐκπυέει· ἢν δὲ σπαδὼν ἐν τῷ φλεβίῳ γένηται, τοῦτο δὲ κατ' ἀρχὰς μὲν ὀδύνας παρέχει καὶ
σφύζει, προϊόντος δὲ τοῦ χρόνου διαδιδοῖ ἡ φλὲψ
τοῦ αἵματος ἐς τὴν σάρκα, καὶ τοῦτο σηπόμενον
ἐν τῇ σαρκὶ πύον γίνεται.

Κατὰ τὸν αὐτὸν δὲ λόγον καὶ ἡ σάρξ, ἢν μὲν
μᾶλλον πονήσῃ, πλέον ἕλκει τοῦ αἵματος ἐς
ἑωυτὴν ἐκ τῶν ἐγγυτάτω φλεβῶν, καὶ παραχρῆμα
ἐκπυέει· ἢν δ' ἧσσον πονήσῃ, σχολαίτερον καὶ
ἕλκει καὶ ἐκπυέει.

Ἐνίοισι δ' ὅταν ἀσθενέα γένηται τὰ σπάσματα
ἢ ἐν τῇ σαρκὶ ἢ ἐν τῇσι φλεψίν, οὐκ ἐκπυΐσκεται,
ἀλλὰ γίνεται ἀλγήματα πολυχρόνια, ἃ καὶ
καλέουσι ῥήγματα.

becomes weaker, and if he is unable to expectorate, but when he is cauterized or incised the pus comes forth, in this case, too, he seems somewhat better for the moment; still, with the passage of time, he perishes from the things indicated in the first case.

20. In the side, tubercles arise from both phlegm and bile, and in the same way as those in the lung. These tubercles also occur from exertions, when one of the small vessels is torn and ruptures, or, although torn, it does not rupture completely, but a tear arises in it. Now, if the vessel ruptures right away, the blood that is poured out of it putrefies and suppurates; if, on the other hand, only a tear occurs in the vessel, at the beginning this produces pains, and throbs; later, the vessel exudes blood into the tissue, and this putrefies in the tissue to become pus.

In the same way, the tissue, too, if seriously affected, attracts more blood from the nearby vessels, and at once suppurates; if it is less affected, it attracts and suppurates more slowly.

In some cases, when small tears occur in the tissue or in vessels, they do not suppurate, but there arise chronic pains, which people also call tears.

ΠΕΡΙ ΝΟΥΣΩΝ Α

Καὶ ὅσα μὲν ἐν τῇ σαρκὶ γίνεται, ὧδε γίνεται·
ὅταν ἡ σὰρξ πονήσῃ τι ἢ σπασθεῖσα ἢ πληγεῖσα
ἢ ἄλλο τι παθοῦσα, γίνεται, ὥσπερ εἴρηται, πελιδνή, πελιδνὴ δὲ[1] οὐκ εἰλικρινεῖ αἵματι, ἀλλὰ
λεπτῷ τε καὶ ὑδαρεῖ καὶ τούτῳ ὀλίγῳ· ὅταν δ'
ὑπερξηρανθῇ μᾶλλον τοῦ εἰωθότος, διαθερμαίνε-
178 ταί τε καὶ ὀδύνην παρέχει, καὶ ἄγει ἐς ἑωυ|τὴν
ἀπὸ τῶν πλησίον καὶ φλεβῶν καὶ σαρκῶν τὸ
ὑγρόν· καὶ ὅταν ὑπερυγρανθῇ καὶ τοῦτο αὐτὸ τὸ
ὑγρὸν διαθερμανθῇ ὑπ' αὐτῆς τῆς σαρκός,
σκίδναται ἀνὰ τὸ σῶμα πᾶν, οἷόν περ εἰρύσθη,
καὶ μᾶλλον δή τι σκίδναται[2] ἐς τὰς φλέβας ἢ ἐς
τὰς σάρκας· ἕλκουσι γὰρ αἱ φλέβες μᾶλλον τῶν
σαρκῶν, ἕλκουσι δὲ καὶ αἱ σάρκες.

Ὅταν δ' ἐς πολλὸν ὑγρόν, τὸ ἐν τῷ σώματι,
ὀλίγον τὸ ἀπὸ τῆς σαρκὸς ἔλθῃ,[3] ἄδηλόν τε γίνεται καὶ ἀνώδυνον, καὶ ἀντὶ νενοσηκότος γίνεται
ὑγιὲς τῷ χρόνῳ.

Ἢν δὲ διαθερμανθῇ τε μᾶλλον ἡ σὰρξ καὶ εἰρύσῃ πλεῖον τὸ ὑγρόν, ὀδύνην παρέχει, καὶ ὅπῃ ἂν
τοῦ σώματος ἀπ' αὐτῆς ὁρμήσῃ καὶ καταστηρίξῃ,
ὀδύνην παρέχει ὀξέην· καὶ δοκέουσιν ἔνιοι ἑωυτοῖσι τὸ ῥῆγμα μεθεστάναι· τὸ δ' οὐκ ἀνυστόν·
ἕλκος γὰρ μεταστῆναι οὐκ ἀνυστόν·[4] ἐγγυτάτω
δ' ἕλκεός ἐστιν ὅσα τοιαῦτα· ἀλλὰ τὸ ἀπὸ τῆς

---

[1] πελιδνὴ δὲ om. M.   [2] ἀνὰ τὸ σῶμα ... σκίδναται om. Θ.
[3] Θ: ἑλκυσθῇ M.   [4] ἕλκος ... ἀνυστόν om. M.

146

DISEASES I

What happens in the tissue happens in the following way: when the tissue is somehow affected, being either torn or struck or suffering some other insult, it becomes, as was stated,[1] livid, and livid not with pure blood, but with thin watery blood of a small amount; this makes the tissue drier than normal, and as a result it becomes hot, produces pain, and attracts moisture from the nearby vessels and tissues; when the tissue then becomes abnormally moist, the moisture that has been attracted becomes hot from the heat of the tissue, and is dispersed throughout the whole body, just as it was attracted; actually, more is dispersed to the vessels than to the tissues, for the vessels attract more than do the tissues, although the tissues do attract some.

When this small amount of moisture from the tissues passes into the large amount of moisture in the body, it becomes inconspicuous, harmless, and, with time, no longer ill but healthy.

However, if the tissue has been more severely heated, and has attracted a greater amount of moisture, this produces pain: wherever in the body the moisture from the tissue rushes and is deposited, at that place there is sharp pain. Some patients believe that the tear in them has moved, but this is not possible, for an ulcer cannot move, and such things as these are most akin to an ulcer;

[1] In chapter 15 above.

σαρκὸς ὑγρὸν ἀΐσσει διὰ τῶν φλεβίων· ὅταν δὲ
διαθερμανθῇ τε καὶ παχυνθῇ καὶ γένηται πλέον,
ὀδύνην παρέχει, ἔστ' ἂν ὅμοιον γένηται τῷ ἄλλῳ
ὑγρῷ κατὰ λεπτότητα καὶ ψυχρότητα.

Ὅσα δ' ἐν τοῖσι φλεβίοισι γίνεται, αὐτὸ μὲν τὸ
φλέβιον, ὅσον ἔσπασται, κατὰ χώρην μένει· ὅταν
δὲ σπασθῇ, σπᾶται δ' ὑπὸ τόνου καὶ βίης, καὶ
ὅταν σπασθῇ, γίνεται οἷον κιρσός· καὶ διαθερμαίνεταί τε καὶ ἕλκει ἐς ἑωυτὸ νοτίδα ὑγρήν· ἡ δὲ
νοτίς ἐστιν ἀπὸ χολῆς καὶ φλέγματος· καὶ ὅταν
μιχθῇ τό τε αἷμα καὶ τὸ ἀπὸ τῆς σαρκὸς ὑγρόν,
παχύνεταί τε τὸ αἷμα πολλαπλασίως αὐτὸ ἑωυτοῦ ταύτῃ, ᾗ ἂν ἡ φλὲψ τυγχάνῃ ἐσπασμένη, καὶ
νοσωδέστερον γίνεται καὶ στασιμώτερόν τε καὶ
πλέον· καὶ ὅταν πλέον γένηται, μετ' οὖν ἔστη τὸ
180 πλήρωμα, ᾗ ἂν | τύχῃ, καὶ ὀδύνην παρέχει ὀξέην
ὥστ' ἐνίοισι δοκέειν τὸ ῥῆγμα ἑωυτοῖσι μεθεστά-
ναι. καὶ ἢν τύχῃ ὥστ' ἐς τὸν ὦμον μεταστῆναι,
βάρος τε τῇ χειρὶ παρέχει καὶ νάρκην καὶ νω-
θρίην· καὶ ἢν μὲν ἐς τὴν φλέβα σκεδασθῇ,[1] ἢ ἐς
τὸν ὦμόν τε καὶ ἐς τὸν νῶτον τείνει, παύεται ἡ
ὀδύνη παραχρῆμα ὡς τὰ πολλά.

Γίνεται δὲ τὰ σπάσματα καὶ ἀπὸ πόνων καὶ
πτωμάτων καὶ πληγῆς, καὶ ἤν τις ἄχθος μέζον
αἴρηται, καὶ ἀπὸ δρόμων καὶ πάλης καὶ τῶν τοιούτων πάντων.

21. Ὅσοι δ' ἀπὸ τρωμάτων ἔμπυοι γίνονται,

rather, it is the moisture from the tissue which darts through the small vessels; and once this has become heated, thickened, and greater in quantity, it will continue to produce pain until it once more becomes as thin and cold as the rest of the moisture in the body.

Whatever else happens in small vessels, the vessel itself, however much it is torn, remains in place. When it is torn, it is torn by tension and violence, and it forms something like a varix; it becomes heated and attracts damp moisture; this moisture is from bile and phlegm. When blood and the moisture from the tissues are mixed where the vessel happens to have been torn, the blood becomes many times thicker than normal, more sickly, more stagnant, and greater in quantity. As it increases in quantity, its fullness moves anywhere it pleases, and produces sharp pain, so that to some patients it seems that their tear has moved. If the blood happens to move to the shoulder, it produces a heaviness in the arm along with numbness and torpor; however, if the blood is dispersed into the vessel that passes to the shoulder and the back, the pain ceases, in most cases, at once.

Tears can also arise from exertions, falls, a blow, if a person lifts some great burden, and from races, wrestling, or anything else of that sort.

21. Persons that suppurate internally as the

---

[1] Potter: σκιμφθῇ ΘΜ.

ἢν ὑπὸ δόρατος ἢ ἐγχειριδίου ἢ τοξεύματος ἐσω-
τέρω τρωθῶσιν, ἕως μὲν ἂν ἔχῃ ἔξω τὸ ἕλκος
ἀναπνοὴν ἀνὰ τὸ ἀρχαῖον τρῶμα, ταύτῃ τε τὸ
ψυχρὸν ἐπάγεται ἐφ᾽ ἑωυτό, καὶ τὸ θερμὸν ἀφ᾽
ἑωυτοῦ ταύτῃ ἀφίησι, καὶ ἀποκαθαίρεται ταύτῃ
τὸ πύον καὶ ἢν δή τι ἄλλο. καί ἢν μὲν ὑγιανθῇ[1]
τό τ᾽ ἔνδον καὶ τὸ ἔξω ὁμοῦ, ὑγιὴς γίνεται παν-
τελῶς· ἢν δὲ τὸ μὲν ἔξω ὑγιανθῇ, τὸ δ᾽ ἔσω μὴ
ὑγιανθῇ, ἔμπυος γίνεται· καὶ ἢν ὑγιανθῇ μὲν
ὁμοῦ τὸ ἔνδον καὶ τὸ ἔξω, ἡ δὲ οὐλὴ ἔσω ἀσθενής
τε γένηται καὶ τρηχέα καὶ πελιδνή, ἀνελκοῦται
ἐνίοτε, καὶ ἔμπυος γίνεται· ἀνελκοῦται δὲ καὶ ἢν
πονήσῃ τι πλέον, καὶ ἢν λεπτυνθῇ, καὶ ἢν
φλέγμα ἢ χολὴ πρὸς τῇ οὐλῇ προσπαγῇ, καὶ ἢν
νούσῳ ἑτέρῃ ληφθεὶς λεπτυνθῇ.

Ὅταν δὲ γένηται ἕλκος, ἤν τε οὕτως ἤν τε
προσυμφυῇ τὸ ἔξω τοῦ ἔσω, ὀδύνην τε παρέχει
ὀξέην καὶ βῆχα καὶ πυρετόν· καὶ τήν τε ψύξιν
ἐπάγεται αὐτὸ ἑωυτῷ τὸ ἕλκος διὰ τὸ πλέον τε
καὶ θερμότερον εἶναι, καὶ αὐτὸ ἀφ᾽ ἑωυτοῦ ἀπο-
πνεῖ τὸ θερμόν· καὶ τὸ πύον ἀποκαθαίρεται διὰ
πλείονος καὶ προσανίητόν τε καὶ σχολαίτερον
ὑγιάζεται πολλῷ, ἐνίοτε δ᾽ οὐδ᾽ | ὑγιάζεται· ἡ
γὰρ σὰρξ ἡ τοῦ ἕλκεος ὑπὸ τοῦ καύματος τοῦ ἐν

182

---

[1] Here and below ὑγιανθῇ ΘΜ: ὑγιασθῇ Wittern, following Van
Brock.

result of being wounded internally by spear, dagger, or arrow: as long as the ulcer maintains a connection to the external air through the original wound, at that point it attracts cold to itself and sends off heat, and pus—along with anything else—is cleaned from it there. If the internal and external parts heal at the same time, the patient recovers completely; however, if the external part heals, but the internal one does not, the person suppurates internally. If the internal and external parts heal at the same time, but the scar becomes weak, rough, and livid within, it sometimes ulcerates afresh, and then the patient suppurates internally. The scar may also ulcerate afresh if the patient exerts himself too much in some way, if he becomes lean, if phlegm or bile becomes fixed in the scar, or if he becomes lean as the result of being attacked by another disease.

When an ulcer has arisen, either in this way or because the external part grew together before the internal one did, it produces sharp pain, coughing, and fever. This ulcer attracts cold to itself, because of its greater magnitude and heat, and exhales heat from itself; pus is cleaned over a longer time, and the patient recovers, but with a tendency to be incurable, and much more slowly, sometimes not at all[1]; for the tissue of the ulcer is boiled by the

---

[1] For a discussion of the Greek conception of health, as a relative rather than an absolute phenomenon, see F. Kudlien, "Gesundheit" in *Reallexikon für Antike und Christentum* (X. 902–45), Stuttgart, 1978, cols. 904–10.

τῷ σώματι ἕψεταί τε καὶ ὑπερυγραίνεται, ὥστε
μὴ δύνασθαι μήτε ξηρανθῆναι μήτε σαρκοφῦσαι
μήτε ὑγιασθῆναι, ἀλλ' ὅταν χρόνος προΐῃ, τελευτᾷ πάσχων ταὐτά, ἃ καὶ ἐν τῇ πρόσθεν εἴρηται.

Ἢν δὲ τύχῃ ὥστε τρωθῆναί τι τῶν φλεβίων
τῶν παχυτέρων καὶ εἴσω ῥυῇ τὸ αἷμα καὶ ἐν-
σαπῇ, ἔμπυος γίνεται· καὶ ἢν μὲν τοῦτο τὸ πύον
πτυσθῇ πᾶν, καὶ ἡ φλὲψ ἡ τετρωμένη στεγνωθῇ,
καὶ τὸ ἕλκος ὑγιασθῇ καὶ τὸ ἔσω καὶ τὸ ἔξω,
ὑγιὴς γίνεται παντελῶς· ἢν δὲ μὴ δύνηται μήτε
τὸ ἕλκος συμφῦναι[1] μήτε ἡ φλὲψ στεγνωθῆναι,
ἀλλ' ἄλλοτε καὶ ἄλλοτε ἀναδιδῷ αἷμα, καὶ ἢ
παραυτίκα ἐμῆται ἢ πτύηται, ἢ καὶ σήπηται καὶ
πύον πτύηται, διαφθείρεται ὡς τὰ πολλά, ἢ
παραυτίκα ἐμέων αἷμα, ἢ ὑστέρῳ χρόνῳ ὑφ' ὧν
καὶ ἐν τῇ πρόσθεν εἴρηται διαφθειρόμενος.

Πολλάκις δὲ ὅσοι τι τῶν ἔσω φλεβίων τιτρώ-
σκονται ἢ ὑπὸ τρωμάτων ἢ πόνων ἢ κατὰ
γυμνασίην ἢ ὑπ' ἄλλου του, ὅταν συμφυῇ καὶ δοκῇ
ὑγιὲς εἶναι τὸ φλέβιον, ἀναρρήγνυνται ὕστερον
χρόνῳ· ἀναρρήγνυνται δ' ὑπὸ τῶν αὐτῶν ὑφ' ὧνπερ καὶ πρόσθεν· ὅταν δ' ἀναρραγῇ, αἱμορροέει,
καὶ παραυτίκα ἀπόλλυνται ἐμέοντες αἷμα πολλόν
τε καὶ πολλάκις, ἢ ἄλλοτε μὲν καὶ ἄλλοτε αἷμα
ἐμέουσι πρόσφατον, πύον δὲ πτύοντες ἀνὰ πᾶσαν
ἡμέρην πολλόν τε καὶ παχύ, δι' οὖν ἐφθάρησαν

[1] M adds τὸ ἔνδον.

152

burning heat in the body, and becomes too moist, so that it can neither be dried out, nor produce new tissue, nor heal; rather, as time passes, the patient meets his end suffering the same things mentioned in the preceding disease.

If it so happens that one of the wider vessels is wounded, and blood flows inward and putrefies, the patient suppurates internally. If this pus is all coughed up, if the wounded vessel closes, and if the ulcer heals both internally and externally, there is complete recovery. However, if the ulcer cannot grow together, nor the vessel close, but from time to time they give off blood, which is either vomited up at once or expectorated, or putrefies and is expectorated in the form of pus, the patient usually perishes, either straightway from vomiting blood, or at some later time from the things mentioned above as being fatal.

It often happens in persons that have been wounded in one of the small internal vessels, either by wounds or by exertions in athletics or otherwise, that after the vessel has grown together and seems to have recovered, it ruptures again at a later time from the same things as before. When it does this, it bleeds, and the patients either die straightway from vomiting blood frequently and in large amounts, or they vomit fresh blood only now and then, but expectorate copious thick pus all day long, and so perish in a way identical, or similar, to

τρόπῳ τοιούτῳ ἢ παραπλησίῳ, ὡς καὶ ἐν τῇσιν
ἄλλησιν[1] εἴρηται.

22. Τοῖσι δὲ ταῦτα τὰ νοσήματα ἴσχουσι καὶ
ὅσα τοιαῦτα διαφέρει εἰς τὸ εὐπετεστέρως τε
ἀπαλλάσσειν καὶ δυσπετεστέρως ἀνήρ τε γυναι-
κός, καὶ νεώτερος γεραιτέρου, καὶ γυνὴ νεωτέρη
184 καὶ παλαιο|τέρη, καὶ πρὸς τούτοισιν ἡ ὥρη τοῦ
ἔτεος, ἐν ᾗ ἂν νοσέωσι, καὶ ἢν ἐξ ἑτέρης νούσου
νοσέωσιν ἤν τε μὴ ἐξ ἑτέρης. διαφέρει δὲ καὶ
πάθημα παθήματος μέζον <ὂν>[2] τε καὶ ἔλασσον
καὶ χρὼς χρωτὸς καὶ θεραπείη θεραπείης.

Τούτων δ' οὕτω διαφερόντων ἀνάγκη διαφέ-
ρειν καὶ τὸν χρόνον, καὶ τοῖσι μὲν πλέω γίνεσθαι,
τοῖσι δ' ἐλάσσω, καὶ ἀπόλλυσθαι ἢ μή, καὶ τοῖσι
μὲν παραμόνιμά τ' εἶναι καὶ μέζω, τοῖσι δ' ἐλάσσω τε καὶ ὀλιγοχρόνια, τοῖσι δὲ παραμένειν ἐς τὸ
γῆρας τὰ νοσήματα καὶ συναποθνήσκειν, τοὺς δ'
ἀπόλλυσθαι δι' ὀλίγου ὑπ' αὐτῶν.

Καὶ ὅσοι μὲν νεώτεροι πάσχουσί τι τούτων,
ὅσα εἴρηται ἀπὸ πόνων παθήματα γίνεσθαι, πάσχουσι πλέω τε καὶ ἰσχυρότερα καὶ ἀλγέουσι μᾶλλον τῶν ἄλλων, καὶ παραυτίκα ἔνδηλα αὐτοῖσιν,
ὥστε ἢ πτύσαι αἷμα ἢ ἐμέσαι, τὰ δὲ καὶ γινόμενα
λανθάνει αὐτοὺς ὑπὸ εὐεξίης τοῦ σώματος.

Οἱ δὲ γεραίτεροι πάσχουσι μὲν ὀλιγάκις, καὶ
ὅταν πάθωσιν, ἀσθενέα πάσχουσιν ἅτε ἀσθενέστε-

---

[1] M adds νούσοισιν.   [2] Potter.

that mentioned in the other diseases.

22. Among persons that have these and similar diseases, a man differs from a woman in the ease or difficulty with which he recovers, a younger man differs from an older man, and a younger woman differs from an older woman; additional factors are the season in which they have fallen ill, and whether or not their disease has followed from another disease. Besides, one affection differs from another, being either greater or less, one body from another, and one treatment from another.

And since these things vary in this way, it necessarily follows that the duration, too, varies, being greater in some instances and less in others, and that patients may or may not die; for such diseases are permanent and more serious in some patients, but of short duration and less serious in others, they last into old age in yet others, clinging to them until death, and still others die from them in a short time.

When younger men are subject to one of the affections that were said to arise from exertions, they suffer in more ways and more severely, and have more pains than do others; diseases usually become apparent in them immediately, so that they either expectorate or vomit blood, although sometimes the disease escapes the patient's notice because of his good bodily condition.

Older men suffer less often and, when they do, more mildly, since they are themselves weaker, and

ροι ἐόντες, καὶ ἐπαίουσι μᾶλλον καὶ ἐπιμέλονται
μᾶλλον τῶν παθημάτων.

Γίνεται οὖν τὴν ἀρχὴν τὸ παράπαν ἧσσον τῷ
γεραιτέρῳ ἢ τῷ νεωτέρῳ· καὶ ὅταν γένηται, τῷ
μὲν γεραιτέρῳ ἀσθενέστερα γίνεται, τῷ δὲ νεωτέρῳ ἰσχυρότερα.

Καὶ τῷ μὲν νεωτέρῳ, ἅτε τοῦ σώματος τόνον
τε ἔχοντος καὶ ξηρασίην καὶ τὴν σάρκα πυκινήν
τε καὶ ἰσχυροτέρην[1] καὶ πρὸς τοῖσι ὀστέοισι
προσκαθημένην καὶ περὶ αὐτὴν τοῦ δέρματος
περιτεταμένου, ὅταν τι πονήσῃ πλέον τοῦ εἰωθότος, μᾶλλον καὶ ἐξαίφνης, σπασμοί τε γίνονται
ἰσχυροὶ καὶ ῥήγματα πολλά τε καὶ παντοῖα τῶν
φλεβῶν καὶ τῶν σαρκῶν· καὶ τούτων τὰ μὲν
παραυτίκα ἔνδηλα[2] γίνεται, τὰ δ' ὕστερον χρόνῳ
ἀναφαίνεται.

Τοῖσι δὲ γεραιτέροισι τόνος τ' ἰσχυρὸς οὐκ ἔνι,
καὶ αἱ σάρκες περὶ τὰ ὀστέα περιρρέουσι, καὶ τὸ
186 δέρμα περὶ τὰς σάρκας, καὶ αὐτὴ ἡ σὰρξ | ἀραιή
τε καὶ ἀσθενής· καὶ οὔτε τι ἂν πάθοι τοιοῦτον
ὁμοίως ὡς καὶ ὁ νεώτερος, καὶ ἢν τι πάθῃ, πάσχει
ἀσθενέα τε καὶ παραυτίκα ἔνδηλα.

Τοσούτῳ μὲν ἐν τῇ ἀρχῇ τῶν παθημάτων
δυσχερέστερον ἀπαλλάσσουσιν οἱ νεώτεροι τῶν
γεραιτέρων.

Ὅταν δ' ἡ νοῦσος ἐμφανὴς γένηται, καὶ ἢ

[1] Θ: ἰσχυρὴν Μ.   [2] Θ: ἔκ- Μ.

156

also they have more understanding and take better care of their affections.

Thus, to begin with, these diseases occur less often, on the whole, in older men than in younger ones, and, when they do occur, they are milder in older men and more violent in younger ones.

In the younger man, inasmuch as his body has tension, dryness, and a tissue that is dense, stronger and adherent to the bones, and inasmuch as his skin is tightly stretched about the tissue, when he exerts himself more than normal, either in greater amounts or violently, severe tears arise, along with many and various ruptures of the vessels and tissues. Of these, some are revealed at once, while others come to light only later.

In older men, strong tension is not present, the tissues are loosely attached to the bones, and the skin to the tissues, and the tissue itself is rarified and weak. Therefore, the older man would never suffer such a thing as the younger man does, and even if he did, his disease would be mild and immediately apparent.

This is how much more difficult it is for younger men to recover at the beginning of affections, than for older ones.

When the disease is revealed, and patients are

πύον ἢ αἷμα πτύσωσιν ἢ ἀμφότερα, ὅσοι μὲν νεώ-
τεροί εἰσιν, ἅτε τοῦ σώματος εὐτόνου τε ἐόντος
καὶ πυκινοῦ, οὐ δύνανται ἀποκαθαίρεσθαι ὁμαλῶς
ἀπὸ τῶν ἑλκέων τῶν ἐν τῇ ἄνω κοιλίῃ τὸ πύον.
ὅ τε γὰρ πλεύμων οὐ κάρτα ἕλκει ἐς τὰς
ἀρτηρίας πυκνότερος ἐών, αἵ τε ἀρτηρίαι λεπταὶ
ἐοῦσαι καὶ στεναὶ οὐκ ἐνδέχονται τὸ πύον, εἰ μὴ
ὀλίγον τε καὶ ὀλιγάκις, ὥστε ἀνάγκη τὸ πύον ἐν
τῷ θώρακί τε καὶ ἐπὶ τῶν ἑλκέων ἀθροίζεσθαί τε
καὶ παχύνεσθαι.

Τῷ δ' ἀφηλικεστέρῳ ὅ τε πλεύμων ἀραιότερος
καὶ κοιλότερος, καὶ αἱ ἀρτηρίαι εὐρύτεραι, ὥστε
μὴ ἐγχρονίζειν τὸ πύον ἐν τῇ κοιλίῃ καὶ ἐπὶ
τῶν ἑλκέων, καὶ ὅ τι ἂν ἐπιγένηται, τοῦτο πᾶν
ἀνάγκη ἀνασπᾶσθαι ἄνω ὑπὸ τοῦ πλεύμονος ἐς
τὰς ἀρτηρίας καὶ παραχρῆμα ἐκπτύεσθαι.

Τῷ μὲν οὖν νεωτέρῳ, ἅτε τῶν παθημάτων
ἰσχυροτέρων ἐόντων καὶ τῆς καθάρσιος οὐ γινομέ-
νης κατὰ λόγον τοῦ πτύσματος, οἵ τε πυρετοὶ
ὀξύτεροι καὶ πυκνότεροι γίνονται, καὶ ὀδύναι ἐμ-
πίπτουσιν ὀξέαι αὐτοῦ τε τοῦ παθήματος καὶ τοῦ
ἄλλου σώματος, ἅτε τῶν φλεβίων ἐντόνων τε
ὄντων καὶ ἐναίμων· ὅταν δὲ ταῦτα διαθερμανθῇ
ἐφ'[1] ἑωυτῶν, ὀδύναι διαΐσσουσιν ἄλλοτε ἄλλῃ τοῦ
σώματος, καὶ οὗτοι μὲν διαφθείρονται ὡς τὰ
πολλὰ δι' ὀλίγου.

[1] Θ: ὑφ' Μ.

expectorating either pus or blood or both, those that are younger, since their body is elastic and dense, are unable to clean the pus adequately from the ulcers in their upper cavity; for the lung, being denser, hardly draws the pus into the bronchial tubes, and the bronchial tubes, being thin and narrow, only accept it rarely and in small amounts; perforce, then, the pus collects in the thorax on the ulcers and becomes thick.

In the elderly man, the lung is rarer and hollower, and the bronchial tubes wider, so that the pus does not delay long in the cavity and on the ulcers, and whatever is added must all be drawn up by the lung into the bronchial tubes and at once expectorated.

Thus, in the younger man, inasmuch as his affections are severer, and cleaning does not proceed properly by expectoration, fevers are higher and more frequent, and sharper pains attack both the affected area itself and the rest of the body, since the small vessels are stretched and charged with blood; when these themselves become heated, pains dart at one time to one part of the body, at another time to another part; such patients generally die in a short time.

Τοῖσι δὲ γεραιτέροισιν, ἅτε τῶν παθημάτων
ἀσθενεστέρων ἐόντων καὶ τοῦ πτύσματος ἀπ'
αὐτῶν καθαιρομένου, οἵ τε πυρετοὶ λεπτότεροι καὶ
ὀλιγάκις γίνονται, καὶ ὀδύναι ἔνεισι μέν, ἔνεισι δὲ
λεπταί· καὶ παντάπασι μὲν τῶν παθημάτων τῶν
τοιούτων οὐκ ἀπαλλάσσονται οὐδ' οἱ γεραίτεροι,
ἀλλ' ἔχοντες αὐτὰ καταφθείρονται πολὺν χρόνον,
188 καὶ ἄλλοτε | πύον πτύουσιν, ἄλλοτε δ' αἷμα,
ἄλλοτε δ' οὐδέτερον, τέλος δὲ συναποθνήσκει
αὐτοῖσιν· ἀποθνήσκουσι δὲ μάλιστα οὕτως, ὅταν
τι αὐτοὺς νόσημα τούτων, ᾧ ἂν ἔχωσι, παρα-
πλήσιον καταλάβῃ, ὥστ' ἔχειν καὶ τοῦτο, καὶ ὃ
ἂν ἔχωσι νόσημα ἰσχυρότερον γίνεται καὶ ὡς τὰ
πολλὰ δι' οὖν ἔφθειρε. ταῦτα δ' ἐστὶ τὰ μάλιστα
ἐξεργαζόμενα τῶν νοσημάτων πλευρῖτίς τε καὶ
περιπλευμονίη.

23. Πυρετὸς δ' ἀπὸ τῶνδε γίνεται· ὅταν χολὴ
ἢ φλέγμα θερμανθῇ, θερμαίνεται πᾶν τὸ ἄλλο
σῶμα ἀπὸ τούτων, καὶ καλέεται πυρετὸς τοῦτο·
θερμαίνεται δὲ ἥ τε χολὴ καὶ τὸ φλέγμα[1] ἔσωθεν
μὲν ἀπὸ σιτίων καὶ ποτῶν, ἀφ' ὧνπερ καὶ τρέφε-
ται καὶ αὔξεται, ἔξωθεν δ' ἀπὸ πόνων καὶ τρωμάτων, καὶ ὑπό τε τοῦ θερμοῦ ὑπερθερμαίνοντος καὶ
τοῦ ψυχροῦ ὑπερψύχοντος· θερμαίνεται δὲ καὶ
ἀπὸ ὄψιος καὶ ἀκοῆς, ἐλάχιστα δ' ἀπὸ τούτων.

[1] M: σῶμα Θ.

In older patients, inasmuch as their affections are milder and the sputum from them is cleaned out, fevers are milder and infrequent, and, although pains are present, they are mild. However, not even older men recover completely from affections like these, but, still retaining them, go down hill over a long period; sometimes they expectorate pus, sometimes blood, sometimes neither, and in the end they die still with the diseases. Generally they die as follows: when some additional disease similar to the one they have befalls them, the disease they first had becomes severer and, in most cases, kills them; the diseases most frequently added in this way are pleurisy and pneumonia.

23. Fever arises from the following: when bile or phlegm becomes heated, from this all the rest of the body, too, is heated, and this is called fever. Both bile and phlegm are heated from inside by the foods and drinks out of which they are nourished and grow, from outside by exertions and wounds, and by heat that makes them too hot, and cold that makes them too cold; they are also heated by seeing and hearing, but least of all by these.

ΠΕΡΙ ΝΟΥΣΩΝ Α

24. Τὸ δὲ ῥῖγος ἐν τῇσι νούσοισι γίνεται μὲν καὶ ἀπὸ τῶν ἔξωθεν ἀνέμων καὶ ὕδατος καὶ αἰθρίης καὶ ἑτέρων τοιούτων, γίνεται δὲ καὶ ἀπὸ τῶν ἐσιόντων σιτίων καὶ ποτῶν· μάλιστα δὲ καὶ ἰσχυρότερον γίνεται, ὅταν χολὴ ἢ φλέγμα συμμιχθῇ ἐς τωὐτὸ τῷ αἵματι, ἢ τὸ ἕτερον ἢ ἀμφότερα· μᾶλλον δέ, ἢν τὸ φλέγμα συμμιχθῇ· ψυχρότατον γὰρ τοῦ ἀνθρώπου φλέγμα, θερμότατον δ' αἷμα, ψυχρότερον[1] δὲ καὶ χολὴ αἵματος· ὅταν οὖν ταῦτα συμμιχθῇ, ἢ ἀμφότερα ἢ τὸ ἕτερον, ἐς τὸ αἷμα, συμπήγνυσι τὸ αἷμα, οὐ παντάπασι δέ, οὐ γὰρ ἂν δύναιτο ζῆν ὥνθρωπος, εἰ τὸ αἷμα πυκνότερόν τε καὶ ψυχρότερον γένοιτο πολλαπλασίως αὐτὸ ἑωυτοῦ.

190 Ψυχομένου | δὲ τοῦ αἵματος, ἀνάγκη ψύχεσθαι καὶ τὸ ἄλλο σῶμα πᾶν, καὶ καλέεται ῥῖγος· ὁκόταν τοῦτο τὸ τοιοῦτον[2] γένηται, ἢν μὲν ἰσχυρῶς γένηται, ῥῖγός τε ἰσχυρὸν καὶ τέτραμος· αἱ γὰρ φλέβες συσπώμεναι καὶ ἐφ' ἑωυτὰς ἰοῦσαι, πηγνυμένου τοῦ αἵματος, συσπῶσί τε τὸ σῶμα καὶ τρέμειν ποιέουσιν· ἢν δὲ ἐπὶ ἧσσον ἡ ξύνοδος γένηται τοῦ αἵματος, τοῦτο δὲ καλέεται ῥῖγος· φρίκη δὲ τὸ ἀσθενέστατον.

Ὅτι δὲ μετὰ τὸ ῥῖγος ἀνάγκη πυρετὸν ἐπιλαβεῖν ἢ πλείω ἢ ἐλάσσω, οὕτως ἔχει· ὅταν τὸ αἷμα διαθερμαίνηταί τε καὶ ἀποβιᾶται καὶ ἀπίῃ πάλιν

[1] M: -τατον Θ. [2] ῥῖγος . . . τοιοῦτον M: τοῦ τοιοῦτο Θ.

24. The chills in diseases arise both from external winds, water, clear air, and other such things, and also from ingested foods and drinks. They occur most frequently and severely when either bile or phlegm or both are mixed together in the same place with blood; in fact, more in the case of phlegm, for phlegm is the coldest part of man, blood the hottest, and bile colder than blood. Accordingly, when either one or both of these are mixed into the blood, they make the blood congeal, not totally, though, for a person could not stay alive if his blood became too many times thicker and colder than normal.

With the chilling of the blood, all the rest of the body must also be cooled, and this is called a chill; when something of this sort occurs, if it is severe, it is called a severe chill and a tremor, for the vessels, being drawn together and closing with the congealing of the blood, draw the body together and make it tremble. If the constriction of the blood is less in degree, it is called a chill; shivering is the name of the mildest form.

The reason why subsequent to the chill fever must occur, to either a greater or a lesser degree, is as follows: when the blood heats up again, regains

163

ἐς τὴν ἑωυτοῦ φύσιν, συνδιαθερμαίνεται καὶ τοῦ
φλέγματος καὶ τῆς χολῆς τὸ ἐν τῷ αἵματι συμ-
μεμιγμένον, καὶ γίνεται τὸ αἷμα θερμότερον αὐτὸ
ἑωυτοῦ πολλαπλασίως· τούτων οὖν διατεθερμα-
σμένων ἀνάγκη πυρετὸν ἐπιγενέσθαι ὑπὸ τῆς
ὑπερθερμασίης τοῦ αἵματος μετὰ τὸ ῥῖγος.

25. Ἱδρὼς δὲ γίνεται διὰ τόδε· οἷσιν ἂν κρί-
νωνται αἱ νοῦσοι ἐν τῇσι κυρίῃσι τῶν ἡμερέων καὶ
τὸ πῦρ μεθίῃ, ἐκτήκεται ἀπὸ τοῦ ἐν τῷ σώματι[1]
φλέγματος καὶ τῆς χολῆς τὸ λεπτότατον καὶ
ἀποκρίνεται καὶ χωρέει τὸ μὲν ἔξω τοῦ σώματος,[2]
τὸ δὲ καὶ αὐτοῦ ἐν τῷ σώματι ὑπολείπεται· τὸ δὲ
ὑπὸ θερμασίης λεπτυνόμενον ἀτμὸς γίνεται, καὶ
σὺν τῷ πνεύματι[3] μισγόμενον ἔξω χωρέει.

Ἔστι μὲν οὖν ταῦτα τοιαῦτα καὶ ἀπὸ τούτων ὁ
ἱδρώς. ὅτι[4] δὲ ὁ[5] μὲν θερμός, ὁ[5] δὲ ψυχρός· ὁ μὲν
192 θερμὸς ἀπὸ δια|τεθερμασμένου τε τοῦ κακοῦ καὶ
ἐκκεκαυμένου καὶ λελεπτυσμένου καὶ ἀσθενέος
καὶ οὐ λίην πολλοῦ ἀποκρίνεται, καὶ ἀνάγκη θερμότερον αὐτὸν ἐκκρίνεσθαι ἐκ τοῦ σώματος· ὁ δὲ
ψυχρὸς ἀπὸ πλέονος τοῦ κακοῦ ἀποκεκριμένος
τοῦ τε ὑπολειπομένου καὶ ἔτι ἰσχύοντος καὶ οὔπω
συσσεσηπότος οὐδὲ λελεπτυσμένου οὐδὲ ἐκκεκαυμένου, ψυχρότερος καὶ παχύτερος καὶ κακωδέστε-
ρος ἐκχωρέει.

[1] Θ: αἵματι Μ.　　　[2] Μ: αἵματος Θ.　　　[3] Μ: αἵματι Θ.
[4] Θ: διότι Μ.　　　[5] Θ: ὁτὲ Μ.

its force, and returns to its normal condition, the phlegm and bile mixed in the blood are heated with it, and as a result the blood becomes many times hotter than normal; that is, when the phlegm and bile become heated, fever follows of necessity because of the overheating of the blood after the chill.

25. Sweating occurs in the following way: in patients whose diseases have their crises on the critical days, and in whom the fever remits, there melts away from the phlegm and bile in the body the finest part, and this is secreted, part of it passing out of the body, and part of it being left behind inside; the part thinned by the heat becomes vapour and, being mixed with the breath, passes out.

Such, then, are the factors, and it is from these that sweat arises. Why the one sweat is hot, and the other one cold: the hot kind of sweat is secreted from peccant material that has been thoroughly heated, burnt up and thinned, that is weak, and that is not all too great in amount; therefore, it must be excreted from the body hotter. The cold sweat, because it is secreted from more copious peccant material, and such as has been left behind and is still strong and not yet brought to maturity, thinned or burnt up, passes out colder, thicker, and more ill-smelling.

ΠΕΡΙ ΝΟΥΣΩΝ Α

Δῆλον δὲ τῷδε τοῦτο· οἱ ψυχρῷ ἱδρῶτι ἱδρῶντες μακρὰς νούσους νοσέουσιν ὡς ἐπὶ τὸ πολύ, ἔτι
ἰσχύοντος τοῦ κακοῦ τοῦ ἐν τῷ σώματι ὑπολειπο-
μένου· οἱ δὲ θερμῷ ἱδρῶτι ἱδρῶντες ταχύτερον
ἀπαλλάσσονται τῶν νοσημάτων.

26. Πλευρῖτις καὶ περιπλευμονίη γίνονται
ὧδε. ἡ μὲν πλευρῖτις· ὅταν πόσιες ἀλέες τε καὶ
ἰσχυραὶ κάρτα λάβωσι, διαθερμαίνεται τὸ σῶμα
πᾶν ὑπὸ τοῦ οἴνου καὶ ὑγραίνεται· μάλιστα δὲ ἥ
τε χολὴ καὶ τὸ φλέγμα διαθερμαίνεταί τε καὶ
ὑγραίνεται· ὅταν οὖν τούτων κεκινημένων τε καὶ
διυγρασμένων ξυγκυρήσῃ, ὥστε ῥιγῶσαι μεθύοντα
ἢ νήφοντα, ἅτε ἐὸν τὸ πλευρὸν ψιλὸν φύσει σαρκὸς μάλιστα τοῦ σώματος καὶ οὐκ ἐόντος αὐτῷ
ἔσωθεν τοῦ ἀντιστηρίζοντος οὐδενός, ἀλλὰ κοιλίης, αἰσθάνεται μάλιστα τοῦ ῥίγεος· καὶ ὅταν ῥι-
γώσῃ τε καὶ ψυχθῇ, ξυνέλκεταί τε καὶ συσπᾶται
ἥ τε σὰρξ ἡ ἐπὶ τῷ πλευρῷ καὶ τὰ φλέβια, καὶ
ὅσον ἐν αὐτῇ τῇ σαρκὶ ἔνι χολῆς ἢ φλέγματος ἢ
ἐν τοῖσιν ἐν αὐτῇ φλεβίοισι, τούτου τὸ πολλὸν ἢ
πᾶν ἀποκρίνεται ἔσω[1] πρὸς τὸ θερμόν, πυκνουμέ-
νης τῆς σαρκὸς ἔξωθεν, καὶ προσπήγνυται πρὸς
τῷ πλευρῷ· καὶ ὀδύνην τε παρέχει ἰσχυρὴν καὶ
διαθερμαίνεται, καὶ ὑπὸ τῆς θερμότητος ἄγει ἐφ'
ἑωυτὸ ἀπὸ τῶν πλησίον καὶ φλεβῶν καὶ σαρκῶν

[1] M adds συνωθεόμενον.

DISEASES I

This is shown by the following: patients that experience cold sweating generally suffer from lengthy illnesses, since the evil left behind in the body is still strong, whereas those with hot sweating recover more quickly from their diseases.

26. Pleurisy and pneumonia arise as follows: first pleurisy: when strong drinks drunk close together have a violent effect, the whole body is heated by the wine and becomes moist—especially the bile and phlegm in it. So when, with these set in motion and greatly moistened, it happens that the person, drunk or sober, has a chill, it is his side that feels the chill most, inasmuch as it is by nature the part of the body most barren of tissue, and since there is nothing inside it to offer any resistance, but only hollowness. When the side has this chill and is cooled, the tissue on the side and the small vessels are drawn together and contracted, and most or all of the bile or phlegm present in the tissue itself or in the small vessels in the tissue is secreted inwards towards the heat— the tissue being condensed from without—and becomes fixed against the side; here it produces severe pains, becomes heated and, because of its heat, attracts phlegm and bile from the nearby

φλέγμα τε καὶ χολήν. γίνεται μὲν οὖν τούτῳ τῷ τρόπῳ.

194 Ὅταν δὲ τὰ πρὸς τῷ πλευρῷ | προσπαγέντα σαπῇ καὶ πτυσθῇ, ὑγιέες γίνονται· ἢν δὲ τό τε ἀρχαῖον πολλὸν προσπαγῇ πρὸς τῷ πλευρῷ καὶ ἄλλο προσεπιγένηται, αὐτίκα ἀπόλλυνται, οὐ δυνάμενοι ἀναπτύσαι ὑπὸ πλήθεος τοῦ σιάλου, ἢ ἔμπυοι γίνονται· καὶ οἱ μὲν ἀπόλλυνται, οἱ δὲ ἐκφεύγουσιν. διαδηλοῖ δὲ ταῦτα ἐν τῇσιν ἑπτὰ ἡμέρῃσιν ἢ ἐννέα ἢ ἕνδεκα ἢ τεσσερεσκαίδεκα.

Ὀδύνην δὲ παρέχει ἐς τὸ ὦμον καὶ ἐς τὴν κληῖδα καὶ ἐς τὴν μασχάλην διὰ τόδε· ἡ φλὲψ ἡ σπληνῖτις καλεομένη τείνει ἀπὸ τοῦ σπληνὸς ἐς τὸ πλευρόν, ἐκ δὲ τοῦ πλευροῦ ἐς τὸν ὦμον καὶ ἐς τὴν χεῖρα[1] τὴν ἀριστερήν· ἡ δὲ ἡπατῖτις ἐς τὰ δεξιὰ ὡσαύτως· καὶ ὅταν ταύτης τὸ ἐπὶ τοῦ πλευροῦ συνειρυσθῇ ὑπὸ τοῦ ῥίγεος, καὶ φρίξῃ τὸ αἷμα ἐν αὐτῇ, ἔς τε τὴν μασχάλην καὶ τὴν κληῖδα καὶ τὸν ὦμον ξυνέρχεταί τε καὶ σπᾷ, καὶ ὀδύνην παρέχει.

Κατὰ δὲ τὸν αὐτὸν λόγον καὶ τὰ περὶ τὸν νῶτον χωρία διαθερμαίνεται ὑπὸ τοῦ προσπεπηγότος πρὸς τῷ πλευρῷ[2] φλέγματός τε καὶ χολῆς.

Παρέχει δ' ὀδύνην ἐνίοτε καὶ τοῖσι τοῦ πλευροῦ κάτωθε χωρίοισι. πολλάκις δέ, ἢν ἐς τὰ κάτω τράπηται ὀδύνη, διαδιδοῖ ἐς τὴν κύστιν διὰ τῶν

[1] Μ: ῥάχιν Θ.   [2] Μ: πλεύμονι Θ.

vessels and tissues. Pleurisy arises, then, in this way.

When what has become fixed on the side reaches maturity and is expectorated, patients recover; but, if the original amount fixed on the side was great, and yet more is added, patients either succumb at once, being unable to clean out their chest because of the amount of the sputum, or they suppurate internally; in the latter case, some die and some escape; these things give a clear indication in seven, nine, eleven, or fourteen days.

Pleurisy produces pain in the shoulder, collar-bone, and axilla in the following way: the vessel called the splenic leads from the spleen to the side, and from the side to the shoulder and left arm; the hepatic vessel does the same on the right. When the part of this vessel along the side is contracted by the chill, and the blood shudders inside it, it closes and contracts as far as the axilla, collar-bone, and shoulder, and so produces pain.

In the same way, the parts in the back, too, become heated by the phlegm and bile fixed on the side.

Sometimes pleurisy also produces pain in the parts below the side. Often, if the pain turns downward, it spreads through the small vessels to the

φλεβίων, καὶ οὐρέει πολλόν τε καὶ χολῶδες. νο-
μίζουσι δὲ ταύτης τῆς νούσου τὸ ῥῖγος αἴτιον
εἶναι καὶ ἀρχήν.

27. Ἡ δὲ περιπλευμονίη· ὅταν κεκινημένου
τε καὶ ὑγραινομένου τοῦ φλέγματος καὶ τῆς χολῆς ἑλκύσῃ ὁ πλεύμων ὑπὸ θερμασίης ἐφ᾽ ἑωυτὸν
ἀπὸ τῶν πλησίον χωρίων πρὸς τοῖσιν ὑπάρχουσιν
ἐν ἑωυτῷ, διαθερμαίνει τε πᾶν τὸ σῶμα καὶ ὀδύνην παρέχει, μάλιστα δὲ τῷ τε νώτῳ καὶ τῇσι
πλευρῇσι καὶ τοῖσιν ὤμοισι καὶ τῇ ῥάχει, ἅτε ἀπὸ
τούτων ἕλκων ἐς ἑωυτὸν τὴν ἰκμάδα τὴν πλεί-
στην καὶ ὑπερξηραίνων τε ταῦτα καὶ ὑπερθερ-
μαίνων· ὅταν δ᾽ εἰρύσῃ ἐς ἑωυτόν, καὶ ἕδρην λάβῃ
ἥ τε χολὴ καὶ τὸ φλέγμα ἐν τῷ | πλεύμονι, σή-
πεται καὶ πτύεται.

Καὶ ἢν μὲν ἐν τῇσι κυρίῃσι τῶν ἡμερέων
σαπέντα πτυσθῇ, ὑγιὴς γίνεται·[1] ἢν δὲ τά τε ἐπ-
ελθόντα τὴν ἀρχὴν δέχηται, καὶ προσεπιγίνηται
ἕτερα, καὶ μήτε πτύων μήτε σήπων κρατῇ ὑπὸ
πλήθεος τῶν ἐπιγινομένων, ἀπογίνονται ὡς τὰ
πολλά· ἢν δὲ πρὸς τὰς ἡμέρας διαγένωνται τὰς
δύο καὶ εἴκοσι, καὶ τὸ πῦρ μεθῇ, καὶ ἐν ταύτῃσι
μὴ ἐκπτυσθῇ, πάντες ἔμπυοι γίνονται, μάλιστα
δὲ τούτων,[2] οἷσιν ἰσχυρόταται ἥ τε πλευρῖτις καὶ
ἡ περιπλευμονίη.

[1] ὑ. γ. Θ: περιγίνονται Μ.   [2] μ. δ. τ. Θ: γίνονται δὲ μάλιστα ἐκ τούτων Μ.

bladder, and the patient passes much bilious urine. People consider the chill to be to blame for this disease, and to be its origin.

27. Pneumonia: when the lung, because of its heat, attracts from the nearby parts phlegm and bile, that have been set in motion and moistened, in addition to the phlegm and bile already present in it, this heats the whole body and produces pain, especially in the back, sides, shoulders and spine, since the lung attracts most of the moisture out of these, and dries and heats them too much. When the lung has drawn the bile and phlegm to itself, and they come to rest in it, they become mature and are coughed up.

If, on becoming mature, they are all expectorated on the critical days, the patient recovers. But, if what arrives at the beginning is taken in, and then more is added, and the patient, on account of the great amount added, cannot gain the upper hand in expectoration and bringing to maturity, he generally dies. If such patients survive for twenty-two days and their fever remits, but during this time expectoration does not take place, they invariably suppurate internally, especially those in whom the pleurisy and pneumonia are severest.

171

ΠΕΡΙ ΝΟΥΣΩΝ Α

28. Γίνεται δὲ καὶ πλευρῖτις ἄπτυστος καὶ περιπλευμονίη, ἄμφω ἀπὸ τοῦ αὐτοῦ, ἀπὸ ξηρασίης· ξηραίνει δὲ καὶ τὰ θερμά, ὅταν ὑπερθερμαίνῃ, καὶ τὰ ψυχρά, ὅταν ὑπερψύχῃ.[1] πήγνυται δὴ τὸ πλευρὸν καὶ τὰ ἐν αὐτῷ τῷ πλευρῷ φλέβια, καὶ ξυσπᾶται, καὶ ὅσον ἐν αὐτῷ ἔνι χολῆς ἢ φλέγματος, τοῦτο ὑπὸ τῆς ξηρασίης[2] ἐνέσκληκέ τε καὶ ὀδύνην παρέχει καὶ ὑπὸ τῆς ὀδύνης πυρετόν.

Τούτου ξυμφέρει τὴν φλέβα ἀποσχάσαι τὴν ἐν τῇ χειρί, τὴν σπληνῖτιν καλεομένην ἢ τὴν ἡπατῖτιν, καθ᾽ ὁπότερον ἂν ᾖ τὸ νόσημα· καὶ οὕτως ἡ ὀδύνη μαλακωτέρη γίνεται τοῦ πλευροῦ τε καὶ τῶν ἄλλων· ἡ γὰρ φλέψ, ὅσον ἔνι ἐν αὐτῇ χολῆς ἢ φλέγματος, αὐτοῦ τοῦ αἵματος νενοσηκότος, μετὰ τούτου μεθίησι τὸ πολὺ ἔξω· τὸ δ᾽ ἐκ τῆς σαρκὸς ὑπό τε φαρμάκων ποτῶν διαχεῖται καὶ ὑπὸ χλιασμάτων προστιθεμένων ἔξωθεν, ὥστε τὴν νοῦσον σκίδνασθαι ἀνὰ πᾶν τὸ σῶμα. καλέεται δ᾽ αὕτη ἡ νοῦσος[3] πλευρῖτις ἄπτυστος.

Ἡ δὲ περιπλευμονίη, ὅταν ὁ πλεύμων ὑπερξηρανθῇ καὶ αὐτός· καὶ ὅσον[4] ἐν αὐτῷ ἔνι χολῆς ἢ φλέγματος, οὔτε σήπει ὁμαλῶς, οὔτε τὸ σίελον ἀναδιδοῖ· ὅσον τε ἐν | αὐτῷ ἐστιν ἰκμάδος, ἢ ἀπὸ ποτοῦ ἢ ἀπὸ ρυφήματος ἢ ἀπὸ τῶν πλησίον

---

[1] Μ: ὑπερθερμανθῇ . . . ὑπερψυχθῇ Θ.   [2] Θ: θερμασίης Μ.
[3] ἡ νοῦσος om. Μ.   [4] Θ: ὁκόταν Μ.

172

DISEASES I

28. Both pleurisy without expectoration and
pneumonia without expectoration arise from the
same thing, from dryness; and both heat, when it
makes anything too hot, and cold, when it makes
anything too cold, dry. The side, then, and the
small vessels in it congeal and draw together, and
whatever bile or phlegm is present in them
becomes hard, because of the dryness, and produces
pain and, because of the pain, fever.

It is of benefit to open the vessel in this patient's
arm, either the one called the splenic or the one
called the hepatic, according to which side the
disease is on. In this way, the pain becomes milder
in both the side and the rest of the body, for the
vessel discharges to the exterior with the blood,
which has itself become diseased, most of the bile
and phlegm that are in it. The bile and phlegm
from the tissue are dispersed by medicinal drinks,
and by fomentations applied externally, so that the
disease is dissipated through the whole body. This
disease is called pleurisy without expectoration.

Pneumonia in which the lung becomes too dry:
in this case, whatever bile or phlegm is in the lung
does not become sufficiently mature, nor does it
give off sputum, but the moisture in it, whether
coming from drink or gruel, or out of the nearby

χωρίων, τοῦτο πᾶν ἐκκαίει ὑπὸ τῆς ὑπερξηρασίης
τε καὶ θερμασίης.

Τούτῳ ξυμφέρει πώματα πίνειν, ὑφ᾽ ὧν
ὑγραίνεται ὁ πλεύμων καὶ πτύσεται· ἢν γὰρ
μὴ πτυσθῇ, σκληρός[1] τε γίνεται ὁ πλεύμων καὶ
συναυαίνεται, καὶ τὸν ἄνθρωπον ἀπόλλυσι.

29. Καῦσος δὲ λαμβάνει μὲν μᾶλλον τοὺς χο-
λώδεας, λαμβάνει δὲ καὶ τοὺς φλεγματίας, λαμβάνει δ᾽ ὧδε· ὅταν χολὴ κινηθῇ κατὰ τὸ σῶμα,
καὶ ξυγκυρήσῃ ὥστε τὰς φλέβας καὶ τὸ αἷμα
εἰρύσαι τῆς χολῆς—εἰρύσαι δὲ [τὸ][2] πλεῖστον ἔκ
τε τῶν σαρκῶν καὶ τῆς κοιλίης τὸ πρόσθεν
ἐνεόν—ἅτε τῇ φύσει θερμότατον ἐὸν ἐν τῷ σώματι, τὸ αἷμα, ὅταν διαθερμανθῇ ἔκ τε τῶν σαρκῶν καὶ τῆς κοιλίης [πρὸς τῷ ἐνεόντι][3] ὑπὸ τῆς
χολῆς, διαθερμαίνει καὶ τὸ ἄλλο σῶμα πᾶν. καὶ
τὰ μὲν ἔνδον[4] ὑπὸ πολλῆς ἰκμάδος οὐ δύναται
ἀποξηραίνεσθαι παντάπασιν· ἢν δ᾽ ἀποξηρανθῇ,
ἀποθνήσκει ὤνθρωπος· τὰ δ᾽ ἐν τοῖσιν ἀκρωτηρίοισι τοῦ σώματος, ἅτε ξηρὰ ἐόντα φύσει, ἀποξηραίνεταί τε καὶ ἐκκαίεται ἐξ αὐτῶν τὸ ὑγρὸν τὸ
πλεῖστον· καὶ εἰ ἐθέλοις ψαύειν, ψυχρά τε αὐτὰ
εὑρήσεις καὶ ξηρά· καὶ διὰ τοῦτο ὅσοι ὑπὸ καύσου
ἁλίσκονται, τὰ μὲν ἔνδον καίονται ὑπὸ τοῦ πυρός,
τὰ δ᾽ ἔξω ψυχροί εἰσιν, ἡ δὲ γλῶσσα καὶ ἡ

---

[1] Θ: -ότερός Μ.   [2] Del. Potter.   [3] Del. Potter: Μ adds
ἔτι μᾶλλον.   [4] ἔνδον om. Μ.

parts, is burnt out completely by the great dryness and heat.

It is beneficial for this patient to drink medications that will moisten the lung and promote expectoration. For, unless expectoration occurs, the lungs become hard and dried up, and kill the person.

29. Ardent fever befalls bilious persons more, although it does also attack phlegmatic ones. It arises as follows: when bile is set in motion through the body, and it happens that the vessels and blood attract some of it—they attract mostly what was previously in the tissues and cavity—the blood, inasmuch as it is by nature the hottest thing in the body, when it has been heated by the bile out of the tissues and the cavity, heats all the rest of the body, too. The internal parts, because of their large proportion of moisture, cannot be completely dried out; if they are, the person dies. But the parts at the extremities of the body, inasmuch as they tend naturally to dryness, are dried up, and most of the moisture is burnt out of them; if you wish to touch them, you will find them cold and dry; for this reason, patients suffering from ardent fever are burnt internally from the heat, but externally are cold.

φάρυγξ τρηχύνεταί τε καὶ ξηραίνεται[1] ὑπὸ τοῦ
πνεύματος τοῦ ἔνδον [καὶ][2] τῆς θερμότητος.

Ὅσον δ᾽ ἂν ἐν τῇ κοιλίῃ ἢ ἐν τῇ κύστι ἐγγέ-
νηται χολῆς, τὸ μὲν ἐν τῇ κοιλίῃ ἐνίοτε μὲν δια-
ταράσσεται κάτω, τὰ δὲ πολλὰ ἐμέεται ἐν τῇσι
πρώτῃσιν ἡμέρῃσιν ἢ τέσσερσιν ἢ πέντε· ἐμέεται
δὲ διὰ τόδε· ὅταν ἡ ἄνω κοιλίη ὑπερθερμανθῇ,
ἕλκει ἐφ᾽ ἑωυτήν, καὶ γίνεται ἔμετος· διὰ τοῦτο |
200 δ᾽ αὐτὸ καὶ ἐς περιπλευμονίην ἐκ καύσου τε καὶ
πλευρίτιδος μάλιστα μεθίσταται τὰ νοσήματα·
ὅταν γὰρ ἡ ἄνω κοιλίη ὑπερθερμανθῇ, ἕλκει ἐφ᾽
ἑωυτήν, καὶ ὑποδέχεται ὁ πλεύμων, καὶ γίνεται
περιπλευμονίη, καὶ ὡς τὰ πολλὰ ἀπόλλυνται,
ἅτε ἐόντες ἤδη ἀσθενέες καί, ἑτέρης νούσου
καινῆς ἐπιγενηθείσης, οὐ δυνάμενοι τὰς ἡμέρας
διατελέειν, ἄχρι οὗ τὸ σίαλον πεπανθῇ ἐν τῷ
πλεύμονι, ἀλλ᾽ ὡς τὰ πολλὰ προαπόλλυνται
ὑπὸ ἀσθενείης· ἔνιοι δὲ περιγίνονται.

Ὅσον[3] δ᾽ ἐς τὴν κύστιν συρρέει χολῆς, οὐρεῖται
παχύ, παχὺ δ᾽ ὑπὸ φλέγματος καὶ χολῆς·
χολῶδες δὲ διαχωρέει, ὅταν διαχωρέῃ,[4] ὑπὸ τοῦ
ξυγκεκαῦσθαι ἐν τῇ κοιλίῃ τὰ ἐνεόντα.

30. Φρενῖτις δ᾽ οὕτως ἔχει· τὸ αἷμα ἐν τῷ ἀν-
θρώπῳ πλεῖστον συμβάλλεται μέρος συνέσιος·

---

[1] Θ: αὐαίνεται Μ.   [2] Del. I, Wittern.   [3] Θ: Ὁκόσοισι Μ.
[4] Wittern: φλέγματος καὶ χολῆς διαχωρέει, ὅταν διαχωρέῃ Θ· φλέγμα-
τος· χολῶδες δὲ διαχωρέει Μ.

176

The tongue and throat become rough and are dried up by the heat of the breath on them.

Any bile that occupies the cavity or bladder: in the case of the cavity, it is sometimes evacuated downwards, but usually vomited up in the first four or five days. It is vomited up for the following reason: when the upper cavity is overheated, it attracts, and vomiting takes place. (For this same reason, progression from ardent fever and pleurisy to pneumonia is also common; for, when the upper cavity becomes very heated, it attracts, the lung accepts, and pneumonia develops; such patients generally die; for, being weak already, when the new disease is added they are not able to make it to the days when the sputum in their lung would reach maturity, but usually die before that due to weakness. Some survive.)

Any bile that flows into the bladder is passed as thick urine, thick because of the phlegm and bile. The stools the patient passes, when he does pass any, are bilious, because the contents in the cavity have been burnt up.

30. Phrenitis is as follows: the blood in man contributes the greatest part to his intelligence, some

ἔνιοι δὲ λέγουσι τὸ πᾶν· ὅταν οὖν χολὴ κινηθεῖσα
ἐς τὰς φλέβας καὶ ἐς τὸ αἷμα ἐσέλθῃ, δι’ οὖν
ἐκίνησε καὶ διώρρωσε[1] τὸ αἷμα ἐκ τῆς ἐωθυίης
συστάσιός τε καὶ κινήσιος, καὶ διεθέρμηνε· διαθερμαίνει δὲ[2] καὶ τὸ ἄλλο σῶμα πᾶν, καὶ παρανοέει
τε ὥνθρωπος καὶ οὐκ ἐν ἑωυτῷ ἐστιν ὑπὸ τοῦ πυρετοῦ τοῦ πλήθεος καὶ τοῦ αἵματος τῆς διορρώ-
σιός[3] τε καὶ κινήσιος γενομένης οὐ τῆς ἐωθυίης.

Προσεοίκασι δὲ μάλιστα οἱ ὑπὸ τῆς φρενίτιδος
ἐχόμενοι τοῖσι μελαγχολώδεσι κατὰ τὴν παρά-
νοιαν· οἵ τε γὰρ μελαγχολώδεις, ὅταν φθαρῇ τὸ
αἷμα ὑπὸ χολῆς καὶ φλέγματος, τὴν νοῦσον ἴσχουσι
καὶ παράνοοι γίνονται, ἔνιοι δὲ καὶ μαίνονται· καὶ
ἐν τῇ φρενίτιδι ὡσαύτως· οὕτω δὲ ἧσσον ἡ μανίη
τε καὶ ἡ παραφρόνησις γίνεται, ὅσῳπερ ἡ χολὴ
τῆς χολῆς ἀσθενεστέρη ἐστίν.

202   31. Ὕφαιμον δὲ καὶ πελιδνὸν ἔν τε τῇ πλευρίτιδι καὶ ἐν τῇ περιπλευμονίῃ διὰ τόδε πτύουσι·
κατ’ ἀρχὰς μὲν ὡς τὸ πολὺ οὐδέτερα πτύουσιν,
οὔτε πελιδνὸν οὔτε ὕφαιμον· εἰδέναι δὲ χρὴ ἰσχυρὴν ἐοῦσαν τὴν νοῦσον, ὅταν τὸ σίαλον ἄρξωνται
ὑπόπαχυ πτύειν, καὶ καθαίρεσθαι μάλιστα τότε.
πτύεται δ’ ἀπὸ διατάσιος τῶν φλεβῶν, τῆς μὲν
πλευρίτιδος ἐκ τῶν ἐν τῷ πλευρῷ, τῆς δὲ περιπλευμονίης ἐκ τῶν ἐν τῷ πλεύμονι [, καὶ θερμα-

---

[1] Foes (*Oeconomia*, s.v. Διορρώσιος)· διούρησε ΘΜ.   [2] Θ: δια-
θερμανθὲν δὲ διαθερμαίνει Μ.   [3] Foes: διουρήσιός ΘΜ.

people say everything; therefore, when bile that has been set in motion enters the vessels and the blood, it stirs the blood up, heats it, and turns it to serum, altering its normal consistency and motion; now the blood heats all the rest of the body, too, and the person, because of the magnitude of his fever, and because his blood has become serous and abnormal in its motion, loses his wits and is no longer himself.

Patients with phrenitis most resemble melancholics in their derangement, for melancholics too, when their blood is disordered by bile and phlegm, have this disease and are deranged—some even rage. In phrenitis it is the same, only here the raging and derangement are less in the same proportion that this bile is weaker than the other one.[1]

31. In pleurisy and pneumonia patients cough up sputum that is bloody and livid because of the following. (Actually, at the beginning they rarely expectorate either livid or bloody sputum. You must recognize that the disease is severe, when patients begin to expectorate sputum that is thickish, and you must clean them right then.) It is coughed up as the result of a dilation of the vessels, in pleurisy, from those of the side, in pneumonia, from those of the lung [, and it draws heat to itself].

[1] The difference in severity, between the mental disturbances of melancholy and those of phrenitis, is related to a difference in strength of the biles producing them.

σίην ἐπάγει ἐφ' ἑωυτήν].[1] ἢν δὲ ῥωγματίης ᾖ ὁ
τὴν νοῦσον ἔχων καὶ σαβακός, ἀπὸ τῆς πρώτης
ἡμέρης καὶ αἷμα καὶ ὕφαιμον καὶ πελιδνὸν σὺν
σιάλῳ πτύει· τὸ δὲ πελιδνὸν ἀπὸ τοῦ αἵματος γίνεται, ἢν ὀλίγον συμμίσγηται ἐς πολὺ σίαλον, καὶ
μὴ παραυτίκα πτύηται, ἀλλ' ἐμμένῃ[2] ἡμισαπὲς
ἐὸν καὶ ἐκτεθηλυσμένον ἐν τῷ σώματι.

32. Ἀποθνήσκουσι δ' ἀπὸ μὲν πλευρίτιδος,
ὅταν πολλὸν μὲν τὴν ἀρχὴν τῷ πλευρῷ προσπαγῇ φλέγμα τε καὶ χολή, πολλὸν δὲ προσεπιρρυῇ
καὶ ἐκ τοῦ ἄλλου σώματος, καὶ μήτε πτύων κρατῇ ὑπὸ πλήθεος τούτων[3] μήτε σήπων, πιμπλῶν-
ται δὲ αἱ ἀρτηρίαι ὑπὸ τῶν ἐνεόντων φλέγματός
τε καὶ πύου· τότε δὲ ῥέγκει καὶ ἀναπνεῖ πυκνόν
τε καὶ αὐτόθεν ἄνωθεν, τέλος δὲ ἀποφράσσεται
πάντα καὶ ἀποθνήσκει. τὸν αὐτὸν δὲ τρόπον[4] καὶ
ἐκ περιπλευμονίης ἀπόλλυνται.

33. Ὅσοι δ' ὑπὸ καύσου ἀποθνήσκουσι, πάντες ὑπὸ ξηρασίης ἀποθνήσκουσιν· ἀποξηραίνεται
204 δ' αὐτῶν πρῶτον μὲν τὰ ἀκρωτήρια, | πόδες τε
καὶ χεῖρες, ἔπειτα δὲ τὰ ἐπιξηρότερα· ὅταν δ' ἐκ-
καυθῇ τε καὶ ἀποξηρανθῇ παντάπασι τὸ ὑγρὸν ἐκ
τοῦ σώματος, τὸ μὲν αἷμα πήγνυταί τε παντελῶς
καὶ ψύχεται, τὸ δ' ἄλλο σῶμα ἀποξηραίνεται, καὶ
οὕτως ἀποθνήσκει.

---

[1] Del. Ermerins.   [2] ἐμμένῃ om. Θ.   [3] τούτων om. M.
[4] M adds τοῦτον.

If the person with the disease is affected by tears and he is feeble, from the first day he expectorates with his sputum blood and material that is bloody and livid. The livid material arises when a small amount of blood is mixed together into much sputum, and this is not expectorated at once, but remains in the body semi-matured and softened.

32. Patients die from pleurisy when much phlegm and bile becomes fixed in the side at the onset, and then much more is added out of the rest of the body, too, so that, because of the great mass of these substances, the patient can succeed neither in coughing them up nor in bringing them to maturity, but his bronchial tubes are filled by the phlegm and pus in them. Then, the patient's breathing becomes stertorous, and he exhales rapidly and only from the upper part of his chest; in the end, he becomes completely blocked up, and dies. Patients also die from pneumonia in this same way.

33. Patients that die from ardent fever all die as the result of dryness. First their extremities become dry, the feet and hands, then the other parts that tend more towards dryness. When the moisture has been altogether burnt out of and dried up from the body, the blood congeals completely and becomes cold, and the rest of the body dries out; this is how the person dies.

34. Ὑπὸ δὲ τῆς φρενίτιδος ἀπόλλυνται[1] οὕτως· παραφρονέουσιν ἐν τῇ νούσῳ διὰ παντός, ἅτε τοῦ αἵματος ἐφθαρμένου τε καὶ κεκινημένου οὐ[2] τὴν ἐωθυῖαν κίνησιν· καὶ ἅτε παραφρονέοντες οὔτε τι τῶν προσφερομένων δέχονται, ὅ τι καὶ ἄξιον λόγου, ὅταν τε προΐῃ ὁ χρόνος, μαραίνονταί τε καὶ μινύθουσιν ὑπό τε τοῦ πυρετοῦ καὶ ὑπὸ τοῦ μηδὲν ἐσιέναι ἐς τὸ σῶμα· καὶ πρῶτα μὲν τὰ ἐν τοῖσιν ἀκρωτηρίοισι μινύθει τε καὶ ψύχεται, ἔπειτα δὲ τὰ ἐπ' ἐγγυτάτω.

Καὶ ψύχεος δὲ καὶ πυρὸς καὶ πόνων ἀρχὴν ταύτην ἴσχει· ὅταν τὸ αἷμα ἐν τῇσι φλεψὶν ὑπὸ τοῦ φλέγματος ψυχθῇ, μεταπίπτει τε καὶ ξυσπᾶται ἁλὲς ἄλλοτε ἄλλῃ καὶ τρέμει· τέλος δὲ ψύχεται πάντα καὶ ἀποθνήσκει.

[1] Θ: -υται Μ.   [2] οὐ om. Μ.

DISEASES I

34. From phrenitis patients die in the following way. Inasmuch as their blood is corrupted and does not move in its normal motion, they are deranged all through the disease. Being deranged, they do not accept anything worth mentioning of what is administered to them, and as time passes they waste away and become emaciated as a result of their fever and of the fact that nothing is entering their body. First the parts at the extremities shrivel up and become cold, then the nearer parts.

This is the origin of the cold, fever and pains: when the blood in the vessels is cooled by the phlegm, it migrates and contracts into a mass at one time in one part, at another time in another part, and trembles. Finally, everything becomes cold and the person dies.

DISEASES II

INTRODUCTION

That one word in Erotian's glossary[1] is present in *Diseases II* but no other treatise of the Hippocratic Collection provides inconclusive evidence of his acquaintance with the work.

Galen includes many words from *Diseases II* in his Hippocratic glossary, and refers to the writing by name five times as *Diseases I the Greater*,[2] once as *Diseases I*,[3] once as *Diseases the Greater*,[4] and once as *Diseases II*.[5] He quotes from the treatise in his commentary on *Epidemics VI*:

Also in *Diseases* ascribed to Hippocrates, which is thought by the followers of Dioscurides to have been written by Hippocrates the son of Thessalus, a livid disease is mentioned. This is the beginning of the book: "When the head becomes overheated, much urine is passed". About the livid disease he writes the following, word for word: "Livid disease: a dry fever comes on, occasional shivering, the patient suffers pain

---

[1] Φ19 φῶδες (ed. E. Nachmanson, Gothenburg, 1918, p. 92).

[2] Under the words ἀμαλῶς, βρῆγμα, κύκλοι προσώπου, <ἐκ>μάσσειν and μελιηδέα (Kühn XIX. 77, 89, 115, 120 and 121).

[3] Under the word κρότωνας (Kühn XIX. 115).

[4] Under the word ἄορτρον (Kühn XIX. 82).

[5] Under the word ἀποσπαρθάζουσι (Kühn XIX. 85).

in his head and inward parts, and he vomits bile; when the pain is present he cannot look up, but feels weighed down. His belly is costive, and his complexion, lips, and the whites of his eyes become livid; he stares as if he were being strangled. Sometimes his colour changes, too, and turns from livid to yellow-green."[1]

Diseases II consists of two parts (chs. 1–11 and 12–75). Chs. 1–11 are devoted primarily to aetiology and pathogenesis; in each chapter, the symptoms and course of one disease are described and then explained according to the author's implicit speculative theories of bodily structure and function. With ch. 12, a fresh beginning is made, and from then on prognosis and treatment predominate. Chs. 1–11 and 12–31 overlap in subject-matter.

The diseases are arranged in the treatise according to anatomy, as follows:

Diseases of the Head: 1–11 and 12–37
General Diseases: Jaundices: 38–39
                     Fevers: 40–43
Diseases centred in the Lungs and Sides: 44–65
Diseases centred in the Abdomen:
                     Vomiting of Bile: 66–70
                     White Phlegm: 71
                     Phrenitis: 72
                     Dark Diseases: 73–75.

[1] Kühn XVII(1). 888 = CMG 10,2,2 pp. 55 f.; the reference is to *Diseases II* 68.

DISEASES II

The disease categories tend to be indefinite and are not clearly reflected in the chapter divisions.[1]

The correspondence in subject-matter between chs. 1–11 and 12–31 is as follows:

Diseases II is included in the standard collected Hippocratic editions and translations, and in the two renaissance works dealing with the Hippocratic

[1] Compare e.g. chapters 17–18; 26–28; 33–37; 44–46; 47–53; 66–70.

DISEASES II

Diseases.[1] In his book cited above (p. 5) Jouanna
provides a new edition of the text of more than half
the chapters of *Diseases II*.

[1] See above, p. 97.

1. Οὐρέεται πολλὸν ὅταν ὑπερθερμανθῇ ἡ κε-
φαλή· τήκεται γὰρ ἐν αὐτῇ τὸ φλέγμα· τηκόμε-
νον δὲ χωρέει τὸ μὲν ἐς τὰς ῥῖνας, τὸ δ᾽ ἐς τὸ
στόμα, τὸ δὲ διὰ τῶν φλεβῶν αἳ ἄγουσιν ἐς τὸ
αἰδοῖον. ὅταν δ᾽ ἐς τὸ αἰδοῖον ἀφίκηται, οὐρέει
καὶ πάσχει οἷά περ ὑπὸ στραγγουρίης. ἀμβλυώσ-
σουσι δ᾽ ὅταν ἐς τὰ ἐν τοῖσιν ὀφθαλμοῖσι φλέβια
ἐσέλθῃ φλέγμα· ὑδαρεστέρη γὰρ γίνεται ἡ ὄψις
καὶ θολερωτέρη, καὶ τὸ λαμπρὸν ἐν τῷ ὀφθαλμῷ
οὐχ ὁμοίως λαμπρόν ἐστιν, οὐδὲ καταφαίνεται ἐν
αὐτῷ, ἢν[1] θέλῃ ὁρᾶν, ὁμοίως ὡς καὶ ὅτε λαμπρὸς
καὶ καθαρὸς[2] ἦν. οὗτος ἐν τεσσεράκοντα ἡμέρῃσι
μάλιστα ὑγιάζεται. ἢν δὲ χρόνῳ ὕστερον πολλῷ
ὑποστρέψῃ ἡ νοῦσος, τὸ δέρμα τῆς κεφαλῆς παχύνεται, καὶ τὸ ἄλλο σῶμα αἴρεται καὶ παχύνεται
καὶ εὐχροέει. τούτῳ[3] τὸ φλέγμα ἐς τὰς σάρκας
τρέπεται, καὶ ὑπὸ τούτου δοκέει παχὺς εἶναι· αἱ
γὰρ σάρκες, ἅτε διάβροχοι ἐοῦσαι καὶ ἠρμέναι καὶ

---

[1] Jouanna (p. 32): ἐὰν later mss., edd.: ἂν ΘΜ.   [2] Μ: -ως
καὶ -ως Θ.   [3] Θ: τοῦτο Μ.

190

DISEASES II

1. When the head becomes overheated, much urine is passed; for the phlegm in it melts, and, as it melts, it goes partly to the nostrils, partly to the mouth, and partly through the vessels that lead to the genital organs. When it arrives in the genital organs, the patient urinates, and suffers the symptoms of strangury. Patients see unclearly, in this condition, when phlegm enters the small vessels of their eyes; for the pupil becomes more watery and turbid, so that the clear part of the eye is no longer as clear as it was, and thus the image does not appear in it, when it wishes to see, the same as when it was clear and pure. This patient generally generally recovers in forty days. If the disease recurs a long time later, the skin on the head becomes thick, and the rest of the body swells, becomes stout, and takes on a good colour. In this patient, phlegm is invading the tissues, and that is why he appears stout; the tissues, being sodden, swollen, and more expanded than normal, draw

ἀραιότεραι, ἕλκουσιν ἐκ τῶν φλεβῶν αἷμα, καὶ
διὰ τοῦτο δοκέουσιν εὕχροοι εἶναι.

2. Ἑτέρη νοῦσος· ἡ κεφαλὴ ἑλκέων καταπίμπλαται, καὶ τὸ σῶμα οἰδέει, καὶ ἡ χροιὴ ἰκτερώδης, καὶ ἄλλοτε ἄλλη τοῦ σώματος ἕλκεα ἐκθύει,
καὶ πυρετὸς λαμβάνει ἄλλοτε καὶ ἄλλοτε, καὶ ἐκ
τῶν ὤτων ὕδωρ ῥεῖ. τούτῳ, ὅταν ἐν τῇ κεφαλῇ
φλέγμα ὑπόχολον ἐντραφῇ,[1] τὰ μὲν ἕλκεα γίνεται, ὅταν τὸ βρέγμα διάβροχον γένηται τῷ φλέγ-
ματι καὶ τῇ χολῇ, καὶ ἀραιὸν ᾖ καὶ ἁλὲς τὸ
φλέγμα καὶ ἡ χολή· ἵσταται γὰρ τοῦτο καὶ σήπεται καὶ ἑλκοῦται· ἐς δὲ τὰ ὦτα λεπτυνόμενον τὸ
φλέγμα διαδιδοῖ. ἐν δὲ τῷ ἄλλῳ σώματι τά τε
ἕλκεα κατὰ τὸν αὐτὸν λόγον τοῖσιν ἐν τῇ κεφαλῇ
10 γίνεται, συσσηπομένου τοῦ αἵματος | καὶ τῆς
χολῆς, ᾗ ἂν τύχῃ ἀλισθέντα· ταύτῃ γὰρ ἡ σὰρξ
σήπεται καὶ ἑλκοῦται, καὶ προσκατασήπει τὸ
ἐσελθὸν τοῦ φλέγματός τε καὶ τῆς χολῆς, καὶ
γίνεται πύον.

3. Ἑτέρη νοῦσος· περιωδυνίη τὴν κεφαλὴν
ἴσχει, καὶ ἐμέει χολήν, καὶ δυσουρέει, καὶ παραφρονέει. οὗτος περιωδυνέει ὑπὸ τῆς ὑπερθερμασίης τῆς κεφαλῆς· παραφρονέει δὲ ὅταν τὸ αἷμα
τὸ ἐν τῇ κεφαλῇ ὑπὸ χολῆς ἢ φλέγματος ὑπερθερμανθῇ καὶ κινηθῇ μᾶλλον τοῦ εἰωθότος· ἐμέει
δὲ χολὴν ἅτε κεκινημένης τῆς χολῆς ἐν τῷ σώ-

---

[1] Μ: ἐντραχῇ Θ.

blood out of the vessels, and it is for this reason that such patients appear to have a good colour.

2. Another disease: the head becomes covered with ulcers, the body swells up, the skin becomes jaundiced, and ulcers break out at one time in one part of the body, at another time in another part; fever is present from time to time, and watery fluid runs out of the ears. The ulcers develop in this patient when his bregma becomes sodden with phlegm and bile that have formed in the head, and these are thin and present in a great quantity; for the bilious phlegm stands, putrefies and ulcerates; on becoming thin, phlegm exudes from the head into the patient's ears. In the rest of the body, the ulcers arise in the same way as those on the head, from blood and bile putrefying wherever they happen to collect; for there the tissue putrefies and ulcerates, and also makes any phlegm and bile that arrive putrefy too, so that they become pus.

3. Another disease: intense pain occupies the head, the patient vomits bile, suffers from dysuria, and becomes deranged. This patient has the intense pain because of overheating of his head. He becomes deranged when the blood in his head is overheated by bile or phlegm, and set in motion more than usual. He vomits bile when his head, in consequence of its being overheated, attracts bile

ματι, καὶ ἡ κεφαλὴ ὑπὸ τῆς ὑπερθερμασίης[1] ἕλκει ἐφ᾽ ἑωυτήν, καὶ τὸ μὲν παχύτατον ἐμέει, τὸ
δὲ λεπτότατον ἐς ἑωυτήν. οὐρέει δὲ καὶ ἐν ταύτῃ
ὑπὸ τῶν αὐτῶν, ὡς καὶ ἐν τῇ πρόσθεν εἴρηται.

4. Ἢν περὶ τὸν ἐγκέφαλον φλέβια ὑπεραιμή-
σῃ[2] (τὸ μὲν οὔνομα οὐκ ὀρθὸν τῇ νούσῳ, οὐ γὰρ
ἀνυστὸν ὑπεραιμῆσαι οὐδὲν τῶν φλεβίων οὔτε
τῶν ἐλασσόνων οὔτε τῶν μειζόνων· ὀνομαίνουσι
δὲ καί φασιν ὑπεραιμέειν. εἰ δ᾽ ὡς μάλιστα
ὑπεραιμήσειε νοῦσος ὑπ᾽ αὐτοῦ οὐκ ἔοικεν ἂν[3]
γίνεσθαι· ἀπὸ ἀγαθοῦ γὰρ κακὸν οὐχ οἷόν τε
γενέσθαι, οὐδ᾽ ἀγαθὸν πλέον τοῦ δέοντος οἷόν τε
γενέσθαι οὐδ᾽ ἀπὸ κακοῦ ἀγαθὸν γένοιτ᾽ ἄν. ἀλλ᾽
ὑπεραιμέειν δοκέει ὅταν ἐς τὰς φλέβας χολὴ ἢ
φλέγμα ἐσέλθῃ), μετεωρίζονταί τε γὰρ αἱ φλέβες
καὶ σφύζουσι, καὶ ὀδύνη κατὰ πᾶσαν τὴν κεφαλὴν ἐγγίνεται, καὶ τὰ ὦτα ἠχέει, καὶ ἀκούει οὐδέν. καὶ ἠχέει μὲν ἅτε τῶν φλεβίων σφυζόντων
καὶ παλλομένων, τηνικαῦτα γὰρ ἦχος ἔνεστιν ἐν
τῇ κεφαλῇ. βαρυηκοεῖ δὲ τὸ μέν τι καὶ ὑπὸ τοῦ
ἔσωθεν ψόφου καὶ ἤχου, τὸ δ᾽ ὅταν ὁ ἐγκέφαλος
καὶ τὰ φλέβια τὰ περὶ αὐτὸν ἐπαρθῇ. ὑπὸ[4] γὰρ
12   τῆς ὑπερθερμασίης ἐμπίμπλησι τὸ κατὰ τὸ | οὖς

---

[1] Θ: θερμασίης M.   [2] Jouanna (p. 46): -εμήσῃ ΘM: -αιμώσῃ
Ermerins. Similarly wherever the verb occurs (see LSJ,
s.v. ὑπερεμέω, where the correction is recommended).
[3] ἂν om. M.   [4] M: τὰ Θ.

that has been set in motion in the body, and the thickest part is vomited up, while the thinnest part stays in the head. He passes urine in this disease, because of the same things mentioned in the disease before.

4. If, around the brain, small vessels overfill with blood (this name is not a correct one for the disease, because no vessel, either one of the lesser ones or one of the greater ones, can actually be overfilled with blood. Still they use this name and say that they overfill with blood. And even if they really did overfill with blood, it does not seem probable that a disease would arise because of it, for bad cannot come from good,[1] nor can good greater than what is fitting, nor would good come from bad. Rather, there *appears* to be an overfilling with blood when bile or phlegm enters the vessels), the vessels are raised up and throb, pain occupies the entire head, the ears ring, and the patient hears nothing. The ears ring because the vessels are throbbing and quivering, for then there is a ringing in the head. The patient is hard of hearing, partly on account of the sound and ringing and partly because the brain and vessels around it swell up; for, owing to the overheating that occurs, the brain

---

[1] For a vessel to be filled with blood is taken to be a condition of health, and thus good; to increase this good, the author argues, could not lead to something bad, disease.

κενεὸν ὁ ἐγκέφαλος ἑωυτοῦ· καὶ ἅτε οὐκ ἐνεόντος
τοῦ ἠέρος ἰσοπληθέος ὡς[1] καὶ ἐν τῷ πρὶν χρόνῳ,
οὐδὲ τὸν ἦχον ἴσον παρέχοντος, οὐκ ἐνσημαίνει οἱ
τὰ λεγόμενα ὁμαλῶς, καὶ ἀπὸ τούτου βαρυηκοέει.

Οὗτος, ἢν μὲν ῥαγῇ αὐτῷ ἐς τὰς ῥῖνας ἢ ἐς τὸ
στόμα ὕδωρ καὶ φλέγμα, ὑγιὴς γίνεται· ἢν δὲ μὴ
ῥαγῇ, ἑβδομαῖος μάλιστα ἀποθνήσκει.

Ἢν δ' αἱ ἐν τῇ κεφαλῇ ὑπεραιμήσωσι φλέβες,
ὑπεραιμέουσι δὲ ὑπὸ τῶν αὐτῶν ἃ καὶ ἐν τῷ
πρόσθεν εἴρηται· σημεῖον δ' ὅτι τοιούτῳ τρόπῳ
ὑπεραιμέουσι τόδε· ὅταν τις ἢ χεῖρα τοῦτο πά-
σχουσαν ἐπιτάμῃ ἢ κεφαλὴν ἢ ἄλλο τι τοῦ σώματος, τὸ αἷμα μέλαν ῥεῖ καὶ θολερὸν καὶ νοσῶδες·
καίτοι οὐ δίκαιον κατὰ τὸ οὔνομα, ἀλλ' ἐρυθρὸν
καὶ εἰλικρινὲς <δεῖ>[2] ῥεῖν.

Ὅταν δ' ὑπεραιμήσωσιν ὑπὸ[3] τῶν αὐτῶν, ἴσχει
ὀδύνη καὶ σκοτοδινίη καὶ βάρος τὴν κεφαλήν·
ὀδύνη μὲν ὑπὸ τῆς ὑπερθερμασίης τοῦ αἵματος,
σκοτοδινίη δ' ὅταν ἁλὲς ἐπὶ τὸ πρόσωπον χωρήσῃ
τὸ αἷμα, βάρος δὲ ὅταν[4] τοῦ αἵματος πλέονος
ἐόντος ἐν τῇ κεφαλῇ καὶ θολερωτέρου[5] καὶ
νοσωδεστέρου ἢ ἔωθεν.

5. Ἢν σφακελίσῃ ὁ ἐγκέφαλος, ὀδύνη ἔχει ἐκ
τῆς κεφαλῆς τὴν ῥάχιν καὶ ἐπὶ τὴν καρδίην φοι-

---

[1] ὡς om. Θ. [2] Potter. [3] Θ: ἀπὸ Μ. [4] Θ: ἅτε Μ.
[5] Later mss: χολερωτέρου Θ: θολωδεστέρου Μ.

by itself fills up the empty space in the direction of the ear; therefore, since the same amount of air as in the time before is no longer present, and thus does not provide the same sound,[1] what is said does not register adequately with the patient, and for this reason he is hard of hearing.

If water and phlegm break through to this patient's nostrils or mouth, he recovers; if not, he generally dies on the seventh day.

If the vessels in the head overfill with blood, they do so on account of the things mentioned before. A sign that overfilling with blood occurs in this way is the following: if you make an incision into an arm suffering from the condition, or into the head or any other part of the body, the blood that flows forth is dark, turbid, and diseased; and yet not rightly so according to the name, but the blood should flow red and pure.

When vessels are overfilled with blood owing to these factors, pain, dizziness and heaviness occupy the head: pain, as a result of the overheating of the blood, dizziness, when blood moves in a mass over the face, heaviness, when the blood in the head is greater in quantity, more turbid, and more diseased than normal.

5. If the brain becomes sphacelous, pain radiating from the head occupies the spine and migrates to the cardia; the patient loses consciousness and

---

[1] The perception of sound is understood as occurring through the agency of air present between the brain and the ear.

τᾷ· καὶ[1] ἀψυχίη καὶ ἱδρώς, καὶ ἄπνοος[2] τελέθει·
καὶ ἐκ τῶν ῥινῶν αἷμα ῥεῖ, πολλάκις δὲ καὶ ἐμέει
αἷμα. σφακελίζει δ' ὁ ἐγκέφαλος ἢν[3] ὑπερθερμανθῇ ἢ ὑπερψυχθῇ, ἢ χολώδης ἢ φλεγματώδης
γένηται μᾶλλον τοῦ ἐωθότος· ὅταν δέ τι τούτων
πάθῃ, ὑπερθερμαίνεται καὶ τὸν νωτιαῖον μυελὸν
διαθερμαίνει, καὶ οὗτος ὀδύνην τῇ ῥάχει παρέχει.
ἀψυχέει δ' ὅταν προσίστηται πρὸς τὴν καρδίην
φλέγμα ἢ χολή· προσίστασθαι δ' ἀνάγκη κεκινη-
14   μένων καὶ ὑγρα|σμένων· ἱδρὼς δὲ ὑπὸ πόνου. τὸ
δὲ αἷμα ἐμέει ὅταν αἱ φλέβες αἱ μὲν ἐν τῇ κεφαλῇ ὑπὸ τοῦ ἐγκεφάλου θερμανθῶσιν, αἱ δὲ παρὰ
τὴν ῥάχιν ὑπὸ τῆς ῥάχιος, ἡ δὲ ῥάχις ὑπὸ τοῦ νω-
τιαίου μυελοῦ, ὁ δὲ νωτιαῖος[4] ὑπὸ τοῦ ἐγκεφάλου,
ὅθεν περ πέφυκεν· ὅταν οὖν θερμανθῶσιν αἱ
φλέβες καὶ τὸ αἷμα ἐν αὐτῇσι ζέσῃ, διαδιδοῦσιν
αἱ μὲν ἀπὸ τῆς κεφαλῆς ἐς τὰς ῥῖνας, αἱ δ' ἀπὸ
τῆς ῥάχιος αἱμορρόοι ἐς τὸ σῶμα. οὗτος τριταῖος
ἀπόλλυται ἢ πεμπταῖος ὡς τὰ πολλά.

6. Ἄλλη ἐξαπίνης ὀδύνη λαμβάνει τὴν κεφαλήν, καὶ παραχρῆμα ἄφωνος γίνεται καὶ ἀκρα-
τὴς ἑωυτοῦ. οὗτος ἀποθνήσκει ἐν ἑπτὰ ἡμέρῃσιν,
ἢν μή μιν πῦρ ἐπιλάβῃ· ἢν δ' ἐπιλάβῃ, ὑγιὴς γίνεται. πάσχει δὲ ταῦτα, ὅταν αὐτῷ μέλαινα χολὴ
ἐν τῇ κεφαλῇ κινηθεῖσα ῥυῇ, καὶ μάλιστα καθ' ὃ

[1] καὶ om. M.   [2] Potter (see ch. 20): ὕπνος Θ: ἄϋπνος M.
[3] Potter: ἢ Θ: ὅταν ἢ M.   [4] Θ: μυελὸς M.

sweats, and becomes lifeless; blood flows from his nostrils, and often he also vomits blood. The brain becomes sphacelous if it is made too hot or too cold, or becomes more bilious or phlegmatic than usual; when it suffers any of these things, it becomes too hot, and heats the spinal marrow, and this produces pain in the spine. The patient loses consciousness when phlegm or bile invades his cardia, and such an invasion is inevitable once these are set in motion and moistened. Sweating is because of the strain. The patient vomits blood, when the vessels in his head are heated by the brain, and those along the spine are heated by the spine; the spine was heated by the spinal marrow, and the spinal marrow by the brain out of which it grows. Now when the vessels are heated and the blood inside them seethes, some exude blood from the head into the nose, while the sanguiferous ones do the same from the back into the body. This patient dies on the third or fifth day, in the majority of cases.

6. Another disease: pain suddenly seizes the head, and the patient immediately becomes speechless and loses power over himself. Unless fever supervenes, this patient dies in seven days; if it does, he recovers. He suffers these things when dark bile is set in motion in his head, and flows

τὰ πλεῖστα ἐν τῷ τραχήλῳ ἐστὶ φλέβια καὶ τοῖσι
στήθεσιν· ἔπειτα δὲ καὶ τῇ ἄλλῃ ἀπόπληκτος γίνεται καὶ ἀκρατής, ἅτε τοῦ αἵματος ἐψυγμένου.
καὶ ἢν κρατήσῃ ὥστε τὸ αἷμα θερμανθῆναι, ἤν τε
ὑπὸ τῶν προσφερομένων ἤν τε ὑφ'[1] ἑωυτοῦ,
μετεωρίζεται καὶ διαχεῖται, καὶ κινέεται, καὶ τὴν
πνοιὴν ἐσάγεταί τε καὶ ἀφρίει καὶ χωρίζεται τῆς
χολῆς, καὶ ὑγιὴς γίνεται. ἢν δὲ μὴ κρατήσῃ, ψύχεται ἐπὶ μᾶλλον· καὶ ὅταν ψυχθῇ παντάπασι καὶ
ἐκλίπῃ ἐξ αὐτοῦ τὸ θερμόν, πήγνυται καὶ κινηθῆ-
ναι οὐ δύναται, ἀλλ' ἀποθνήσκει. ἢν δ' ἐκ θωρηξίων ταῦτα πάθῃ, πάσχει τε ὑπὸ τῶν αὐτῶν, καὶ
ἀπόλλυται ὑπὸ τῶν αὐτῶν, καὶ διαφεύγει ὑπὸ
τῶν αὐτῶν.

7. Ὅταν τερηδὼν γένηται ἐν τῷ ὀστέῳ, ὀδύνη
λαμβάνει ἐκ τοῦ ὀστέου, χρόνῳ δ' ἀφίσταται τὸ
δέρμα ἀπὸ τῆς κεφαλῆς ἄλλῃ καὶ ἄλλῃ. οὗτος
ταῦτα πάσχει, ὅταν ἐν τῇ διπλόῃ τοῦ | ὀστέου
ὑπογενόμενον φλέγμα ἐναποξηρανθῇ ταύτῃ
ἀραιὸν γίνεται, καὶ ἐκλείπει ἐξ αὐτοῦ ἡ ἰκμὰς
πᾶσα, καὶ ἅτε ξηροῦ ἐόντος ἀφίσταται ἀπ'[2]
αὐτοῦ τὸ δέρμα. αὕτη ἡ νοῦσος[3] θανάσιμός ἐστιν.

8. Ἢν βλητὸς γένηται, ἀλγέει τῆς κεφαλῆς
τὸ πρόσθεν, καὶ τοῖσιν ὀφθαλμοῖσιν οὐχ ὁμαλῶς
ὁρᾷ, καὶ κωμαίνει, καὶ αἱ φλέβες σφύζουσι, καὶ
πυρετὸς ἴσχει βληχρὸς καὶ τοῦ σώματος ἀκρασίη.

---

[1] Θ: ἐφ' Μ.   [2] Θ: ὑπ' Μ.   [3] M adds οὐ.

16

mainly to where most of the vessels in the neck and chest are; then, owing to a cooling of the blood, he becomes paralysed in his other parts, and powerless. If this patient gains the upper hand, so that his blood is warmed either as the result of what is administered or by itself, the blood is lifted, dispersed, and set in motion, it takes in vapour, foams, and separates itself from the bile, and he recovers. But if he does not gain the upper hand, the blood is cooled even more; when it has been cooled completely and given up its heat, it congeals and can no longer move, and the patient dies. If a person suffers this condition subsequent to drunkenness, he suffers it because of the same things, and he dies or escapes because of the same things.

7. When a teredo forms in the skull, a pain originating in the bone arises, and, as time passes, skin separates from the head in one place after another. The patient suffers these things when phlegm that has formed in the diploe of the bone becomes dried up inside it; where this happens, the bone becomes loose in texture, all the moisture leaves it, and, because it is dry, the skin separates from it. This disease is usually mortal.

8. If a person is stricken, he has pain in the front of his head, he does not see properly, and he is drowsy; the vessels throb, and there are mild fevers and powerlessness of the body. The patient suffers

οὗτος ταῦτα πάσχει, ὅταν αἱ ἐν τῇ κεφαλῇ φλέβες θερμανθῶσιν, θερμανθεῖσαι δὲ εἰρύωσι[1] φλέγμα ἐς ἑωυτάς. ἡ μέν νυν ἀρχὴ τῆς νούσου ἐκ
τούτου γίνεται· τὸ δ' ἔμπροσθεν τῆς κεφαλῆς διὰ
τόδε ἀλγέει, ὅτι αἱ φλέβες ταύτῃ εἰσὶν αἱ παχύ-
ταται, καὶ ὁ ἐγκέφαλος εἰς τὸ πρόσθεν μᾶλλον
κεῖται τῆς κεφαλῆς ἢ ἐς τοὔπισθεν· καὶ τοῖσιν
ὀφθαλμοῖσι διὰ τοῦτο οὐχ ὁρᾷ προκειμένου τοῦ
ἐγκεφάλου καὶ φλεγμαίνοντος. τὸ δὲ σῶμα διὰ
τοῦτο δὲ[2] ἀκρασίαι ἔχουσιν· αἱ φλέβες ἐπὴν εἰς
ἑωυτὰς εἰρύσωσι φλέγμα, ἀνάγκη ὑπὸ ψυχρότη-
τος τοῦ φλέγματος τὸ αἷμα ἑστάναι μᾶλλον ἢ ἐν
τῷ πρὶν χρόνῳ καὶ ἐψῦχθαι· μὴ κινεομένου δὲ τοῦ
αἵματος, οὐχ οἷόν τε μὴ οὐ καὶ τὸ σῶμα ἀτρεμί-
ζειν καὶ κεκωφῶσθαι. καὶ ἢν μὲν τὸ αἷμα καὶ τὸ
ἄλλο σῶμα κρατήσῃ ὥστε διαθερμανθῆναι, δια-
φεύγει· ἢν δὲ τὸ φλέγμα κρατήσῃ, ἐπιψύχεται
μᾶλλον τὸ αἷμα καὶ πήγνυται· καὶ ἢν ἐς τοῦτο
ἐπιδῷ[3] ψυχόμενον καὶ πηγνύμενον, πήγνυται
παντελῶς καὶ ἐκψύχεται ὤνθρωπος καὶ ἀποθνήσκει.

9. Κυνάγχη δὲ γίνεται ὅταν ἐν τῇ κεφαλῇ
φλέγμα κινηθὲν ῥυῇ ἁλὲς κάτω καὶ στῇ ἐν τῇσι
σιαγόσι καὶ περὶ τὸν τράχηλον. οὗτος οὔτε τὸ
σίαλον δύναται καταπίνειν, ἀναπνεῖ δὲ[4] βιαίως

---

[1] Θ: εἰρύσωσι Μ.   [2] Θ: τόδε Μ.   [3] Potter: ἐπιδοι Θ: ἐπιδιδοῖ Μ.
[4] Θ: τε Μ.

these things when the vessels in his head become heated and, being heated, attract phlegm. Now the disease takes its origin from the following: the patient suffers pain in the front of his head because the vessels there are widest, and because the brain lies more towards the front of the head than towards the back; he loses the sight from his eyes because his brain projects and is swollen; powerlessness befalls his body on account of the following: when the vessels draw phlegm into themselves, the blood must, on account of the coldness of the phlegm, stand more still than before and be cooled, and so, with the blood immobile, it is impossible for the body not to become still and numb. Now if the blood and the rest of the body gain the upper hand, so that they are warmed, the patient escapes. But if the phlegm predominates, the blood is cooled and congeals more, and if it reaches a certain stage of cooling and congelation, it congeals completely, the person becomes cold, and he dies.

9. Angina arises when phlegm that has been set in motion in the head flows downward in a large mass, and comes to rest in the jaws and about the neck. This patient is unable to swallow his saliva,

καὶ ῥέγκει, καὶ ἔστιν ὅτε καὶ πυρετὸς αὐτὸν ἴσχει.
18 τὸ μὲν οὖν νό|σημα ἀπὸ τούτου γίνεται καὶ ἄλλο-
τε ὑπ᾽ αὐτὴν τὴν γλῶσσαν, ἄλλοτε ὑπὲρ τῶν
στηθέων ὀλίγον.

10. Σταφυλὴ δὲ γίνεται ὅταν ἐς τὸν γαργαρεῶνα καταβῇ φλέγμα ἐκ τῆς κεφαλῆς· κατα-
κρίμαται καὶ γίνεται ἐρυθρός. ἢν δὲ πλείων
χρόνος ἐγγίνηται,[1] μελαίνεται· μελαίνεται δ᾽ ὧδε·
ἐπὶ φλεβὸς πέφυκεν ὁ γαργαρεὼν παχέης, καὶ
ἐπὴν φλεγμήνῃ, θερμαίνεται, καὶ ὑπὸ τῆς θερμασίης ἕλκει καὶ ἐκ τῆς φλεβὸς τοῦ αἵματος, καὶ
μελαίνεται ὑπ᾽ αὐτοῦ. διὰ τοῦτο δὲ καὶ ἢν μὴ
ὀργῶντα τάμνῃς, παραχρῆμα ἀποσπαρθάζουσιν· ἡ
γὰρ φλὲψ θερμαίνει καὶ ὑπὸ τῆς θερμασίης καὶ
φλεγμασίης[2] ἐμπιμπλεῖ τὰ περὶ τὸν γαργαρεῶνα
αἵματος, καὶ δι᾽ ὀλίγου[3] ἀποπνίγονται.

11. Ἀντιάδες δὲ καὶ ὑπογλωσσίδες καὶ οὖλα
καὶ γλῶσσα καὶ ὅσα τοιαῦτα ταύτῃ πεφυκότα,
ταῦτα πάντα νοσέει ἀπὸ φλέγματος· τὸ δὲ φλέγμα ἀπὸ τῆς κεφαλῆς καταβαίνει. ἡ δὲ κεφαλὴ
ἐκ τοῦ σώματος ἕλκει· ἕλκει δ᾽ ὅταν διαθερμανθῇ·
διαθερμαίνεται δὲ καὶ ὑπὸ σιτίων καὶ ὑπὸ ποσίων[4]
καὶ ἡλίου καὶ ψύχεος καὶ πόνων καὶ πυρός.
ὅταν δὲ διαθερμανθῇ, ἕλκει[5] ἐς ἑωυτὴν ἐκ τοῦ

---

[1] Θ: γένηται Μ.   [2] καὶ φλεγμασίης om. Μ.   [3] Θ: ὅλου Μ.
[4] καὶ ὑπὸ ποσίων om. Μ.   [5] Μ adds τὸ λεπτότατον.

204

and he respires laboriously and stertorously; sometimes fever, too, befalls him. The condition arises in this way, then, sometimes beneath the tongue itself, and sometimes a little above the chest.

10. Staphylitis occurs when phlegm descends out of the head into the uvula, and it hangs down and becomes red. As more time passes, the uvula becomes dark, and in the following way: at the base of the uvula is a wide vessel; now when the uvula swells, it becomes hot, because of its heat it draws blood out of the vessel, and from this blood it becomes dark. Thus, if you do not incise it when it is turgid, patients immediately begin to gasp for breath; for the vessel heats and, because of the heat and swelling, fills the region about the uvula with blood, so that in a short time patients choke.

11. The tonsils, the area beneath the tongue, the gums, the tongue, and other such structures growing in the region all become ill as the result of phlegm that comes down from the head. First, the head draws phlegm out of the body; it does this on becoming heated, and it becomes heated from foods, drinks, sun, cold, exertions and fire. When it becomes heated, then, it draws phlegm to itself out

σώματος· ὅταν δὲ εἰρύσῃ, καταβαίνει[1] πάλιν ἐς τὸ
σῶμα, ὅταν πλήρης γένηται ἡ κεφαλὴ καὶ
τύχῃ ὑπό τινος τούτων διαθερμανθεῖσα.[2]

12. Νοῦσοι αἱ ἀπὸ τῶν κεφαλῶν· νάρκα ἴσχει
τὴν κεφαλήν, καὶ οὐρέει θαμινά, καὶ τἆλλα πάσχει ἅπερ ὑπὸ στραγγουρίης. οὗτος ἡμέρας ἐννέα
ταῦτα πάσχει, καὶ ἢν μὲν ῥαγῇ | κατὰ τὰς ῥῖνας
ἢ κατὰ τὰ ὦτα ὕδωρ καὶ πλέννα, ἀπαλλάσσεται
τῆς νούσου, καὶ παύεται τῆς στραγγουρίης· οὐρέει
δὲ ἀπόνως[3] πολὺ καὶ λευκὸν ἐς τὰς εἴκοσιν ἡμέ-
ρας· καὶ ἐκ τῆς κεφαλῆς ἡ ὀδύνη ἐκλείπει, καὶ ἐκ
τῶν ὀφθαλμῶν ἐσορῶντι κλέπταί οἱ αὐγή,[4]
καὶ δοκέει τὸ ἥμισυ τῶν προσώπων ὁρᾶν. οὗτος
τεσσερακοσταῖος παντάπασιν ὑγιὴς γίνεται.

Ἐνίοτε δὲ ὑπ᾽ οὖν ἔστρεψεν ἡ νοῦσος ἑβδόμῳ
ἔτει ἢ τεσσερεσκαιδεκάτῳ· καὶ τὸ δέρμα οἱ παχύνεται τῆς κεφαλῆς, καὶ ψαυόμενον ὑπείκει· καὶ
ἀπ᾽ ὀλίγων σιτίων ἁπαλὸς καὶ εὔχρως φαίνεται,
καὶ ἀκούει οὐκ ὀξέα.

Ὅταν οὕτως ἔχοντι ἐπιτύχῃς[5] ἀρχομένῳ τῆς
νούσου, πρόσθεν ἢ ῥαγῆναι κατὰ τὰς ῥῖνας τὸ
ὕδωρ καὶ κατὰ τὰ ὦτα, καὶ ἔχῃ αὐτὸν ἡ περιωδυνίη, ἀποξυρήσαντα χρὴ αὐτὸν τὴν κεφαλήν, περι-

---

[1] M adds καί.   [2] So ΘM, Jouanna (p. 80): ὅταν ...
διαθ. placed after κεφαλῶν (next line) by later mss, edd.
[3] M adds καί.   [4] This paragraph (νάρκα ἴσχει τὴν κεφαλήν
... οἱ αὐγή) recurs in *Regimen in Health* 8 (Loeb vol. IV,
56).   [5] Θ adds ἤ.

of the body but, after this attraction has taken place, the phlegm descends back into the body, after the head has become full, on happening to be heated by one of the things mentioned.

12. Diseases of the head: the head becomes numb, and the patient urinates frequently and suffers the rest of the symptoms of strangury. He experiences these things for nine days, and then, if fluid and mucus break out through his nostrils or ears, he is relieved of the disease, and the strangury stops. He passes copious white urine without pain for twenty days, and the pain goes away out of his head, but when he looks at any-thing, the sight is snatched from his eyes, and he seems to see only the half of faces. This patient recovers completely about the fortieth day.

Sometimes the disease recurs in the seventh or fourteenth year. In that case, the skin on the patient's head becomes thick, and on being touched gives way; he takes on a delicate appearance and a good colour from little food; he does not hear keenly.

When you happen upon a patient in this state, at the beginning of his illness before fluid has bro-ken out through his nostrils or ears, and he is suffering intense pain, you must shave his head

δέοντα περὶ τὸ μέτωπον τὸν ἀσκὸν τὸν σκύτινον,
ὕδατος ἐμπιμπλάντα ὡς ἂν ἀνέχηται θερμοτά-
του, ἐᾶν αὐτὸν χλιαίνεσθαι, καὶ ἐπὴν ἀποψυχθῇ,
ἕτερον ἐγχεῖν. ἢν δ' ἀσθενέῃ, παύεσθαι, καὶ διαλιπὼν αὖτις ποιέειν ταὐτὰ ἔστ' ἂν χαλάσῃ ἡ
περιωδυνίη. καὶ ἢν ἡ κοιλίη μὴ ὑποχωρέῃ, ὑποκλύσαι αὐτόν, καὶ πιπίσκοντα τῶν οὐρητικῶν μελίκρητα διδόναι πίνειν[1] ὑδαρέα· καὶ θαλπέσθω ὡς
μάλιστα· ῥυφανέτω δὲ τὸν χυλὸν τῆς πτισάνης
λεπτόν. ἢν δέ οἱ[2] ἡ γαστὴρ μὴ ὑποχωρέῃ, λινό-
ζωστιν ἑψήσας ἐν ὕδατι, τρίβων, διηθέων τὸν
χυλόν, συμμίσγειν ἴσον[3] τοῦ ἀπὸ τῆς πτισάνης
χυλοῦ καὶ ἀπὸ τῆς λινοζώστιος, καὶ μέλι ὀλίγον
παραμίσγειν ἐς τὸν χυλόν· τοῦτο ῥυφάνειν τρὶς
τῆς ἡμέρης, καὶ ἐπιπίνειν οἶνον μελιχρόν, ὑδαρέα,
λευκόν, ὀλίγον ἐπὶ τῷ ῥυφήματι.

Ἐπὴν δέ οἱ ῥαγῇ κατὰ τὰς ῥῖνας τὰ βλεννώ-
δεα, καὶ οὐρέῃ παχύ, καὶ τῆς ὀδύνης ἀπηλλαγμέ-
22 νος ᾖ, | τῷ ἀσκῷ τῆς κεφαλῆς μηκέτι χρήσθω,
ἀλλὰ λουόμενος πολλῷ καὶ θερμῷ πινέτω τὰ διουρητικὰ καὶ μελίκρητα ὑδαρέα. καὶ τὰς μὲν
πρώτας ἡμέρας κέγχρον λειχέτω, καὶ κολοκύντην
ἐσθιέτω ἢ τεῦτλα τρεῖς ἡμέρας· ἔπειτα σιτίοισι
χρήσθω ὡς μαλθακωτάτοισι καὶ διαχωρητικω-
τάτοισι,[4] προστιθεὶς ὀλίγον αἰεὶ τῶν σιτίων. ἐπὴν

---

[1] Θ: ἐπιπ- Μ.   [2] οἱ om. Μ.   [3] Cornarius: ἴσου ΘΜ.
[4] Later mss, edd.: -χωρηκω- Θ: -φορητικω- Μ.

clean, attach a leather skin about his forehead, fill
it up with water as hot as he can stand, and leave
him to be warmed; when the water becomes cold,
pour in fresh hot water. If the patient becomes
weak, stop treatment for a while, and then resume
it again until the pain slackens. If the cavity does
not pass anything downwards, administer an
enema; have the patient drink diuretics, and give
him dilute melicrat to drink; let him be warmed as
thoroughly as possible, and drink as gruel thin
barley-water. If his belly does not pass anything
downwards, boil the herb mercury in water, mash
it, filter the juice, mix together equal amounts of
barley-water and the mercury juice, and add a little
honey. Let him take this as gruel three times
daily, and after the gruel drink a little dilute white
honeyed wine.

When mucus breaks out through the patient's
nostrils, when he passes thick urine, and when his
pain goes away, let him stop applying the skin to
his head, but bathe in copious hot water, and drink
diuretics and dilute melicrat. On the first days, let
him take millet, and eat gourd or beets for three
days. Then let him have foods that are as soft and
laxative as possible, gradually increasing the

δὲ αἱ[1] τεσσεράκοντα ἡμέραι ἐξέλθωσι—
καθίσταται γὰρ μάλιστα ἡ νοῦσος ἐν τοσούτῳ
χρόνῳ—καθήρας αὐτοῦ τὴν κεφαλὴν πρότερόν οἱ
φάρμακον δοὺς κάτω καθαρόν[2] ἢν ὥρη ᾖ τοῦ
ἔτεος, ἔπειτα ὀρὸν μεταπῖσαι ἑπτὰ ἡμέρας, ἢν δ'
ἀσθενὴς ᾖ,[3] ἐλάσσονας.

Ἢν δ' ὑποστρέψῃ ἡ νοῦσος, πυριάσας αὐτὸν
ὅλον, ἐς αὔριον δοῦναι ἐλλέβορον πιεῖν· κἄπειτα
διαλείπειν ὁπόσον ἄν σοι δοκέῃ χρόνον, καὶ τότε
τὴν κεφαλὴν καθήρας, κατωτερικὸν δοὺς φάρμακον, καῦσον τὴν κεφαλὴν ἐσχάρας ὀκτώ, δύο μὲν
παρὰ τὰ ὦτα, δύο δ' ἐν τοῖσι κροτάφοισι,[4] δύο δ'
ὄπισθεν τῆς κεφαλῆς ἔνθεν καὶ ἔνθεν ἐν τῇ
κοτίδι, δύο δ'[5] ἐν τῇ ῥινὶ παρὰ τοὺς κανθούς. τὰς
φλέβας καίειν δὲ τὰς μὲν παρὰ τὰ ὦτα, ἔστ' ἂν
παύσωνται σφύζουσαι· τοῖσι δὲ σιδηρίοισι σφηνί-
σκους ποιησάμενος, διακαίειν πλαγίας τὰς φλέβας. ταῦτα ποιήσαντι ὑγιείη ἐγγίνεται.

13. Ἑτέρη νοῦσος· ἑλκέων καταπίμπλαται
τὴν κεφαλήν, καὶ τὰ σκέλεα οἰδίσκεται ὥσπερ
ἀπὸ ὕδατος, καὶ ἐν τῇσι κνήμῃσιν ἐμπλάσσεται,
ἢν πιέσῃς, καὶ[6] ἡ χροιὴ ἰκτερώδης. καὶ ἐκθύει
ἕλκεα ἄλλοτε ἄλλῃ, μάλιστα δ' ἐς[7] τὰς κνήμας,
καὶ φαίνεται πονηρὰ προσιδεῖν, ἀποφλεγμήναντα

---

[1] αἱ om. M.   [2] Θ: κάθηρον M.   [3] Potter (cf. e.g. chs. 40,
50, 72): ἀσθενήσῃ ΘΜ.   [4] δύο δ' ἐν τοῖσι κροτάφοισι om. M.
[5] δ' om. M.   [6] ἢν πιέσῃς, καὶ Jouanna (p. 36): καὶ ἢν
πιέσῃς ΘΜ.   [7] ἐς om. M.

amount. When forty days have expired—for the disease generally subsides in that length of time—first clean out the patient's head, and then give him a medication that cleans downwards; if it is the right season, have him then drink whey for seven days; if he is too weak, though, for fewer.

If the disease recurs, apply a vapour-bath to the whole body and, on the morrow, give the patient hellebore to drink. Then, leaving whatever period of time you think correct, clean out the head, give a medication to act downwards, and burn eight eschars on the head: two beside the ears, two on the temples, two behind the head at different places on the occiput, and two on the nose by the corners of the eyes. Also burn the vessels beside the ears until they no longer throb; make the irons wedge-shaped, and cauterize across the oblique vessels. If a person does these things, he recovers.

13. Another disease: the head becomes covered with ulcers, the legs swell up as though from dropsy—below the knee, if you apply pressure, an indentation is left—and the skin becomes jaundiced. Ulcers break out at one time in one part of the body, at another time in another part, but mainly on the legs below the knee; on inspection, these ulcers appear to be very bad, but when they

δὲ ταχέως ὑγιᾶ γίνονται. καὶ πυρετὸς ἄλλοτε
καὶ ἄλλοτε λαμβάνει· ἡ δὲ κεφαλὴ αἰεὶ θερμὴ
γίνεται, καὶ ἐκ τῶν ὤτων ὕδωρ ῥεῖ.

24   Ὅταν οὕτως ἔχῃ, φάρμακόν οἱ δοῦναι, ὑφ' | οὗ
φλέγμα καὶ χολὴν καθαρεῖται ἄνω· ἢν μὲν ψύχη
ᾖ, προπυριήσας, λούσας θερμῷ, ἔπειτα διαλείπων
ἡμέρας τρεῖς, τὴν κεφαλὴν καθῆραι. μετὰ δὲ
κάτω φάρμακον πῖσαι· ἢν δὲ ὥρη ᾖ, καὶ ὀρὸν μετα-
πιέτω· εἰ δὲ μή, γάλα ὄνου. μετὰ δὲ τὰς καθάρ-
σιας σιτίοισιν ὡς ἐλαχίστοισι χρήσθω καὶ δια-
χωρητικωτάτοισι, καὶ ἀλουτείτω. ἢν δὲ ἡ κεφαλὴ
ἑλκῶται, τρύγα κατακαίων οἰνηρήν, σμῆγμα
ποιέων, σύμμισγε τῆς βαλάνου τὸ ἔκλεμμα λεῖον
τρίβων, λίτρον συμμίσγων ἴσον, ἀποσμήξας τούτοισι, λούσθω θερμῷ. χριέσθω δὲ τὴν κεφαλήν,
δαφνίδας τρίψας καὶ κηκίδας καὶ σμύρναν καὶ
λιβανωτὸν καὶ ἀργύρου[1] ἄνθος καὶ ὕειον ἄλειφα
καὶ δάφνινον ἔλαιον· ταῦτα μίξας χρίειν. τὸν δ'
ἔπειτα χρόνον ἐμέτοισι χρήσθω τρὶς τοῦ μηνός,
καὶ γυμναζέτω[2] καὶ θερμολουτείτω καὶ τὴν ὥρην
ὀροποτείτω.[3]

    Ἢν δέ σοι ταῦτα[4] ποιέοντι ἐκ μὲν τοῦ ἄλλου
σώματος ἡ νοῦσος ἐξεληλύθῃ, ἐν δὲ τῇ κεφαλῇ
ἕλκεά οἱ γίνηται, καθήρας τὴν κεφαλὴν αὖτις,

---

[1] Θ: ἀργυρίου Μ.   [2] Θ: γυμναζέσθω Μ.   [3] καὶ τὴν ὥρην ὀροποτείτω om. Μ.   [4] Θ: τάδε Μ.

get over their swelling, they heal quickly. There is intermittent fever, the head becomes permanently warm, and watery fluid runs out of the ears.

When the case is such, give the patient a medication that will clean upwards of phlegm and bile. If the head is cold, first apply a vapour-bath and wash with hot water, and then, leaving an interval of three days, clean out the head. Afterwards, have the patient drink a medication to act downwards; if it is the right season, let him also drink whey, if not that, then ass's milk. After these cleanings, give a very little food, and that of the most laxative kind; let the patient go without bathing. If the head ulcerates, burn wine lees, make these into a paste, add finely ground acorn shell, and mix in an equal amount of soda; smear the patient with this, and have him bathe in hot water. Let him anoint his head with ground bayberry, galls, myrrh, frankincense, flower of silver, lard, and bay oil: mix these together, and smear them on. Then, for the time being, let the patient employ vomiting three times a month, take exercises, bathe in hot water, and in season drink whey.

If, when you do these things, the disease passes out of the rest of the body, but ulcers still arise on the head, clean out the head again, and afterwards

φάρμακον κάτω μεταπῖσαι. ἔπειτα ξυρήσας τὴν
κεφαλήν, καταταμεῖν τομὰς ἀραιάς, καὶ ἐπὴν
ἀπορρυῇ τὸ αἷμα, ἀνατρῖψαι. ἔπειτα εἴρια πι-
νόεντα οἴνῳ ῥαίνων ἐπιδεῖν, καὶ ἐπὴν ἀπολύσῃς,
περισπογγίζειν καὶ μὴ βρέχειν. ἔπειτα κυπάρισ-
σον ἐπιπάσσειν ἐλαίῳ ὑποχρίων· τοῖσι δὲ εἰρίοισιν
ἐπιδέσμοισι χρήσθω, ἔστ' ἂν ὑγιὴς γένηται.

14. Ἑτέρη νοῦσος· περιωδυνίη λαμβάνει τὴν κεφαλήν, καὶ ἐπὴν κινήσῃ τις ἧσσον, ἐμέει χολὴν·
ἐνίοτε δὲ καὶ δυσουρέει καὶ παραφρονέει. ἐπὴν
δ' ἑβδομαῖος γένηται, ἐνίοτε καὶ[1] ἀποθνήσκει·
ἢν δὲ ταύτην ἐκφύγῃ, ἐναταῖος ἢ ἑνδεκαταῖος,
ἢν μή οἱ ῥαγῇ κατὰ τὰς ῥῖνας ἢ κατὰ τὰ ὦτα.
26 ἢν δὲ ῥαγῇ, ὑπεκφυγγάνει· ῥεῖ δὲ | ὑπόχολον
ὕδωρ, ἔπειτα τῷ χρόνῳ πύον γίνεται ἐπὴν σαπῇ.

Ὅταν οὕτως ἔχῃ, ἕως μὲν ἂν ἡ περιωδυνίη
ἔχῃ κατ' ἀρχὰς πρὶν ῥαγῆναι ἐκ τῶν ῥινῶν καὶ
τῶν ὤτων, σπόγγους ἐν ὕδατι θερμῷ βρέχων,
ἆσσον προστιθέναι πρὸς τὴν κεφαλήν· ἢν δὲ μὴ
τούτοισι χαλᾷ, τῷ ἀσκῷ χρῆσθαι τὸν αὐτὸν
τρόπον, ὅνπερ ἐπὶ τῆς προτέρης. πινέτω δὲ μελίκρητα ὑδαρέα· ἢν δὲ μηδ' ἀπὸ τοῦ μελικρήτου,
ἀπὸ τῶν κρίμνων ὕδωρ πινέτω. ῥυφανέτω δὲ τὸν
χυλὸν τῆς πτισάνης, καὶ ἐπιπινέτω λευκὸν οἶνον
ὑδαρέα ὀλίγον.[2] ἐπὴν δὲ ῥαγῇ κατὰ τὰ ὦτα καὶ ὁ
πυρετὸς ἀνῇ καὶ ἡ ὀδύνη, σιτίοισι χρήσθω διαχω-

---

[1] καὶ om. M.   [2] ὀλίγον om. M.

have the patient drink a medication that acts downwards. Then shave the head and cut narrow incisions; when blood flows out, rub the incisions clean. Then sprinkle unwashed wool with wine, and tie this on; when you remove it, sponge the head all round, without making it wet. Then anoint it with oil, and sprinkle on cypress; employ these woollen bandages until the patient recovers.

14. Another disease: intense pain occupies the head, and whenever anyone moves the patient even a bit, he vomits bile; sometimes he also suffers from dysuria, and becomes deranged. By the seventh day the patient may sometimes even die; if he escapes that day, then he dies on the ninth or the eleventh day, unless a flux breaks out through his nostrils or ears; if this occurs, the patient escapes. The fluid is somewhat bilious, and then in time becomes pus through putrefaction.

When the case is such, as long as the intense pain obtains at the beginning of the illness before any flux has broken out through the nostrils or ears, soak sponges in hot water and apply them tightly against the head; if, with these, the pain does not slacken, employ a leather skin as in the preceding disease. Have the patient drink melicrat diluted with water; if not melicrat, then water made from groats. As gruel let him drink barley-water, and afterwards a little dilute white wine. When a flux does break out through the ears, and the fever and pain remit, give laxative foods,

ρητικοῖσιν, ἀρξάμενος ἐξ ὀλίγων, προστιθεὶς αἰεί,
καὶ λούσθω θερμῷ κατὰ κεφαλῆς· καὶ τὰ ὦτα δια-
κλύζειν ὕδατι καθαρῷ, καὶ ἐντιθέναι[1] σπογγία
μέλιτι ἐμβάπτων. ἢν δέ οἱ[2] μὴ ξηραίνηται
οὕτως, ἀλλὰ χρόνιον γίνηται τὸ ῥεῦμα, διακλύ-
σας, ἐμβάλλειν ἀργύρου ἄνθος, σανδαράκην, ψι-
μύθιον, ἴσον ἑκάστου, λεῖα τρίβων,[3] ἐμπιμπλεὶς
τὸ οὖς σάσσειν, καὶ ἢν παραρρέῃ, ἐπεμβάλλειν
τοῦ φαρμάκου. ἐπὴν[4] ξηρὸν γένηται τὸ οὖς, ἐκ-
καθήρας, ἐκκλύσαι τὸ φάρμακον· ἔπειτα, κωφὸν
γὰρ γίνεται τὸ πρῶτον ἀποξηρανθέν, πυριᾶν αὐτὸν βληχρῇσι πυρίῃσι τὰ ὦτα· καταστήσεται γάρ
οἱ τῷ[5] χρόνῳ. ἀποθνήσκουσι δὲ καὶ ἢν ἐς τὸ οὖς
περιωδυνίη γενομένη μὴ ῥαγῇ ἐν τῇσιν ἑπτὰ
ἡμέρῃσι. τοῦτον λούειν θερμῷ πολλῷ, καὶ σπόγγους ἐν ὕδατι θερμῷ βρέχων, ἐκμάσσων, χλιαροὺς
προστιθέναι πρὸς τὸ οὖς. ἢν δὲ μηδ' οὕτω ῥηγνύηται, πυριᾶν αὐτὸ τὸ οὖς. ῥυφήμασι δὲ καὶ
πώμασι τοῖσιν αὐτοῖσι χρῆσθαι οἷσί περ ἐπὶ τοῖσι
πρόσθεν.

15. Ἢν ὕδωρ ἐπὶ τῷ ἐγκεφάλῳ γένηται, ὀδύ-
νη | ὀξέη ἴσχει διὰ τοῦ βρέγματος καὶ τῶν κροτά-
φων ἄλλοτε ἄλλῃ, καὶ ῥῖγος καὶ πυρετὸς ἄλλοτε
καὶ ἄλλοτε· καὶ τὰς χώρας τῶν ὀφθαλμῶν ἀλγέει,
καὶ ἀμβλυώσσει, καὶ ἡ κόρη σχίζεται, καὶ δοκέει

28

---

[1] Θ: ἐντιθέτω Μ.   [2] Potter: τοι ΘΜ.   [3] λεῖα τρίβων Μ: δια-
τρίβων Θ.   [4] Μ adds δέ.   [5] οἱ τῷ Θ: οὕτω Μ.

beginning with little and then adding more and more, and wash down over the head with hot water; wash out the ears with clean water, and insert small sponges dipped in honey. If the flux is not dried up with this treatment, but becomes chronic, wash the ear out and insert equal amounts of finely ground flower of silver, red arsenic and white lead; fill the ear right up and pack it tight; if any of the medication slips out, add more. When the ear becomes dry, clean it out and rinse out the medication. Then, since on being dried out the ear is deaf at first, apply mild vapour-baths to the ears; for in time the patient's ears will get better.

Patients also die if, when intense pain has occupied their ear, no break occurs for seven days. Wash this patient with copious hot water, soak sponges in hot water, squeeze them dry, and apply them warm against the ear. If, with this treatment, no break occurs, apply a vapour-bath to the ear itself. Give the same gruels and drinks as to the patients above.

15. If fluid forms on the brain, violent pain is present between the bregma and the temples, at one time in one place, at another time in another place, and from time to time there are chills and fevers. The patient suffers pain in the sockets of his eyes, he sees unclearly, his pupil is divided, and

ἐκ τοῦ ἑνὸς δύο ὁρᾶν, καὶ ἢν ἀναστῇ, σκοτοδινίη
μιν λαμβάνει. καὶ τὸν ἄνεμον οὐκ ἀνέχεται οὐδὲ
τὸν ἥλιον. καὶ τὰ ὦτα τέτριγε, καὶ τῷ ψόφῳ
ἄχθεται,[1] καὶ ἐμέει σίαλα καὶ λάπην, ἐνίοτε δὲ
καὶ τὰ σιτία. καὶ τὸ δέρμα λεπτύνεται τῆς
κεφαλῆς, καὶ ἥδεται ψαυόμενος.

Ὅταν οὕτως ἔχῃ, πρῶτον μέν οἱ πιεῖν φάρμακον δοῦναι ἄνω, ὅ τι φλέγμα ἄξει, καὶ μετὰ τοῦτο
τὴν κεφαλὴν καθῆραι. ἔπειτα διαλείπων φάρμακον πῖσαι κάτω· ἔπειτα σιτίοισιν ἀνακομίζειν
αὐτὸν ὡς ὑποχωρητικωτάτοισιν, ὀλίγα αἰεὶ προσ-
τιθείς. ἐπὴν δὲ κατεσθίῃ ἤδη τὰ σιτία ἀρκοῦντα,
ἐμέτοισι χρήσθω νῆστις, τῷ φακίῳ ξυμμίσγων
μέλι καὶ ὄξος, λάχανα προτρώγων. καὶ τῇ ἡμέρῃ
ταύτῃ ᾗ[2] ἂν ἐμέσῃ, πρῶτον μὲν κυκεῶνα πινέτω
λεπτόν· ἔπειτα ἐς ἑσπέρην σιτίοισιν ὀλίγοισι
χρήσθω. καὶ ἀλουτείτω καὶ περιπατείτω[3] ἀπὸ
τῶν σιτίων καὶ ὄρθρου, φυλασσόμενος τὸν ἄνεμον
καὶ τὸν ἥλιον, καὶ πρὸς πῦρ μὴ προσίτω.[4] καὶ ἢν
μέντοι ταῦτα[5] ποιήσαντι ὑγιὴς γένηται· εἰ δὲ μή,
προκαθήρας αὐτὸν τοῦ ὕδατος[6] πρῶτον μὲν ἐλλεβόρῳ, ἔπειτα ἐς τὰς ῥῖνας ἐγχέαι φάρμακον· καὶ
διαλιπὼν ὀλίγον χρόνον κάτω καθῆραι. ἔπειτα
ἀνακομίσας σιτίοισιν, εἶτα ταμὼν τὴν κεφαλὴν

---

[1] M adds ἀκούων.   [2] ᾗ om. M.   [3] καὶ περιπατείτω om. Θ.
[4] Θ: προσιέτω Μ.   [5] Θ: μὲν τοιαῦτα Μ.   [6] Potter: ἧρος ΘΜ.

218

he seems to see two things instead of one; if he gets up, dizziness comes over him; he can tolerate neither wind nor sun; his ears ring, he is vexed by any noise, and he vomits saliva and scum, sometimes food as well. The skin on his head becomes thin, and he feels pleasure on being touched there.

When the case is such, first give the patient a medication to drink that will draw phlegm upwards, and after that clean out his head. Then, leaving a space of time, have him drink a medication to act downwards; next, restore him with foods of the most laxative kind, continually adding a little more. When he has reached the stage where he is eating an adequate amount of food, have him employ vomiting in the fasting state, by first eating vegetables, and then drinking a decoction of lentils into which honey and vinegar have been mixed. On the same day he vomits, let him first drink a thin cyceon, and then, towards evening, eat a little food; let him go without bathing, take walks after meals and early in the morning—but out of the wind and sun—and not go near any fire. If, when he does these things, he becomes well, fine; if not, first clean him of water by using hellebore, and then pour a medication into his nostrils; after a short time, clean downwards. Next, restore with foods, and then incise the head at the bregma; bore

κατὰ τὸ βρέγμα, τρυπῆσαι πρὸς τὸν ἐγκέφαλον,
καὶ ἰᾶσθαι ὡς πρίσμα.

16. Ἑτέρη νοῦσος· ῥῖγος καὶ ὀδύνη καὶ πυρετὸς διὰ τῆς κεφαλῆς, μάλιστα δ' ἐς τὸ οὖς καὶ ἐς
τοὺς κροτάφους καὶ ἐς τὸ βρέγμα. καὶ τὰς χώρας
τῶν ὀφθαλμῶν ἀλγέει, καὶ αἱ ὀφρύες δοκέουσίν οἱ
ἐπικεῖσθαι, καὶ τὴν κεφαλὴν βάρος ἔχει. καὶ ἢν
30 τίς μιν κινήσῃ, ἐμέσει, | καὶ ἐμέει[1] πολὺ καὶ ῥηϊ-
δίως, καὶ τοὺς ὀδόντας αἱμωδίη καὶ νάρκα ἔχει.
καὶ αἱ φλέβες αἴρονται καὶ σφύζουσιν αἱ ἐν τῇ
κεφαλῇ, καὶ οὐκ ἀνέχεται ἠρεμέων, ἀλλὰ ἀλύει
καὶ ἀλλοφρονέει ὑπὸ τῆς ὀδύνης. τούτῳ ἢν μὲν
κατὰ τὰς ῥῖνας ἢ κατὰ τὰ ὦτα ῥαγῇ ὕδρωψ, ῥεῖ
ὑπόπυος, καὶ ὑγιὴς γίνεται· εἰ[2] δὲ μή, ἀποθνήσκει
ἐν ἑπτὰ ἡμέρῃσιν ὡς τὰ πολλά. αὕτη ἡ νοῦσος
γίνεται μάλιστα ἐκ λιπυρίης, ἐπὴν ἀπαλλαγῇ
τοῦ πυρός, ἀκάθαρτος ἐών, ἢ σιτίων ἐμπιμπλῆται,
ἢ θωρήσσηται, ἢ ἐν ἡλίῳ κάμῃ.

Ὅταν οὕτως ἔχῃ, πρῶτον μὲν ἀφιέναι ἀπὸ
τῆς κεφαλῆς τοῦ αἵματος ὁπόθεν ἄν σοι δοκέῃ.
ἐπὴν δὲ ἀφῇς, τὴν κεφαλὴν ξυρήσας, ψύγματά οἱ
προσφέρειν· καὶ ἢν μὴ ὑποχωρέῃ ἡ γαστήρ, ὑποκλύσαι. πίνειν δὲ διδόναι τὸ ἀπὸ τῶν κρίμνων
ὕδωρ, ῥυφάνειν δὲ διδόναι[3] τὸν ἀπὸ τῆς πτισάνης
χυλὸν ψυχρὸν καὶ ἐπιπίνειν ὕδωρ. ἢν δέ οἱ πρὸς

---

[1] ἐμέσει, καὶ ἐμέει Littré: οὐρήσειει καὶ οὐρήσει Θ: οὐρήσῃ, εἰ καὶ οὐρέοι
Μ.   [2] Θ: ἢν Μ.   [3] τὸ . . . διδόναι om. Μ.

right through to the brain, and heal the wound as you would one made by sawing.

16. Another disease: chills, pain and fever throughout the head, especially in the ear, temples and bregma. The patient feels pain in the sockets of his eyes, his eyebrows seem to press down on him, and heaviness befalls his head. If anyone moves him, he vomits copiously and easily; his teeth are set on edge, and he is numb. The vessels in his head are raised up and throb, and he cannot bear to be still, but is beside himself and frenzied from the pain. If, in this patient, a watery discharge breaks out through the nostrils or ears, it runs out mixed with pus, and he recovers; if not, he usually dies in seven days. The disease generally arises from a remittent fever, when, during a remission of the fever-heat, the patient, in an unclean state, fills himself with food, becomes drunk, or toils in the sun.

When the case is such, first draw blood from the head, from wherever you think appropriate. After you have drawn this, shave the head and apply cold compresses; if the belly does not pass anything downwards, clean it out with an enema. Give the patient water from barley-meal to drink, as gruel cold barley-water, and after that water. If the

τὰ ψύγματα μὴ χαλᾷ, μεταβαλών, τῷ ἀσκῷ χρῆσθαι καὶ θερμαίνειν. ἐπὴν δὲ παύσηται ἡ ὀδύνη,
σιτίοισι χρήσθω διαχωρητικοῖσι, καὶ μὴ ἐμπιμ-
πλάσθω. ἐπὴν δὲ γένηται εἰκοσταῖος πεπαυμέ-
νος τῆς ὀδύνης, πυριήσας αὐτοῦ τὴν κεφαλήν,
πρὸς τὰς ῥῖνας φάρμακον προστίθει, καὶ διαλιπὼν ἡμέρας τρεῖς φάρμακον πῖσαι κάτω.

17. Ἑτέρη νοῦσος· ἢν ὑπεραιμήσαντα[1] τὰ
φλέβια τὰ ἔναιμα τὰ περὶ τὸν ἐγκέφαλον θερμή-
νῃ τὸν ἐγκέφαλον, πυρετὸς ἴσχει ἰσχυρός, καὶ
ὀδύνη ἐς τοὺς κροτάφους καὶ τὸ βρέγμα καὶ ἐς
τοὔπισθεν τῆς κεφαλῆς· καὶ τὰ ὦτα ἠχέει καὶ
πνεύματος ἐμπίμπλαται καὶ ἀκούει οὐδέν· καὶ
ἀλύει, καὶ ῥιπτάζει αὐτὸς ἑωυτὸν ὑπὸ τῆς ὀδύνης. οὗτος ἀποθνήσκει πεμπταῖος ἢ ἑκταῖος.

Ὅταν οὕτως ἔχῃ, θερμαίνειν αὐτοῦ τὴν κεφα-
λήν· ἢν γὰρ ῥαγῇ διὰ τῶν ὤτων ἢ διὰ τῶν ῥινῶν
ὕδωρ, οὕτως ἐκφυγγάνει. ἢν δ᾽ ἐκφύγῃ τὰς ἡμέρας τὰς ἕξ, διαιτᾶν ὥσπερ τὴν προτέρην.

18. Ἢν ὑπεραιμήσωσιν[2] αἱ φλέβες ἐν τῇ
κεφαλῇ, ὀδύνη ἴσχει βραχέη τὴν κεφαλὴν πᾶσαν,
καὶ ἐς τὸν τράχηλον, καὶ μεταβάλλει ἄλλοτε ἄλλῃ τῆς κεφαλῆς· καὶ ἐπειδὰν ἀναστῇ, σκοτοδινίη
μιν ἴσχει· πυρετὸς δ᾽ οὐ λαμβάνει.

Ὅταν οὕτως ἔχῃ, ξυρήσας αὐτοῦ[3] τὴν κεφα-

---

[1] Potter: -εμήσαντα ΘΜ: -αιμώσαντα Ermerins.   [2] Jouanna
(p. 48): -εμήσωσιν ΘΜ: -αιμώσωσι Ermerins.   [3] αὐτοῦ om. M.

disease does not slacken when treated with cold compresses, switch over and use the leather skin to warm. When the pain stops, let the patient take laxative foods, and not fill himself. On the twentieth day after the cessation of pain, apply a vapour-bath to his head, administer a medication to his nostrils, and, leaving an interval of three days, have him drink a medication to act downwards.

17. Another disease: if the small blood-vessels around the brain overfill with blood, they heat the brain, and there is violent fever, and pain in the temples, bregma and back of the head. The ears ring, they are filled with air, and they hear nothing; the patient is distraught and casts himself about from the pain. He dies on the fifth or sixth day.

When the case is such, warm the patient's head; for if water breaks out through his ears or nostrils, he escapes. If he survives the six days, employ the same regimen as in the preceding disease.

18. If the vessels in the head overfill with blood, a brief pain occupies the entire head, radiating to the neck, and moves at one time to one part of the head, at another time to another part; when the patient gets up, dizziness comes over him; there is no fever.

When the case is such, if the condition does not

λήν, ἢν μὴ τοῖσι χλιάσμασιν ὑπακούῃ, σχίσαι ἀπὸ
τῆς κεφαλῆς τὸ μέτωπον, ᾗ ἀπολήγει τὸ δασύ.
ἐπὴν δὲ τάμῃς, διαστείλας τὸ δέρμα, ὅταν ἀπορρυῇ τὸ αἷμα, ἁλσὶ λεπτοῖσι διαπάσαι· ἐπὴν δέ σοι
τὸ αἷμα ἀπορρυῇ, συνθεὶς τὴν τομήν, κρόκῃ δι-
πλῇ κατελίξαι πᾶσαν τὴν τομήν. ἔπειτα περι-
χρίσας τῇ κηροπίσσῳ σπληνίσκον, ἐπιθεὶς κάτω
ἐπὶ τῷ ἕλκει, εἴριον πινόεν ἐπιθείς, καταδῆσαι,
καὶ μὴ ἐπιλῦσαι[1] ἑπτὰ ἡμερέων, ἢν μὴ ὀδύνη ἔχῃ·
ἢν δ' ἔχῃ, ἀπολύσασθαι. διδόναι δ'[2] ἔστ' ἂν ὑγιὴς
γένηται, πίνειν μὲν τὸ ἀπὸ τοῦ κρίμνου, ῥυφάνειν
δὲ τὸν χυλὸν τῆς πτισάνης καὶ ἐπιπίνειν ὕδωρ.

19. Ἢν δὲ χολᾷ ὁ ἐγκέφαλος, πυρετὸς ἴσχει
βληχρὸς καὶ ῥῖγος καὶ ὀδύνη διὰ τῆς κεφαλῆς
πάσης, μάλιστα δὲ[3] τοὺς κροτάφους καὶ ἐς τὸ
βρέγμα καὶ ἐς τὰς χώρας τῶν ὀφθαλμῶν. καὶ αἱ
ὀφρύες ἐπικρέμασθαι δοκέουσι, καὶ ἐς τὰ ὦτα
ὀδύνη ἐσφοιτᾷ ἐνίοτε, καὶ κατὰ τὰς ῥῖνας χολὴ
ῥεῖ, καὶ ἀμβλυώσσει[4] τοῖς ὀφθαλμοῖσι. καὶ τοῖσι
μὲν πλείστοισιν ἐς τὸ ἥμισυ τῆς κεφαλῆς ὀδύνη
φοιτᾷ, γίνεται δὲ καὶ ἐν πάσῃ τῇ κεφαλῇ.

Ὅταν οὕτως ἔχῃ, ψύγματά οἱ προσθιθέναι
πρὸς τὴν κεφαλήν, καὶ ἐπὴν ἡ[5] ὀδύνη καὶ τὸ
ῥεῦμα παύηται, σελίνου χυλὸν ἐς τὰς ῥῖνας ἐν-

---

[1] Θ adds ἐπ'.   [2] δ' om. M.   [3] M adds ἐς.   [4] Θ: ἀμ-
βλυώσσουσι M.   [5] ἡ om. M.

yield to fomentations, shave the patient's head, and make an incision in the forehead between the eyes, where the hairs stop; when you have made this cut, separate the skin, and, when blood flows out, sprinkle with fine salt. When blood has flowed out to your satisfaction, close the incision, and wrap the whole of it with a bandage of double threads. Next, anoint a linen pad with wax-pitch ointment, place it directly over the incision, apply unwashed wool, and bind it fast; do not remove this for seven days, unless there is pain; if there is, remove it. Until the patient is well, give him water made from groats to drink, as gruel barley-water, and after that water.

19. If the brain suffers from bile, a mild fever is present, chills, and pain through the whole head, especially in the temples, bregma, and the sockets of the eyes. The eyebrows seem to overhang, pain sometimes migrates to the ears, bile runs out through the nostrils, and the patient sees unclearly. In most patients, pain occupies one half of the head, but it can also arise in the whole head.

When the case is such, apply cold compresses to the patient's head, and, when the pain and flux cease, instill celery juice into his nostrils. Let him

στάζειν. καὶ ἀλουτείτω ἕως ἂν ἡ ὀδύνη ἔχῃ, καὶ
ῥυφανέτω κέγχρον λεπτόν, μέλι ὀλίγον παραχέων,
καὶ πινέτω ὕδωρ. ἢν δὲ μὴ ὑποχωρέῃ, κράμβας
34 τρωγέτω | καὶ τὸν χυλὸν ῥυφανέτω· εἰ δὲ μή, τῆς
ἀκτῆς τῶν φύλλων τὸν αὐτὸν τρόπον. καὶ ἐπὴν
σοι δοκέῃ καιρὸς εἶναι, σιτία οἱ προσφέρειν ὡς
ὑποχωρητικώτατα. καὶ ἢν, ἀπηλλαγμένου τοῦ
ῥεύματος καὶ τῆς ὀδύνης, ὑπὲρ τῆς ὀφρύος¹ αὐτῷ
βάρος ἐγγίνηται κατὰ τὸν μυξωτῆρα ἢ² μύξα πα-
χέη³ καὶ σαπρή, πυριήσας αὐτὸν ὄξει καὶ ὕδατι
καὶ ὀριγάνῳ, ἔπειτα λούσας θερμῷ ὕδατι, προσθεῖναι τὸ ἄνθος τοῦ χαλκοῦ καὶ τὴν σμύρναν πρὸς
τὰς ῥῖνας. ταῦτα ποιήσας, ὡς τὰ πολλὰ ὑγιὴς
γίνεται· ἡ δὲ νοῦσος οὐ θανατώδης.

20. Ἢν σφακελίσῃ ὁ ἐγκέφαλος, ὀδύνη λάζεται ἐκ τῆς κοτίδος ἐς τὴν ῥάχιν, καὶ ἐπὶ τὴν καρδίην καταφοιτᾷ ψῦχος· καὶ ἱδρὼς ἐξαπίνης καὶ
ἄπνοος τελέθει, καὶ διὰ τῶν ῥινῶν αἷμα ῥεῖ· πολλοὶ δὲ καὶ ἐμέουσιν. οὗτος ἐν τρισὶν ἡμέρῃσιν
ἀποθνήσκει· ἢν δὲ τὰς ἑπτὰ ἡμέρας ὑπερφύγῃ·
οὐχ ὑπεκφεύγουσι⁴ δὲ πολλοί.

Οὗτος ἢν μὲν τὸ αἷμα ἐμέῃ ἢ ἐκ τῶν ῥινῶν
ῥέῃ, μήτε λούειν αὐτὸν θερμῷ, μήτε χλιάσματα
προσφέρειν, πίνειν δὲ⁵ ὄξος λευκὸν ὑδαρὲς κιρνάς,
καὶ ἢν ἀσθενέῃ, τῆς πτισάνης ῥυφάνειν. ἢν δὲ

¹ Θ: ὀσφύος Μ.　　² Θ: ἢν Μ.　　³ Μ adds ἢ.　　⁴ Θ: ὑπερ-
φεύγουσι Μ.　　⁵ Μ adds διδόναι.
226

avoid bathing, as long as the pain is present, take as gruel thin millet to which a little honey has been added, and drink water. If nothing passes off below, have him eat cabbage, and drink the juice as gruel; if not that, then the juice of elder leaves in the same way. When you think it is the right moment, give foods of the most laxative kind. If, when the patient is relieved of his flux and pain, a heaviness sets in from above the eyebrow down to the nostril, or there is thick purulent mucus, apply a vapour-bath of vinegar, water and marjoram; then wash him with hot water, and apply flower of copper and myrrh to his nostrils. If the patient does these things, he usually recovers. The disease is seldom mortal.

20. If the brain becomes sphacelous, pain radiates from the occiput to the spine, and coldness moves down over the cardia. The patient suddenly sweats and becomes lifeless, and blood runs out through his nostrils; many vomit it as well. This person generally dies in three days, but if he escapes for seven, he recovers; not many survive.

If this patient vomits blood, or blood flows from his nostrils, do not wash him with hot water or apply fomentations, but have him drink white vinegar, mixing it dilutely, or, if he is weak, barley-

πλέον σοι δοκέῃ τοῦ δικαίου ἐμέειν τὸ αἷμα ἢ ἐκ
τῶν ῥινῶν οἱ ῥέῃ, τοῦ μὲν ἐμέτου πινέτω ἄλητον
σητάνιον ἐπὶ ὕδωρ ἐπιπάσσων· ἢν δ᾽ ἐκ τῶν ῥινῶν ῥέῃ, καὶ ἀποδείτω τὰς φλέβας τὰς ἐν τοῖσι
βραχίοσι καὶ τὰς ἐν τοῖσι κροτάφοισι, σπλῆνας
ὑποτιθείς. ἢν δὲ τούτων οἱ μηδ᾽ ἕτερον ᾖ, ἀλγέῃ
δὲ τὴν κοτίδα καὶ τὸν τράχηλον καὶ τὴν ῥάχιν,
καὶ ἐπὶ τὴν καρδίην ἴῃ τὸ ψῦχος, χλιαίνειν
τοῖσιν ὀρόβοισι τὰ στέρνα καὶ τὸν νῶτον καὶ τὴν
κοτίδα καὶ τὸν τράχηλον. ταῦτα ποιέων μάλιστ᾽
ἂν ὠφελέοις·[1] ἐκφεύγουσι δὲ ὀλίγοι.

36   21. Ἄλλη νοῦσος· ἐξαπίνης ὑγιαίνοντα ὀδύνη
ἔλαβε τὴν κεφαλήν, καὶ παραχρῆμα ἄφωνος
γίνεται, καὶ ῥέγκει, καὶ τὸ στόμα κέχηνε· καὶ ἤν
τις αὐτὸν καλέῃ ἢ κινήσῃ, στενάζει, ξυνίει δ᾽
οὐδέν, καὶ οὐρέει πολύ, καὶ οὐκ ἐπαΐει οὐρέων.
οὗτος, ἢν μιν μὴ πυρετὸς λάβῃ, ἐν τῇσιν ἑπτὰ
ἡμέρῃσιν ἀποθνήσκει· ἢν δὲ λάβῃ, ὡς τὰ πολλὰ
ὑγιὴς γίνεται. ἡ δὲ νοῦσος πρεσβυτέροισι μᾶλλον
γίνεται ἢ νεωτέροισι.

Τοῦτον, ὅταν οὕτως ἔχῃ, λούειν χρὴ θερμῷ καὶ
πολλῷ, καὶ θάλπειν ὡς μάλιστα, καὶ ἐνστάζειν
μελίκρητον χλιαίνων ἐς τὸ στόμα. ἢν δ᾽ ἔμφρων
γένηται καὶ ἐκφεύγῃ τὴν νοῦσον, ἀνακομίσας
αὐτὸν σιτίοισιν, ἐπήν σοι δοκέῃ ἰσχύειν, ἐς τὰς
ῥῖνας ἐνθεὶς αὐτῷ φάρμακον, καὶ διαλείπων ὀλί-

[1] Θ: ὠφελέοι Μ.

gruel. If the amount of blood the patient vomits or that runs out of his nostrils is greater than you think it should be, in the case of vomiting, have him drink this year's flour sprinkled over water, or in the case of blood flowing from the nostrils, let him bind up the vessels in his arms and temples, placing compresses over them. If neither of these things happens to the patient, but he suffers pain in his occiput, neck and spine, and coldness goes to his cardia, foment his chest, back, occiput and neck with vetches. With these measures, you will be most helpful; few escape.

21. Another disease: pain suddenly seizes the head in a healthy person, and he at once becomes speechless, breathes stertorously, and gapes with his mouth; if anyone calls to him or moves him, he moans; he comprehends nothing; he passes copious urine, but is not aware of it when he does. Unless fever occurs in this patient, he dies in seven days; if it does, he usually recovers. The disease is more frequent in older persons than in younger ones.

When the case is such, wash the patient with copious hot water, and warm him as much as possible; heat melicrat and instil it into his mouth. If he regains his senses and escapes from the disease, restore him with foods; when you think he is strong, introduce a medication into his nostrils,

γας ἡμέρας, κατωτερικὸν δὸς φάρμακον[1] πιεῖν·
ἢν γὰρ μὴ καθάρῃς, κίνδυνος αὖτις τὴν νοῦσον
ὑποστρέψαι· ἐκφυγγάνουσι δὲ οὐ μάλα ἐκ τῆς
πρώτης.

22. Ἢν δ' ἐκ θωρήξιος ἄφωνος γένηται, ἢν μὲν
αὐτίκα καὶ παραχρῆμα λάβῃ μιν πυρετός, ὑγιὴς
γίνεται· ἢν δὲ μὴ λάβῃ, τριταῖος ἀποθνήσκει.

Ἢν δὲ μὲν[2] οὕτως ἔχοντι ἐπιτύχῃς, λούειν
πολλῷ καὶ θερμῷ, καὶ πρὸς τὴν κεφαλὴν σπόγγους ἐν ὕδατι βάπτων θερμῷ προστιθέναι, καὶ ἐς
τὰς ῥῖνας κρόμμυα ἀπολέπων ἐντιθέναι. οὗτος
ἢν μὲν ἀνατείνας τοὺς ὀφθαλμοὺς καὶ φθεγξάμε-
νος παρ' ἑωυτῷ γένηται καὶ μὴ φλυηρῇ, τὴν μὲν
ἡμέρην ταύτην κεῖται κωμαίνων, τῇ δ' ὑστεραίῃ
ὑγιὴς γίνεται. ἢν δ' ἀνιστάμενος χολὴν ἐμέῃ,
μαίνεται, καὶ ἀποθνήσκει μάλιστα ἐν πέντε ἡμέρῃσιν, ἢν μὴ κατακοιμηθῇ. τοῦτον οὖν τάδε χρὴ
ποιεῖν· λούειν πολλῷ καὶ θερμῷ, ἔστ' ἂν αὐτὸς
<ἐς>[3] ἑωυτὸν παρῇ[4] ἔπειτα ἀλείψας ἀλείφατι
πολλῷ, κατακλῖναι ἐς στρώματα μαλθακῶς, καὶ
ἐπιβαλεῖν ἱμάτια, καὶ μήτε λύχνον καίειν παρ'
αὐτῷ, μήτε φθέγγεσθαι· ὡς γὰρ ἐπὶ τὸ πολὺ ἐκ
τοῦ λουτροῦ κατακοιμᾶται, καὶ ἢν τοῦτο ποιήσῃ,
ὑγιὴς γίνεται. ἢν δὲ παρ' ἑωυτῷ[5] γένηται, τὰς
πρώτας ἡμέρας τῶν σιτίων ἐρύκειν αὐτόν, ἡμέ-

38

---

[1] δὸς φάρμακον om. M.   [2] Foes (n. 47): μὴ ΘM.   [3] Later
mss, edd.   [4] M: παρίῃ Θ.   [5] Θ: ἑωυτοῦ M.

wait for a few days, and then give him a medication to drink that acts downwards; for if you do not clean the patient out, there is a danger that the disease will return. Not many survive the first bout.

22. In a person that has lost his speech as the result of drunkenness, if fever comes on immediately at that moment, he recovers, but if it does not, he dies on the third day.

If you happen upon a patient with this condition, wash him in copious hot water, and soak sponges in hot water and apply them to his head; peel onions and insert them into his nostrils. If this patient raises his eyes, recovers his normal speech, and ceases to talk nonsense, on that day he still lies in a drowsy state, but on the next one he recovers. If, on getting up, he vomits bile, then he is likely to rage, and usually dies in five days, unless he falls asleep. Now, you must do the following for him: wash him in copious hot water until he comes to himself; then anoint him generously with oil, put him to bed in soft bed-clothes, covering him with blankets, and neither kindle a lamp beside him, nor speak to him; for after his bath he is very likely to fall asleep, and if he does, he recovers. If the patient comes to himself, on the first days prohibit

ρας[1] τρεῖς ἢ τέσσερας διδόναι δὲ κέγχρον λεπτὸν
ῥυφάνειν ἢ πτισάνης χυλόν, καὶ οἶνον μελιηδέα[2]
πίνειν· ἔπειτα σιτίοισι χρῆσθαι ὡς μαλθακωτάτοισι καὶ ὀλίγοισι τὸ πρῶτον.

23. Ἢν σφάκελος λάβῃ,[3] ὀδύνη ἴσχει μάλιστα
τὸ πρόσθεν τῆς κεφαλῆς κατὰ σμικρόν· καὶ ἀνοιδέει, καὶ πελιδνὸν γίνεται, καὶ πυρετὸς καὶ ῥῖγος
λαμβάνει. ὅταν οὕτως ἔχῃ, ταμόντα χρή, ᾗ ἂν[4]
ἐξοιδέῃ,[5] διακαθήραντα τὸ ὀστέον ξῦσαι ἔστ' ἂν
ἀφίκηται πρὸς τὴν διπλοΐδα· ἔπειτα ἰᾶσθαι ὡς
κάτηγμα.

24. Ὅταν τερηδὼν γένηται ἐν τῷ ὀστέῳ, ὀδύνη λαμβάνει ἀπὸ τούτου τοῦ ὀστέου. τῷ δὲ χρόνῳ τὸ δέρμα[6] λεπτὸν γίνεται καὶ ἀναφυσᾶται, καὶ
γίνεται ἐν αὐτῷ κάτηγμα, καὶ ἢν τοῦτο ἀνα-
τάμῃς, εὑρήσεις ἀνατεῖνον ὀστέον καὶ τρηχὺ καὶ
πυρρόν, ἐνίοισι δὲ διαβεβρωμένον πρὸς τὸν ἐγκέφαλον. ὅταν οὕτως ἔχοντι ἐπιτύχῃς, ἢν μὲν ᾖ
πέρην διαβεβρωμένον, ἐὰν ἄριστον, καὶ ἰᾶσθαι ὡς
τάχιστα τὸ ἕλκος. ἢν δὲ τετρημένον[7] μὲν μὴ ᾖ,
τρηχὺ δέ, ξύσας ἐς[8] τὴν διπλοΐδα, ἰᾶσθαι ὥσπερ
τὴν πρόσθεν.

---

[1] τῶν σιτίων ... ἡμέρας om. M.   [2] Θ: μελιτιηδέα M.
[3] M adds ἡ.   [4] ᾗ ἂν Foes (n. 49): ἢν ΘM.
[5] Θ: ἐξοιδέει ηι M.   [6] τὸ δέρμα om. M.   [7] Potter (perfora-
tum Foes): τετρωμένον ΘM.   [8] ἐς om. Θ.

food, and for three or four days give him thin millet or barley-water and honey-sweet wine to drink; then give very soft foods, beginning with small amounts.

23. If a sphacelus occurs,[1] pain gradually occupies, in particular, the front of the head; the patient swells up and becomes livid, and fever and chills are present. When this happens, you must incise wherever the swelling is, and clean and scrape the bone until you arrive at the diploe. Then treat as you would in the case of a fracture.

24. When a teredo forms in the skull, a pain originating in the substance of the bone arises; in time, the skin over the teredo becomes thin and puffed up, and a break develops in it; if you cut this open, you will discover the bone to be raised, jagged, reddish and, in some cases, eroded through to the brain. When you happen upon a patient like this, if the bone is eroded right through, it is best to leave it alone, and to heal up the ulcer as fast as possible; however, if the bone is not perforated, but only jagged, scrape into the diploe, and treat as in the preceding disease.

[1] This disease is to be distinguished from sphacelus of the brain, described in chapters 5 and 20 above; here the organ affected is the skull.

25. Ἢν βλητὸς γένηται, ἀλγέει τὸ πρόσθεν τῆς κεφαλῆς, καὶ τοῖσιν ὀφθαλμοῖσι οὐ δύναται ὁρᾶν, ἀλλὰ κῶμά μιν ἔχει καὶ αἱ φλέβες ἐν τοῖσι κροτάφοισι σφύζουσι, καὶ πυρετὸς βλη|χρὸς ἔχει καὶ τοῦ σώματος παντὸς ἀκρασίη, καὶ μινύθει.[1] ὅταν οὕτως ἔχῃ, λούειν[2] αὐτὸν θερμῷ πολλῷ, καὶ χλιάσματα πρὸς τὴν κεφαλὴν προστιθέναι· ἐκ δὲ τῆς πυρίης ἐς τὰς ῥῖνας σμύρναν καὶ ἄνθος χαλκοῦ. ῥυφάνειν δὲ τὸν χυλὸν τῆς πτισάνης, καὶ πίνειν ὕδωρ. καὶ ἢν μὲν ταῦτα ποιέοντι ῥᾷον[3] γίνηται· εἰ δὲ μή, αὕτη[4] γὰρ μόνη ἐλπίς, σχίσαι αὐτοῦ τὸ βρέγμα, καὶ ἢν[5] ἀπορρυῇ τὸ αἷμα, συνθεὶς τὰ χείλεα, ἰᾶσθαι καὶ καταδῆσαι· ἢν δὲ μὴ σχίσῃς, ἀποθνῄσκει ὀκτωκαιδεκαταῖος, ἢ εἰκοσταῖος ὡς ἐπὶ[6] τὰ πολλά.

26. Κυνάγχη· πυρετὸς λαμβάνει καὶ ῥῖγος καὶ ὀδύνη τὴν κεφαλήν· καὶ τὰ σιηγόνια οἰδίσκεται, καὶ τὸ πτύαλον χαλεπῶς καταπίνει, καὶ ἀποπτύει τὰ σίελα σκληρὰ κατ᾽ ὀλίγα, καὶ τῇ[7] φάρυγγι κάτω ῥέγκει. καὶ ἢν καταλαβὼν τὴν γλῶσσαν σκέπτῃ, ὁ γαργαρεὼν οὐ μέγας, ἀλλὰ λαπαρός. ἡ δὲ φάρυξ ἔσωθεν σιάλου γλίσχρου ἔμπλεως, καὶ οὐ δύναται ἐκχρέμπτεσθαι· καὶ οὐκ

---

[1] Indicatives (ἀλγέει . . . μινύθει) later mss., edd.: subjunctives (incl. μὴ δύνηται) ΘΜ.   [2] Θ: καίειν Μ.   [3] Θ: ῥάων Μ.
[4] Potter: αὐτὴ Θ: ταύτῃ Μ.   [5] Θ: ἐπὴν Μ.   [6] ἐπὶ om. Μ.
[7] Θ: ἐν Μ.

DISEASES II

25. If a person is stricken, he has pain in the front of his head, he cannot see but he is drowsy, the vessels in his temples throb, there are mild fever and powerlessness of the whole body, and he wastes away. When the case is such, wash the patient in copious hot water, and apply fomentations to his head; after a vapour-bath, insert myrrh and flower of copper into his nostrils; have him take barley-water as gruel, and also drink water. If he feels better when he does these things, fine; if not, this is his only hope: cut open the bregma, and, if blood flows out, set the lips of the incision together, treat and bind them. If you do not make this incision, the patient usually dies on the eighteenth or twentieth day.

26. Angina: there are fever, chills, and pain in the head; the jaws swell up, the patient has difficulty swallowing his saliva, he expectorates thick sputum a little at a time, and he breathes stertorously, low down in his throat. If you hold his tongue down and look carefully, the uvula is not enlarged or swollen. Inside, the throat is quite full of sticky sputum, and the person cannot cough it

ἀνέχεται κείμενος, ἀλλ᾽ ἢν κατακέηται,[1] πνίγεται.

Τοῦτον ἢν οὕτως ἐπιτύχῃς ἔχοντα, ποιέειν τάδε· πρῶτον μὲν σικύην[2] προσβαλεῖν πρὸς τὸν σφόνδυλον τὸν ἐν τῷ τραχήλῳ τὸν πρῶτον, [ἔπειτα][3] παραξυρήσας ἐν τῇ κεφαλῇ παρὰ τὸ οὖς ἔνθεν καὶ ἔνθεν, καὶ[4] ἐπὴν ἀποσφίγξῃ, τὴν σικυώνην ἐᾶν προσκεῖσθαι ὡς πλεῖστον χρόνον. ἔπειτα πυριᾶν αὐτὸν ὄξει καὶ λίτρῳ καὶ καρδάμου σπέρματι καὶ ὀριγάνῳ, τρίψας λεῖα, κεράσας τὸ ὄξος ἰσόχοον ὕδατι, καὶ ἄλειφα ὀλίγον ἐπι|στάξας, διεῖναι τούτῳ· ἔπειτα ἐς χυτρίδα ἐγχέας, ἐπιθεὶς ἐπίθημα καὶ κατασκεπάσας, τρυπήσας τὸ ἐπίθημα,[5] κάλαμον ἐνθεῖναι κοῖλον. ἔπειτα ἐπιθεὶς ἐπ᾽ ἄνθρακας, ἀναζέσαι, καὶ ἐπὴν διὰ τοῦ καλάμου ἡ ἀτμὶς ἴῃ, περιχάσκων ἑλκέτω ἔσω τὴν ἀτμίδα, φυλασσόμενος μὴ κατακαύσῃ τὴν φάρυγγα. καὶ ἔξωθεν σπόγγους βάπτων ἐς ὕδωρ θερμόν, προστιθέσθω πρὸς τὰς γνάθους καὶ τὰ σιηγόνια. ἀναγαργάριστον δ᾽ αὐτῷ ποιέειν ὀρίγανον καὶ πήγανον[6] καὶ θύμβραν καὶ σέλινον καὶ μίνθην καὶ λίτρον ὀλίγον, μελίκρητον κεράσας ὑδαρές, ὄξος ὀλίγον ἐπιστάξαι· λεῖα τρίψας τὰ φύλλα καὶ τὸ λίτρον, τούτῳ[7] διείς, χλιήνας, ἀναγαργαριζέτω.

---

[1] Μ: -καίηται Θ. [2] Θ: σικύης Μ. [3] Del. Potter: ἐπὶ τὰ καὶ ἐπὶ τὰ Littré. [4] καὶ om. Μ. [5] καὶ ... ἐπίθημα om. Μ. [6] καὶ πήγανον om. Μ. [7] Θ: τοῦτο Μ.

up; he will not lie down, for if he does, he chokes.

If you happen upon a patient in this state, do the following: first, apply a cupping instrument to the first vertebra of his neck, after shaving his head beside the ear on each side, and when the cupping-vessel presses tightly, leave it in place as long as possible. Then apply a vapour-bath of vinegar, soda, cress seed and marjoram: grind these fine, mix the vinegar into an equal amount of water, instil a little oil, and then dissolve the soda, cress seed and marjoram into it; pour into a pot, set on a lid that covers it completely, bore a hole through the lid, and insert a hollow reed. Then set the pot on coals to boil, and, when vapour passes up through the reed, have the patient open his mouth wide and draw in the vapour, taking care not to burn his throat. Soak sponges in hot water, and have the patient apply these externally to his upper and lower jaws. Make a gargle for him of marjoram, rue, savory, celery, mint, and a little soda: prepare dilute melicrat, and instil a little vinegar into it; grind the leaves and soda fine, dissolve them in the liquid, warm it, and have the patient

ἢν δὲ τὸ σίαλον ἴσχηται, μύρτου λαβὼν ῥάβδον,
λείην ποιήσας αὐτήν, ἐπικάμψας τὸ ἄκρον τὸ
ἁπαλὸν τῆς ῥάβδου, κατελίξας εἰρίῳ μαλθακῷ,
καθορῶν ἐς τὴν φάρυγα, τὸ σίελον ἐκκαθαίρειν. καὶ
ἢν ἡ γαστὴρ μὴ ὑποχωρέῃ, βάλανον προστιθέναι
ἢ ὑποκλύζειν. ῥυφανέτω δὲ τὸν χυλὸν τῆς πτισάνης καὶ ὕδωρ ἐπιπινέτω. ἢν δέ οἱ οἴδημα ἐκ-
θύῃ καὶ οἰδίσκηται πρὸς τὰ στήθεα καὶ ἐρυθρὸν ᾖ
καὶ καίηται, ἐλπίδες πλέονες σωτηρίης. ποιέειν
δέ οἱ τάδε· ἐπὴν ἔξω τράπηται τὸ φλέγμα, τεῦτλα ἐμβάπτων ἐς ὕδωρ ψυχρόν, προστιθέτω[1]
ἀναγαργαριζέτω δὲ χλιαροῖσι, καὶ ἀλουτείτω.
ταῦτα ποιέων μάλιστ' ἂν ἐκφυγγάνοι· ἡ δὲ νοῦσος
θανατώδης, καὶ ἐκφυγγάνουσιν ὀλίγοι.

27. Ἑτέρη κυνάγχη πυρετὸς καὶ ὀδύνη λαμβάνει τὴν κεφαλήν, καὶ ἡ φάρυγξ φλεγμαίνει καὶ
τὰ σιηγόνια· καὶ τὸ σίαλον καταπίνειν οὐ δύναται, πτύει δὲ παχὺ καὶ πολλόν, καὶ φθέγγεται
χαλεπῶς.

Ὅταν οὕτως ἔχῃ, πρῶτον μὲν σικύην προσ-
βαλεῖν τὸν αὐτὸν τρόπον ὥσπερ τῷ πρόσθεν. ἔπει-
τα προσίσχειν σπόγγον βρέχων | ἐν ὕδατι θερμῷ
πρὸς τὸν τράχηλον καὶ τὰ σιηγόνια. ἀναγαργαρίζειν δὲ διδόναι τόδε ὅ[2] τῶν φύλλων ἐλειθερές·
πίνειν δὲ διδόναι[3] μελίκρητον ὑδαρές· ῥυφάνειν δὲ

---

[1] Θ: -τιθέναι Μ.   [2] τοδεο Θ: τὸ ἀπὸ Littré (cf. chs. 29, 30).
[3] τόδε ὅ . . . διδόναι om. Μ.

gargle. If sputum is still held back in the throat, take a twig of myrtle, smoothe it off, bend back the flexible end of the twig, and wrap it round with soft wool; then look down into the throat, and clean away the sputum. If the belly does not pass anything downwards, administer a suppository or an enema. Have the patient drink as gruel barley-water, and then water. If he suffers an outbreak of swelling so that he swells up in the chest, becomes red, and burns, he has more hope of being saved. Do the following for this patient: when his phlegm turns upwards, have him soak beets in cold water, and apply them; also let him gargle with warming agents, and abstain from the bath. If he does these things, he will have the greatest chance of surviving; the disease is often mortal, and few escape it.

27. Another angina: there are fever and pain in the head, the throat and jaws swell up, and the patient cannot swallow his saliva, but spits it out thick and plentiful; he speaks with difficulty.

When the case is such, first apply a cupping instrument in the same way you would for the preceding patient; then soak a sponge in hot water, and apply it against the neck and jaws. As gargle give the one made from herbs warmed in the sun; to drink give dilute melicrat; compel the patient to

ἀναγκάζειν τὸν χυλὸν τῆς πτισάνης. ἢν δέ οἱ τὸ
σίαλον ταῦτα ποιέοντι μὴ ᾖ, πυριᾶν τὸν αὐτὸν
τρόπον ὥσπερ ἐν τῇ πρόσθεν. ἢν δέ οἱ ἐς τὰ
στήθεα τρέπηται ἢ ἐς τὸν τράχηλον τὸ φλέγμα,
τεῦτλα ἢ κολοκύντας καταταμών, ἐσβαλὼν ἐς
ὕδωρ χλιαρὸν ἐπιρυφανέτω, καὶ πινέτω ψυχρόν,
ὅπως τὸ σίαλον εὐπετέστατα ἀποχρέμπτηται.
ὅταν δ' ἐξοιδήσῃ ἐς τὰ στήθεα, πλεῦνες ἐκ-
φεύγουσιν. ἢν δέ, τῆς φάρυγος καθεστηκυίης καὶ
τῶν οἰδημάτων, τρεφθῇ ἐς τὸν πλεύμονα ἡ
νοῦσος, πυρετὸς ἐπ' οὖν ἔλαβε, καὶ ὀδύνη τοῦ
πλευροῦ, καὶ ὡς ἐπὶ τὸ πολὺ ἀπ' οὖν ἔθανεν[1] ἐπὴν
τοῦτο γένηται· ἢν δ' ὑπεκφύγῃ[2] ἡμέρας πέντε,
ἔμπυος[3] γίνεται. ἢν μή μιν βὴξ λάβῃ[4] αὐτίκα,
ἢν δ' ἐπιλάβῃ,[5] ὑποχρεμψάμενος καὶ ἀποκαθαρ-
θείς, ὑγιὴς γίνεται. τοῦτον, ἔστ' ἂν μὲν ἡ ὀδύνη
τὸ πλευρὸν ἔχῃ, χλιαίνειν τὸ πλευρόν, καὶ προσφέρειν ὅσα περ εἰ[6] περιπλευμονίη ἔχοιτο. ἢν δὲ
ὑπεκφύγῃ[7] τὰς πέντε ἡμέρας καὶ ὁ πυρετὸς ἀνῇ,
ἡ δὲ βὴξ ἔχῃ, τὰς μὲν πρώτας ἡμέρας ῥυφήμασι
διαχρῆσθαι· ἐπὴν δὲ τῶν σιτίων ἄρξηται, ὡς
λιπαρώτατα καὶ ἁλυκώτατα ἐσθίειν. ἢν δέ οἱ
ἡ[8] βὴξ μὴ ᾖ, ἀλλὰ γινώσκῃς ἔμπυον γινόμενον,

---

[1] ἀπ' οὖν ἔθανεν Θ: ἀπέθανεν Μ.   [2] Θ: ὑπερφύγῃ Μ.   [3] Θ: ἔμπυρος Μ.   [4] Θ: ἐπι- Μ.   [5] Θ: ὑπο- Μ.   [6] περ εἰ om. Μ.
[7] Jouanna (p. 74): ὑπεκφύγοι Θ: ὑπερφύγῃ Μ.   [8] οἱ ἡ Θ: τοι Μ.

drink barley-water. If he does these things, but
there is still no sputum, apply a vapour-bath as in
the disease before. If the phlegm turns towards the
chest or neck, cut up beets or gourds, put them in
water, and have the patient drink them warm as
gruel; let him also drink this cold, in order to cough
up his sputum as easily as possible. When swelling
occurs in the chest, most patients survive. If, when
the throat returns to normal and the swellings go
down, the disease turns towards the lung, then
fever resumes together with pain in the side. Generally, the patient dies when this happens; if he
survives for five days, he suppurates internally. If
coughing is not present immediately, but comes on
later, the sputum is coughed up and cleaned out,
and the patient recovers. As long as the pain in the
patient's side remains, warm the side and make the
same applications as if he were suffering from
pneumonia. If he survives for five days and his
fever remits, but coughing is still present, on the
first days give gruels. When the patient begins
with foods again, let him eat the richest and salti-
est ones. If he does not have a cough, but you know
he is suppurating internally, after his main meal,

δειπνήσαντα, ἐπὴν μέλλῃ καθεύδειν, σκόροδα
ὠμὰ τρωγέτω ὡς πλεῖστα, καὶ ἐπιπινέτω οἶνον
οἰνώδεα ἀκρητέστερον. ἢν μὲν οὕτως οἱ ῥαγῇ τὸ
πύον· εἰ[1] δὲ μή, τῇ ὑστεραίῃ λούσας θερμῷ θυμι-
ᾶσαι, καὶ ἢν ῥαγῇ, ἰᾶσθαι ὥσπερ ἔμπυον.

46   28. Ἑτέρη κυνάγχῃ φλεγμαίνει τοὔπισθεν
τῆς γλώσσης καὶ τὸ κλήϊθρον τὸ ὑπὸ τῷ βρόγχῳ,
καὶ οὐ δύναται καταπίνειν τὸ σίαλον, οὐδ' ἄλλο
οὐδέν· ἢν δ' ἀναγκασθῇ, διὰ τῶν ῥινῶν οἱ ῥεῖ.[2]

Ὅταν οὕτως ἔχῃ, τρίψας μίνθην χλωρὴν καὶ
σέλινον καὶ ὀρίγανον καὶ λίτρον καὶ τῆς ῥόου τῆς
ἐρυθρῆς, μέλιτι διείς, παχὺ ποιέων, ἐγχρίειν τὴν
γλῶσσαν ἔσωθεν ᾗ[3] ἂν οἰδέῃ. ἔπειτα ἀναζέσας
σῦκα, ἀποχέας[4] τὸ ὕδωρ, τρίψας τῆς ῥόου ὀλίγην
διεῖναι[5] τῷ συκίῳ τούτῳ, ἐὰν δύνηται ἀναγαργαρίζειν, ἢν δὲ μή, διακλύζεσθαι· πίνειν δὲ διδόναι
τὸ ἀπὸ τῶν κρίμνων ὕδωρ. ἔξωθεν δὲ τὸν αὐχένα
καὶ τὰ σιηγόνια καταπλάσσειν ἀλήτῳ, ἐν οἴνῳ
καὶ ἐλαίῳ ἕψοντα, χλιαρῷ, καὶ ἄρτους προστιθέναι θερμούς. ἀποπυΐσκεται γὰρ ὡς τὰ πολλὰ ἐν
τῷ κλήϊθρῳ καὶ ἢν μὲν ῥαγῇ αὐτόματον, ὑγιὴς
γίνεται· ἢν δὲ μὴ ῥηγνύηται, ψηλαφήσας τῷ
δακτύλῳ ἢν μαλακὸν ᾖ, σιδήριον ὀξὺ προσδησά-
μενος πρὸς[6] δάκτυλον τύψαι[7] τοῖσι πλείστοισι.

---

[1] Θ: ἢν Μ.   [2] οἱ ῥεῖ Θ: οἰδεῖ Μ.   [3] ᾗ om. Μ.   [4] Θ: ἀπο-
ζέσας Μ.   [5] Θ: ὑπο- Μ.   [6] πρὸς om. Θ.   [7] Θ: τρίψαι Μ.

when he is about to go to bed, have him eat a generous amount of raw garlic, and after that drink a strong unmixed wine. If, with this, the pus breaks out, fine; if not, on the next day wash the patient in hot water and apply a fumigation; if the pus breaks out, treat as you would in a case of internal suppuration.

28. Another angina: the back of the tongue and the epiglottis under the wind-pipe swell up, so that the patient can swallow neither his saliva nor anything else; if he is forced to, it runs out through his nostrils.

When the case is such, grind green mint, celery, marjoram, soda, and red sumach, soak them in honey, let this thicken, and anoint the tongue inside where it is swollen. Then, if the patient is able to gargle, boil figs, pour off the water, grind sumach, and soak a little of this in the fig-juice; if he is not able to gargle, have him wash his mouth out, and give him water made from groats to drink. Externally, plaster the neck and jaws with flour boiled in wine and oil, and still warm, and apply warm loaves of bread; for suppuration often occurs in the epiglottis, and if pus breaks out spontaneously, the patient recovers. If not, feel with a finger whether the epiglottis has become soft, and then attach an iron blade to your finger and strike the

ταῦτα ποιήσαντες[1] ὑγιέες γίνονται· ἡ δὲ νοῦσος
αὕτη ἥκιστα θανατώδης.

29. Ἢν σταφυλὴ γένηται ἐν τῇ φάρυγγι,
ἐμπίμπλαται ἄκρος ὁ γαργαρεὼν ὕδατος, καὶ
γίνεται στρογγύλος τὸ ἄκρον καὶ διαφανής, καὶ
ἐπιλαμβάνει τὴν πνοιήν. καὶ ἢν φλεγμήνῃ τὰ σιηγόνια ἔνθεν καὶ ἔνθεν, ἀποπνίγεται· ἢν δὲ αὐτὸς[2]
ἐφ' ἑωυτοῦ γένηται, τούτων μὴ φλεγμαινόντων,
ἧσσον ἀποθνήσκει. ὅταν οὕτως ἔχῃ, λαβὼν τῷ
δακτύλῳ τὸν γαργαρεῶνα, ἄνω ἐς τὴν ὑπερῴην
ἀποπιέσας, διατα|μεῖν ἄκρον· ἔπειτα διδόναι ἀνα-
γαργαρίζειν τὸ ἀπὸ τῶν φύλλων· λείχειν δὲ ἄλητον ψυχρόν, καὶ ὕδωρ ἐπιπίνειν, καὶ μὴ λούεσθαι.

30. Ἢν ἀντιάδες γένωνται, συνοιδέει ὑπὸ τὴν
γνάθον ἔνθεν καὶ ἔνθεν, καὶ ψαυόμενον σκληρόν
ἐστιν ἔξωθεν, καὶ ὁ γαργαρεὼν ὅλος φλεγμαίνει.
ὅταν οὕτως ἔχῃ, καθεὶς τὸν δάκτυλον, διωθέειν
τὰς ἀντιάδας· πρὸς δὲ τὸν γαργαρεῶνα προσ-
χρίειν ἄνθος χαλκοῦ ξηρόν, καὶ ἀναγαργαρίζειν
τῷ ἀπὸ τῶν φύλλων ἐλειθερεῖ. ἔξωθεν δὲ καταπλάσσειν, ᾗ ἂν ἀποιδέῃ, ὠμήλυσιν, ἐν οἴνῳ καὶ
ἐλαίῳ ἑψῶν, χλιαρήν. ἐπὴν δέ σοι δοκέωσι τὰ
φύματα μαλακὰ[3] ἔσωθεν ἀφασσόμενα, ὑποτύψαι
μαχαιρίῳ· ἔνια δὲ καὶ αὐτόματα καθίσταται.

31. Ἢν ὑπογλωσσὶς γένηται, ἡ γλῶσσα οἰδίσκεται, καὶ τὸ ὑποκάτω, καὶ τὸ ἔξω ψαυόμενον

---

[1] Θ: -σαντι M.   [2] Θ: αὐτὸ M.   [3] M adds εἶναι.

epiglottis several times. Patients that do these things recover; the disease is very seldom mortal.

29. If staphylitis arises in the throat, the extremity of the uvula fills up with fluid, becomes spherical and translucent, and stops the breath. If the jaws swell up on both sides, the patient suffocates, but if it is the uvula alone and not the other parts that swell up, the patient has less chance of dying. When the case is such, take the uvula with a finger, press it upwards against the palate, and cut away its extremity. Then give water prepared from herbs, to gargle; have the patient take cold flour, and afterwards drink water; the bath is to be avoided.

30. If tonsillitis occurs, there are swellings beneath the jaws on both sides which, if felt from the outside, are hard, and the whole uvula swells up. When the case is such, put a finger down the throat, and tear away the tonsils. On the uvula anoint dry flower of copper, and have the patient gargle with water made from herbs, warmed in the sun. Plaster the patient on the outside, wherever there is swelling, with bruised meal of raw grain boiled in wine and oil, and still warm. Feel the tubercles from inside the mouth, and when they seem to be soft, strike them off with a knife; some subside spontaneously, too.

31. If an affection occurs in the area beneath the tongue, the tongue and the area below it swell and from the outside feel hard; the patient cannot

σκληρόν ἐστι· καὶ τὸ σίαλον καταπίνειν οὐ δύναται. ὅταν οὕτως ἔχῃ, σπόγγον ἐς ὕδωρ θερμὸν
ἐμβάπτων προστιθέναι, καὶ τὴν ὠμήλυσιν ἑψῶν
ἐν οἴνῳ καὶ ἐλαίῳ καταπλάσσειν ἔξω ᾗ ἂν ἀποιδέῃ. ἀναγαργαρίζειν δὲ τῷ συκίῳ, καὶ μὴ λούεσθαι. ἐπὴν δὲ διάπυον γένηται, τάμνειν· ἐνίοτε
δὲ αὐτόματον ῥήγνυται,[1] καὶ καθίσταται οὐ τμη-
θέν· ἐπὴν δ' ἔξω ἀποπυήσῃ, διακαῦσαι.

32. Ἢν[2] φλέγμα συστῇ ἐς τὴν ὑπερῴην, ὑποι-
δέει[3] καὶ ἐμπυΐσκεται. ὅταν οὕτως ἔχῃ, καίειν τὸ
φῦμα. ἐπὴν δ' ἐξίῃ τὸ πύον, κλύζειν τὸ λοιπόν,
πρῶτον μὲν λίτρῳ καὶ ὕδατι χλιαρῷ, ἔπειτα οἴνῳ·
ἐπὴν δὲ κλύσῃς, ἀσταφίδα τρίψας λευκήν, ἐξελὼν τὸ γίγαρτον, ἐντιθέναι ἐς τὸ καῦμα· ἐπὴν δ'
ἐκρυῇ, οἴνῳ ἀκρήτῳ χλιαρῷ διακλυζέσθω, καὶ
ἐπὴν μέλλῃ ἐσθίειν τι ἢ ῥυφάνειν, σπόγγιον ἐντιθέναι. ταῦτα ποιέειν ἔστ' ἂν ὑγιὴς γένηται.

33. Ἢν πώλυπος γένηται ἐν τῇ ῥινί, ἐκ μέσου
τοῦ χόνδρου κατακρέμαται, οἷον γαργαρεών· καὶ
ἐπὴν ὤσῃ τὴν πνοιήν, προέρχεται[4] ἔξω, καί ἐστι
μαλθακόν· καὶ ἐπὴν ἀναπνεύσῃ, οἴχεται ὀπίσω
καὶ φθέγγεται σομφόν, καὶ ἐπὴν καθεύδῃ, ῥέγκει.

Ὅταν οὕτως ἔχῃ, σπόγγον[5] κατατάμων
στρογγύλον ποιήσας οἷον σπεῖραν, κατελίξαι λίνῳ
Αἰγυπτίῳ καὶ ποιῆσαι[6] σκληρόν· εἶναι δὲ μέγεθος

---

[1] Θ: ἐκ- Μ.   [2] Μ adds δὲ.   [3] Θ: ἀπ- Μ.   [4] Θ: προσ- Μ
[5] Θ: -ιον Μ.   [6] Μ: -σας Θ.

swallow his saliva. When the case is such, soak a sponge in hot water, and apply it to the affected area; boil bruised meal of grain in wine and oil, and plaster it on from the outside wherever there is swelling. Have the patient gargle with fig-juice, and abstain from the bath. When the swelling reaches maturity, incise it; sometimes, it ruptures spontaneously and goes down without being incised. When the swelling suppurates towards the outside, cauterize it.

32. If phlegm collects in the palate, the palate swells slightly and suppurates. When the case is such, cauterize the tubercle, and when its pus comes out, rinse away any that is left, first with soda and warm water, and then with wine. After you have rinsed in this way, mash a white raisin, remove its seed, and insert it into the burn; when the raisin falls out, rinse the mouth with warm unmixed wine. Whenever the person is about to eat anything or to take gruel, put a small sponge in the burn. Do these things until he recovers.

33. If a polyp forms in the nose, it hangs down from the central cartilage like a uvula. When the patient breathes out, the polyp moves outside, and it has a soft consistency; when the patient breathes in, the polyp moves back inside. The patient's voice lacks resonance, and, when he sleeps, he snores.

When the case is such, cut down a sponge to make it spherical like a ball, and wind it around with a cord of Egyptian linen so that it is hard;

ὥστ' ἐσαρτίζειν ἐς τὸν μυκτῆρα· καὶ δῆσαι τὸ
σπόγγιον λίνῳ τετραχόθι· μῆκος δ' ἔστω ὅσον
πυγονιαῖον ἕκαστον. ἔπειτα ποιήσας αὐτῶν μίαν
ἀρχήν, ῥάβδον λαβὼν κασσιτερίνην λεπτὴν[1] ἐκ
τοῦ ἑτέρου κύαρ ἔχουσαν, †διείρειν ἐς τὸ στόμα
τὴν ῥάβδον ἐπὶ τὸ ὀξύ, καὶ ἐπὴν λάβῃς,[2] διέρσας
διὰ τοῦ κύαρος,[3] ἕλκειν ἔστ' ἂν λάβῃς τὴν
ἀρχήν.† ἔπειτα χηλὴν ὑποθεὶς ὑπὸ τὸν γαργαρεῶνα, ἀντερείδων, ἕλκειν ἔστ' ἂν ἐξειρύσῃς τὸν
πώλυπον. ἐπὴν δ' ἐκσπάσῃς καὶ παύσηται τὸ
αἷμα ῥέον, περιθεὶς περὶ τὴν μήλην ξηρὸν ὀθόνιον
μοτῶσαι· καὶ τὸ λοιπὸν ἀναζέσαι τοῦ ἄνθους ἐν
μέλιτι, καὶ χρίων τὸν μοτὸν ἐντιθέναι ἐς τὴν ῥῖνα.
καὶ ἐπὴν ἤδη τὸ ἕλκος ἀλθαίνηται, μόλιβδον
ποιησάμενος ὥς τοι καθίκῃ πρὸς τὸ ἕλκος,[4] μέλιτι χρίων προστιθέναι ἔστ' ἂν ὑγιὴς γένηται.

34. Ἕτερος πώλυπος· ἐμπίμπλαται ἡ ῥὶς
κρέασι, καὶ ψαυόμενον τὸ κρέας σκληρὸν γίνεται,
καὶ διαπνεῖν οὐ δύναται διὰ τῆς ῥινός. ὅταν
οὕτως ἔχῃ, ἐνθέντα χρὴ συρίγγια[5] καῦσαι σιδη-
ρίοισιν ἢ τρισὶν | ἢ τέσσερσιν. ἐπὴν δὲ καύσῃς,
ἐμβάλλειν τοῦ ἐλλεβόρου τοῦ μέλανος τρίψας
λεῖον,[6] καὶ ἐπὴν ἐκσαπῇ καὶ ἐκπέσῃ τὸ κρέας,
μοτοὺς τοὺς λινέους χρίων τῷ μέλιτι καὶ τῷ ἄνθει

---

[1] M: λαβών Θ.　　[2] Ermerins: λάβῃ ΘM.　　[3] M adds τὸ λίνον.
[4] M adds τοῦτον.　　[5] Θ: σύριγγα M.　　[6] Mack: λίον Θ: om. M.

let the sponge be of a size to fit into the nostril; bind it with threads in four places, the length of each being a cubit, and twist these into a single beginning. Take a light tin rod with an eye at one end, and draw the rod into the mouth at an acute angle; when you have hold of it, draw through the eye, and pull until you have the beginning.[1] Then, placing a forked probe under the uvula and using this as a fulcrum, pull until you tear the polyp out. When you have removed the polyp, and the blood has stopped flowing, place dry linen around a probe and use this as a tent; after that boil up flower of copper in honey, anoint the tent, and insert it into the nostril. When the ulcer is already healing, make a lead sound that reaches the ulcer; smear it with honey, and insert it until the patient recovers.

34. Another polyp: the nostril is filled with flesh that is hard to the touch, and the patient cannot breathe through that nostril. When the case is such, you must insert a protective tube, and cauterize with three or four irons. After you have cauterized, put in finely ground black hellebore and, when the flesh becomes putrid and falls away, insert linen tents smeared with honey and flower of

[1] The details of this procedure are so obscure as to suggest that the text is corrupt; divergent explanations are given by Fuchs (II. 427 f.), E. Gurlt (*Geschichte der Chirurgie*, Berlin, 1898, I. 284) and J. S. Milne (*Surgical Instruments in Greek and Roman Times*, Oxford, 1907, 83).

ἐστιθέναι. ἐπὴν δ᾽ ἀλθαίνηται, τοὺς μολίβδους
χρίων τῷ μέλιτι ἐστίθει, ἔστ᾽ ἂν ὑγιὴς γένηται.

35. Ἕτερος πώλυπος· ἔσωθεν[1] τοῦ χόνδρου
προέχει κρέας στρογγύλον· ψαυόμενον δὲ μαλθακόν ἐστιν. ὅταν οὕτως ἔχῃ, χορδὴν λαβὼν νευρί-
νην, βρόχον αὐτῆς[2] σμικρὸν ποιήσας, κατειλίξαι
λίνῳ λεπτῷ· ἔπειτα τὴν ἀρχὴν τὴν ἑτέρην διεῖναι
διὰ τοῦ βρόχου, μέζονα ποιήσας τὸν βρόχον· ἔπειτα τὴν ἀρχὴν διεῖραι διὰ τῆς ῥάβδου τῆς κασσι-
τερίνης· ἔπειτα ἐνθεὶς τὸν βρόχον ἐς τὴν ῥῖνα, τῇ
μήλῃ τῇ ἐντετμημένῃ περιτείνας περὶ τὸν πώλυπον τὸν βρόχον, ἐπὴν περικέηται, διείρειν[3] ἐς τὸ
στόμα, καὶ λαβὼν ἕλκειν τὸν αὐτὸν τρόπον, τῆς
χηλῆς ὑπερειδούσης. ἐπὴν δ᾽ ἐξελκύσῃς, ἰᾶσθαι
ὥσπερ τὸν πρόσθεν.

36. Ἕτερος πώλυπος· ἔσωθεν παρὰ τὸν χόνδρον ἀπό τευ σκληρὸν φύεται· δοκέει μὲν εἶναι
κρέας· ἢν δὲ ψαύσῃς αὐτοῦ, ψοφέει οἷον λίθος.
ὅταν οὕτως ἔχῃ, σχίσαντα τὴν ῥῖνα σμίλῃ ἐκκα-
θῆραι, ἔπειτα ἐπικαῦσαι. τοῦτο δὲ ποιήσας, συρ-
ράψαι τὴν ῥῖνα, καὶ ἰᾶσθαι τὸ ἕλκος τῷ χρίσματι
ἐναλείφων, ῥάκος ἐντιθέναι, καὶ ἐπὴν περισαπῇ
ἐγχρίειν τὸ ἄνθος τὸ ἐν τῷ μέλιτι· ἀλθίσκειν δὲ
τῷ μολίβδῳ.

37. Ἕτερος· φύεται ἐκ πλαγίου τοῦ χόνδρου
ἐν ἄκρῳ οἷον καρκίνια. πάντα ταῦτα καίειν χρή·

---

[1] M adds ἐκ.   [2] Θ: αὐτῇ M.   [3] M adds τὴν ῥάβδον.

copper. When the wound is healing, smear lead sounds with honey, and insert them until the patient recovers.

35. Another polyp: from within the cartilage a spherical mass of flesh projects, and is soft to the touch. When the case is such, take a fibrous cord, make a small loop in it, and wrap it in fine linen; then put the opposite end of the cord into the loop, thereby making a larger noose, and draw the end through a tin probe. Place the noose into the nostril, stretch it over the polyp by means of a notched probe and, when the noose lies in place, draw the end of the cord into the mouth; catch hold of this, and pull it the same way as above, using a forked probe as support. When you have torn the polyp out, treat in the same way as the preceding patient.

36. Another polyp: from inside near the cartilage for some reason a hardness forms; it appears to be flesh but, if you touch it, it makes a sound like stone. When the case is such, divide the nostril with a scalpel, clean the polyp out, and then apply cautery. After you do this, stitch the nostril together, and heal the ulcer by anointing it with ointment; insert a rag and, when this putrefies all around, smear on flower of copper in honey; promote healing with the lead sound.

37. Another polyp: out of the oblique cartilage at the extremity grow certain cancers; these must all

ὅταν δὲ καύσῃς, ἐμπάσαι τοῦ ἐλλεβόρου· ἐπὴν δὲ
σαπῇ, καθαίρειν τῷ ἄνθει τῷ σὺν τῷ μέλιτι· ἀλθίσκειν δὲ τῷ μολίβδῳ.

54   38. Ἵκτερος· ἡ χροιὴ μέλαινα γίνεται καὶ τὸ
πρόσωπον, μάλιστα δὲ τὰ ἐσκιασμένα, καὶ οἱ
ὀφθαλμοὶ χλωροὶ καὶ ἡ γλῶσσα κάτωθεν, καὶ αἱ
φλέβες αἱ ὑπὸ τῇ γλώσσῃ παχεῖαι καὶ μέλαιναι·
καὶ ἄπυρος γίνεται, καὶ οὐρέει παχὺ χολῶδες.

Ὅταν οὕτως ἔχῃ, πρῶτον μὲν τὰς φλέβας τὰς
ὑπὸ τῇ γλώσσῃ ἀποσχᾶν· ἔπειτα λούοντα πολλῷ
καὶ θερμῷ, διδόναι πίνειν νήστει τοῦ ἀσφοδέλου
τὰς ῥίζας, ἀποκαθαίρων, ἑψῶν ἐν οἴνῳ ὅσον[1] πέντε ῥίζας, καὶ σέλινα συμμίξαι ὅσον χεῖρα πλήρη
τῶν φύλλων· ἐπιχεῖν δὲ οἴνου γλυκέος τρία ἡμι-
κοτύλια Αἰγιναῖα, καὶ λιπεῖν ἡμικοτύλιον· τοῦτο
κιρνάς, ἕκτον αὐτῷ διδόναι πίνειν. ἐπὴν δὲ οὐρήσῃ, σιτίοισι χρήσθω διαχωρητικοῖσι· καὶ μετὰ τὸ
σιτίον ἐρεβίνθους λευκοὺς τρωγέτω, καὶ πινέτω
οἶνον λευκόν, ὑδαρέα, πολύν, καὶ σέλινα τρωγέτω
ἐπὶ τῷ σιτίῳ καὶ πράσα. ποιείτω δὲ ταῦτα ἑπτὰ
ἡμέρας, καὶ ἢν μέν οἱ δοκέῃ ἐν ταύτῃσιν ἡ χροιὴ
κεκαθάρθαι ἐπιεικῶς· ἢν δὲ μή, ἑτέρας τρεῖς ταῦτα ποιείτω. μετὰ δέ, ἐπισχὼν μίαν ἢ δύο ἡμέρας, πρόσθες φάρμακον πρὸς τὰς ῥῖνας· μετὰ δέ,
φάρμακον πῖσαι[2] κάτω, ὑφ' οὗ χολὴν καθαρεῖται,
καὶ ἢν μὲν σπληνώδης ᾖ, ὄνου γάλα ἢ ὀρὸν μετά-

_____
[1] Θ: ἴσον Μ.   [2] Θ: πῖσον Μ.

be cauterized. When you have done so, sprinkle on hellebore and, when the wound putrefies, clean it with flower of copper in honey; promote healing with the lead sound.

38. Jaundice: the skin and the face become dark, especially the part that is normally shaded, the eyes and the undersurface of the tongue are yellow-green, and the vessels beneath the tongue are wide and dark; the patient is without fever, and he passes thick bilious urine.

When the case is such, first lance the vessels beneath the tongue; then wash the patient in copious hot water, and give him, in the fasting state, asphodel roots to drink: clean five roots well and boil them in wine; mix in celery leaves to the amount of one handful; pour in three Aeginetan half cotylai of sweet wine, and boil until one half cotyle remains; then mix this with water, and give the patient one sixth of it at a time to drink. When he has passed urine, let him take foods that are laxative; after his meals, have him eat white chickpeas, and drink a generous amount of dilute white wine; also let him eat celery and leeks after the meal. Have him do these things for seven days; if, in this time, his skin seems to be cleaned quite well, fine; if not, continue the treatment for three more days. After that, hold back for a day or two, and then administer a medication to the nostrils; afterwards, have the patient drink a medication that will clean bile downwards and, if he is suffering from an affection of the spleen, after that ass's

πισον. ταῦτα ποιέων ὑγιὴς γίνεται.

39. Ἕτερος ἴκτερος· πυρετὸς λαμβάνει βληχρὸς καὶ τὴν κεφαλὴν βάρος· καὶ οἱ πυρετοὶ ἐπαύσαντο οὖν ἐνίοις. αὐτὸς δὲ γίνεται χλωρός, οἵ τε
ὀφθαλμοὶ μάλιστα, καὶ ἀσθενείη καὶ ἀκρασίη τοῦ
σώματος· καὶ οὐρέει παχὺ καὶ χλωρόν. τοῦτον
θερμῷ λούειν, καὶ διδόναι πίνειν διουρητικά.
ἐπὴν δέ σοι δοκέῃ καθαρώτερος εἶναι καὶ ἡ χροιὴ
βελτίων, πρόσθες φάρμακον πρὸς τὰς ῥῖνας, καὶ |
56   μετάπισον κάτω. σιτίοισιν[1] ὡς μαλθακωτάτοισι
χρήσθω[2] οἶνον δὲ πινέτω λευκόν, γλυκύν, ὑδαρέα.
ταῦτα ποιέων ὑγιὴς γίνεται.

40. Ἢν χολᾷ ὁ[3] ἄνθρωπος, πυρετὸς αὐτὸν
λαμβάνει καθ' ἡμέρην καὶ ἀφίει, ἔχει δὲ μάλιστα
τὸ μέσον τῆς ἡμέρης. καὶ τὸ στόμα πικρόν, καὶ
ὅταν ἄσιτος ᾖ, λυπέει αὐτόν· ἐπὴν δὲ[4] φάγῃ, πνίγεται, καὶ ὑπ' ὀλίγων τινῶν ἐμπίμπλαται, καὶ
βδελύττεται, καὶ ἐμεσίαι μιν λαμβάνουσι. καὶ ἐς
τὴν ὀσφῦν βάρος ἐμπίπτει καὶ ἐς τὰ σκέλεα, καὶ
ὑπνώσσει. τοῦτον, ἢν μὲν μετὰ τὸ πῦρ ἐξιδρῷ,
καί οἱ ψυχρὸς καὶ πολὺς ᾖ, καὶ τοῦ πυρετοῦ μὴ
ἀπαλλάσσηται, ἡ νοῦσος χρονίη γίνεται· ἢν δὲ μὴ
ἱδρῷ, θᾶσσον κρίνεται.

Ὅταν οὕτως ἔχῃ, ἐπὴν γένηται ἐναταῖος,
φάρμακον δοῦναι· ἢν γὰρ αὐτίκα ἀρχομένου τοῦ

[1] M adds δέ.   [2] Θ: χρῆσθαι M.   [3] ὁ om. Θ.   [4] Θ adds
καί.

milk or whey. If a patient does these things, he
recovers.

39. Another jaundice: there are mild fevers and
a heaviness of the head—in some cases, the fevers
actually go away; the person becomes yellow-green,
especially his eyes, he experiences weakness and
powerlessness of the body, and he passes thick
yellow-green urine. Wash this patient in hot water,
and give him diuretic drinks. When he seems to
you to be cleaner, and his colour to be better,
administer a medication to his nostrils; afterwards,
have him drink a medication that acts downwards.
Let him eat very soft foods, and drink dilute sweet
white wine. If he does these things, he recovers.

40. If a person has an affection due to bile, each
day fever attacks and then remits, being greatest at
the middle of the day. His mouth is bitter, and
when he goes without food he feels pain; when he
eats, though, he chokes, becomes full on very little,
has nausea, and is seized by retching; he has a
heaviness in his loins and legs, and is sleepy. If,
after the fever heat, the patient breaks out in a
sweat that is cold and abundant, and is not relieved
of the fever, the disease becomes chronic; if he does
not sweat, the condition reaches its crisis sooner.

This being the case, give a medication, but only
on the ninth day; for if you give one when the fever

πυρετοῦ διδῷς, ἐπὴν καθαρθῇ, ἐπ' οὖν[1] ἔλαβε
πυρετός, καὶ αὖτις φαρμάκου δεῖται. ἐπὴν δὲ τὸ
μὲν στόμα πικρὸν ᾖ,[2] ἐς[3] δὲ τὴν νείαιραν γαστέρα
στρόφος ἐμπίπτῃ, φάρμακον πῖσαι κάτω, καὶ
μεταπῖσαι γάλα ὄνου ἢ ὀρὸν ἢ τῶν χυλῶν τινά·
ἢν δ' ἀσθενὴς ᾖ, ὑποκλύσαι. πρὸ δὲ τοῦ φαρμάκου τῆς πόσιος, ᾧ ἂν[4] πυρεταίνῃ διδόναι μὲν ἔωθεν μελίκρητον ὑδαρές· τὴν δ' ἄλλην ἡμέρην
ἐπὴν ὁ πυρετὸς ἔχῃ, ὕδωρ ὁπόσον ἂν θέλῃ[5] διδόναι πίνειν ψυχρόν· ἐπὴν δὲ ἀνῇ ὁ πυρετός, ῥυφεῖν
διδόναι πτισάνης χυλὸν ἢ κέγχρον λεπτόν, καὶ
ἐπιπίνειν οἶνον λευκόν, οἰνώδεα, ὑδαρέα. ἢν δ'
ἔμπυρος ᾖ καὶ μὴ ἀνίῃ μήτε τῆς νυκτὸς μήτε τῆς
ἡμέρης, ψαυόμενος δὲ ᾖ τὰ μὲν ἄνω | θερμὰ καὶ
ἡ κοιλίη, οἱ πόδες δὲ[6] ψυχροὶ καὶ ἡ γλῶσσα τρηχέη, τούτῳ μὴ δῶς φάρμακον, ἀλλ' ὑποκλύζειν
μαλθακῷ κλύσματι, καὶ διδόναι ῥυφάνειν τὸν χυλὸν τῆς πτισάνης ψυχρὸν δὶς τῆς ἡμέρης, καὶ
ἐπιπίνειν οἶνον ὑδαρέα, τὸν δ' ἄλλον χρόνον ὕδωρ
ὡς ψυχρότατον. οὗτος ἢν μὲν ἑβδομαῖος ἐξιδρώσῃ καὶ τὸ πῦρ αὐτὸν μεθῇ· εἰ[7] δὲ μή, τεσσερεσκαι-
δεκαταῖος ἀποθνῄσκει ὡς τὰ πολλά.

41. Ἄλλος πυρετός· ἔξωθεν ἀφασσόμενος
βληχρός, ἔσωθεν δὲ καίεται· καὶ ἡ γλῶσσα τρη-

---

[1] ἐπ' οὖν M: ἐπὴν Θ.　　[2] πικρὸν ᾖ Θ: μὴ πονῇ M.　　[3] M: ἐπὴν Θ.
[4] ᾧ ἂν Potter: ὃς ἂν ΘΜ: ἢν Littré.　　[5] ὕδωρ ... θέλῃ om. Θ.
[6] δὲ om. M.　　[7] Θ: ἢν M.

first begins, then after the patient has been cleaned out the fever will resume, and another medication will be required. When his mouth is bitter, and colic is present in his lower belly, have the patient drink a medication to act downwards, and afterwards ass's milk or whey, or some juice; if, however, he is weak, administer an enema. To any patient with fever give dilute melicrat early in the morning, before he drinks the medication; the rest of the day, when fever is present give him as much cold water to drink as he wants, when it is not present give barley-water or dilute millet as gruel, and after that dilute strong white wine. If the patient has a fever that remits neither by night nor by day, and the upper part of his body is hot to the touch, and also his cavity, but his feet are cold and his tongue is rough, do not give a medication, but administer a mild enema, and give cold barley-water gruel twice a day, after that dilute wine, and from then on water as cold as possible. If this patient breaks out in a sweat on the seventh day and the fever heat releases him, fine; if not, he generally dies on the fourteenth day.

41. Another fever: felt from the outside, it is mild, inside, it blazes; the tongue is rough, and the

χέη, καὶ πνεῖ διὰ τῶν ῥινῶν καὶ διὰ τοῦ στόματος
θερμόν. ὅταν δὲ πεμπταῖος γένηται, τὰ ὑποχόνδρια σκληρά, καὶ ὀδύνη ἔνεστι, καὶ ἡ χροιὴ οἷον
ὑπὸ ἰκτέρου ἐχομένῳ[1] φαίνεται, καὶ οὐρέει παχὺ[2]
χολῶδες. τοῦτον ἢν μὲν ἑβδομαῖον ὄντα ῥῖγος
λάβῃ καὶ πυρετὸς ἰσχυρὸς καὶ ἐξιδρώσῃ· εἰ δὲ μή,
ἀποθνήσκει ἑβδομαῖος ἢ ἐναταῖος. λαμβάνει δὲ
μάλιστα, ἢν μὴ τὸ ἔτος αὐχμηρὸν γένηται, αὕτη
ἡ νοῦσος.

Ὅταν οὕτως ἔχῃ, λούειν θερμῷ ἑκάστης ἡμέρης, καὶ πίνειν διδόναι μελίκρητον ὑδαρὲς πολλόν, καὶ ῥυφάνειν τὸν χυλὸν τῆς πτισάνης ψυχρὸν
δὶς τῆς ἡμέρης· ἐπὶ δὲ τῷ ῥυφήματι πίνειν οἶνον
ὑδαρέα, λευκόν, ὀλίγον. ἢν δὲ ἡ γαστὴρ μὴ ὑποχωρέῃ, ὑποκλύσαι, ἢ βάλανον προσθεῖναι. σιτίον
δὲ μὴ προσφέρειν, ἔστ' ἂν ὁ πυρετὸς ἀνῇ· ἐπὴν δὲ
παύσηται, φάρμακον πῖσαι κάτω· ὑποστρέφει γὰρ
ἔστιν ὅτε ἡ νοῦσος, ἢν ἀκάθαρτος διαφέρηται. ἡ
δὲ[3] νοῦσος λαμβάνει, ἢν ὑπερχολήσῃ τὸ αἷμα.

42. Ἢν τριταῖος πυρετὸς ἔχῃ ἢν μὲν μὴ
παρεὶς τρεῖς λήψιας τῇ τετάρτῃ λάβῃ, φάρμακον
πῖσαι κάτω· ἢν δέ σοι δοκέῃ φαρμάκου μὴ δεῖσθαι, τρίψας τοῦ πενταφύλλου τῶν ῥιζῶν ὅσον
ὀξύβαφον ἐν ὕδατι, δοῦναι πιεῖν. ἢν δὲ μηδὲ τού-
τῳ παύηται, λούσας αὐτὸν θερμῷ | πολλῷ, πῖσαι
τὸ τρίφυλλον καὶ ὀπὸν σιλφίου ἐν οἴνῳ ἰσοκρατέϊ,
καὶ κατακλίνας ἐπιβάλλειν ἱμάτια πολλὰ ἕως

patient exhales hot breath through his nostrils and
mouth. On the fifth day, the hypochondrium
becomes hard, there is pain, the skin takes on the
appearance of a person with jaundice, and thick
bilious urine is passed. If, on the seventh day,
chills seize this person, together with a violent
fever, and he breaks out in a sweat, fine; if not, he
dies on the seventh or ninth day. This disease usually occurs in years that are not dry.

When the case is such, wash the patient each
day in hot water, give him a generous amount of
dilute melicrat to drink, and as gruel have him
take cold barley-water twice daily; after he takes
gruel, let him drink a little dilute white wine. If
the belly does not pass anything downwards,
administer an enema or a suppository. Do not give
food until the fever remits; when it stops, have the
patient drink a medication that acts downwards;
for sometimes this disease recurs if the patient goes
on in an unclean state. The disease occurs when
the blood becomes over-charged with bile.

42. When a tertian fever occurs, if, after three
accesses in a row, it attacks for a fourth time, let
the patient drink a medication to act downwards; if
you think he does not require a medication, then
grind into water cinquefoil roots to the amount of
one oxybaphon, and give this to drink. If, with this,
the fever does not stop, wash the patient in copious
hot water, have him drink clover and silphium juice
in wine diluted with an equal amount of water, and
put him to bed, covering him with many blankets,

---

[1] Θ: -μένου M.   [2] M adds καὶ.   [3] δὲ om. M.

ἱδρώσῃ. ἐπὴν δ᾽ ἐξιδρώσῃ, ἢν διψῇ,[1] δοῦναι
πιεῖν ἄλφιτον καὶ ὕδωρ ἐς[2] ἑσπέρην δὲ κέγχρον
ἑψήσας λεπτόν, ῥυφησάτω, καὶ οἶνον ἐπιπιέτω.
ἕως δ᾽ ἂν διαλείπῃ, σιτίοισιν ὡς μαλθακωτάτοισι
χρήσθω.

43. Τεταρταῖος πυρετὸς ὅταν ἔχῃ, ἢν μὲν ἐξ
ἄλλης νούσου λάβῃ ἀκάθαρτον, φάρμακον πῖσαι
κάτω· ἔπειτα τὴν κεφαλὴν καθῆραι, ἔπειτα φάρμακον πῖσαι κάτω. ἢν δὲ μηδὲ[3] ταῦτα ποιήσαντι
παύηται, διαλείπων δύο λήψιας μετὰ τὴν κάτω
κάθαρσιν, λούσας αὐτὸν πολλῷ θερμῷ, πῖσον τοῦ
καρποῦ τοῦ ὑοσκυάμου ὅσον κέγχρον, καὶ μανδρα-
γόρου ἴσον, καὶ ὁποῦ τρεῖς κυάμους, καὶ τριφύλλου ἴσον, ἐν οἴνῳ ἀκράτῳ πιεῖν. ἢν δ᾽ ἐρρωμένος
καὶ ὑγιαίνειν δοκέων, ἐκ κόπου ἢ ἐξ ὁδοιπορίης
πυρετήνας, καταστῇ αὐτῷ ἐς τεταρταῖον, πυριήσας αὐτόν, σκόροδα δοῦναι ἐς μέλι βάπτων· ἔπειτα ἐπιπινέτω φάκιον, μέλι καὶ ὄξος μίξας· ἐπὴν
δ᾽ ἐμπλησθῇ, ἐμεσάτω· ἔπειτα λουσάμενος θερμῷ, ἐπὴν ψυχθῇ, πιέτω κυκεῶνα ἐφ᾽ ὕδατι· ἐς
ἑσπέρην δὲ σιτίοισι μαλθακοῖσι καὶ μὴ πολλοῖσι
διαχρήσθω. τῇ δ᾽ ἑτέρῃ λήψει λούσας θερμῷ,
ἱμάτια ἐπιβαλὼν ἕως ἐξιδρώσῃ, πῖσαι παραχρῆμα
λευκοῦ ἑλλεβόρου τῶν ῥιζῶν ὅσον τριῶν δακτύλων
μῆκος, καὶ τοῦ τριφύλλου ὅσον δραχμὴν μέγεθος,

[1] ἢν διψῇ om. Θ.   [2] ἐς om. Μ.   [3] Θ: μὴ Μ.

until he sweats. When he breaks out sweating, if he is thirsty give him meal and water to drink; towards evening, boil thin millet for him to take as gruel, and afterwards let him drink wine. Until the disease goes away, give foods of the softest kinds.

43. When a quartan fever occurs, if it attacks a person that is in an unclean state subsequent to another disease, have him drink a medication to clean downwards; then clean out his head, and after that have him drink another medication that acts downwards; if the fever does not stop when the patient does these things, leave an interval of two accesses after the downward cleaning, wash him in copious hot water, and give him henbane seed equal in quantity to a millet-seed, the same amount of mandrake, mandrake juice to the amount of three beans, and the same quantity of clover juice; have him drink these in unmixed wine. If, on the other hand, a person that is in good health and seems to be sound falls ill with a fever, as the result of weariness or walking, and the fever becomes a quartan, apply a vapour-bath and give garlic heads soaked in honey; then let the patient drink a decoction of lentils to which honey and vinegar have been added and, when he is full, let him vomit. Next, wash him in hot water and, when he becomes cold, let him drink a cyceon in water; towards evening let him eat soft foods in small amounts. At the next access, wash the patient in hot water, cover him with blankets until he breaks out sweating, and then have him immediately drink the following in unmixed wine: roots of white hellebore three fingers in length, a drachma of clover, and clover

261

καὶ ὁποῦ δύο κυάμους, ἐν οἴνῳ ἀκρήτῳ. καὶ ἢν
ἐμεσίαι μιν ἔχωσιν, ἐμεσάτω· ἢν δὲ μή, ὁμοίως,
μετὰ δὲ τὴν κεφαλὴν καθῆραι· σιτίοισι δὲ χρήσθω
ὡς μαλθακωτάτοισι καὶ δριμυτάτοισιν· ὅταν δὲ
ἡ ληψίς μιν ἔχῃ, μὴ νῆστις ἐὼν τὸ φάρμακον
πινέτω.

62   44. Περὶ πλευρίτιδος[1] πλευρῖτις ὅταν λάβῃ,
πυρετὸς καὶ ῥῖγος ἔχει, καὶ ὀδύνη διὰ τῆς ῥάχιος
ἐς τὸ στῆθος· καὶ ὀρθοπνοίη, καὶ βήξ, καὶ τὸ σίαλον λευκὸν καὶ ὑπόχολον, καὶ ἀποβήσσεται οὐ
ῥηϊδίως, καὶ διὰ τῶν βουβώνων ὀδύνη, καὶ οὐρέει
αἱματῶδες. ὅταν οὕτως ἔχῃ, ἢν μὲν τὸ πῦρ ἀνῇ
ἑβδομαῖον ὄντα, ὑγιὴς γίνεται· ἢν δὲ μὴ ἀνῇ,
ἀφικνεῖται ἡ νοῦσος ἐς τὰς ἕνδεκα ἡμέρας ἢ τεσ-
σερεσκαίδεκα· οἱ μὲν οὖν πολλοὶ ἐν ταύτῃσιν
ἀπόλλυνται· ἢν δὲ ὑπερβάλῃ τὴν τεσσερεσκαι-
δεκάτην, ἐκφυγγάνει.

Ὅταν οὕτως ἡ ὀδύνη ἔχῃ, χλιάσματα προστι-
θέναι· πινέτω δὲ μέλι χύτρῃ[2] ἀναζέσας, ἐπιχέας[3]
ὄξος ἴσον τῷ μέλιτι, ἔπειτα ὁπόσον ἂν γένηται
μέτρον τοῦ ἑφθοῦ μέλιτος καὶ τοῦ ὄξους, ἐπιχέαι
ὕδατος ἑνὸς δέοντα εἴκοσι· τοῦτο διδόναι πίνειν
κατ' ὀλίγον πυκινά, καὶ μεταμίσγειν ὕδωρ, ὄξος
ὀλίγον παραχέων. ῥυφείτω δὲ[4] κέγχρου χυλόν,
μέλι ὀλίγον παραστάζων, ψυχρόν, ὅσον τεταρτη-

---

[1] Θ: Πλευρῖτις M.   [2] Potter: μέτρῳ ΘM: om. later mss,
edd.   [3] Θ adds ὕδωρ.   [4] M adds καί.

juice to the amount of two beans. If retching comes over the patient, let him vomit; if it does not, still let him vomit, but after you have cleaned out his head; let him take foods that are very soft and very sharp. During an access, the patient should not drink the medication in the fasting state.

44. On pleurisy: when pleurisy occurs, fevers and chills are present, and pain along the spine and in the chest; there are orthopnoea and coughing, the sputum is white, slightly bilious and not easily coughed up, pain is present in the groins, and bloody urine is passed. When the case is such, if the fever heat remits on the seventh day, the patient recovers; if not, the disease continues on to the eleventh or fourteenth day; now many die by that time, but if a person gets beyond the fourteenth day, he escapes.

When the pains are such, apply fomentations; also have the patient drink honey prepared as follows: boil the honey in a pot, and add an equal amount of vinegar; then, whatever the measure of boiled honey and vinegar is, to this add nineteen measures of water; give this to drink frequently, a little at a time, mixing into it water and a little vinegar. Let the patient take as gruel cold millet-juice into which a little honey has been instilled, this to the amount of one quarter cotyle after every

μόριον κοτύλης ἐφ' ἑκατέρῳ σιτίῳ· πινέτω δὲ[1]
οἶνον λευκόν, οἰνώδεα, ὑδαρέα, ὀλίγον· ὁ δὲ οἶνος
ἔστω ὡς μαλθακώτατος ὀδμὴν μὴ ἔχων. ὅταν δ'
ὁ πυρετὸς ἀφῇ, ἡμέρας μὲν δύο τὸν κέγχρον
ῥυφείτω δὶς τῆς ἡμέρης, καὶ τεῦτλα ἡδυντὰ[2]
ἐσθιέτω· ἔπειτα μετὰ ταῦτα σκύλακα ἢ ὄρνιθα
κάθεφθον ποιήσας, τοῦ ζωμοῦ ῥυφείτω, καὶ τῶν
κρεῶν φαγέτω ὀλίγα. τὸν δὲ λοιπὸν χρόνον
μάλιστα ὅσον ὑπὸ τῆς νούσου ἔχεται, ἀριστιζέσθω
μὲν τὸν κέγχρον, ἐς ἑσπέρην δὲ σιτίοισιν ὡς ἐλαχίστοισι χρήσθω καὶ μαλθακωτάτοισιν.

45. Ἑτέρη πλευρῖτις· πυρετὸς ἔχει καὶ βὴξ
καὶ ῥῖγος καὶ ὀδύνη ἐς τὸ πλευρὸν καὶ ἐς τὴν
κληῖδα ἐνίοτε· καὶ τὸ σίαλον πτύει ὑπό|χολον καὶ
ὕφαιμον ὅταν τύχῃ ῥηγματίας ὤν. τούτῳ ᾗ ἂν
ᾖ[3] ὀδύνη ἔχῃ μάλιστα, προστιθέναι χλιάσματα·
καὶ λούειν θερμῷ, ἢν μὴ ὁ πυρετὸς πολὺς ἔχῃ· ἢν
δὲ μή, μή. πίνειν δὲ διδόναι κηρίον ἐν ὕδατι
ἀποβρέχων, ἄρτι ὑπόγλυκυ ποιέων, καὶ μεταμίσγειν ὕδωρ, ῥυφάνειν δὲ τὸν χυλὸν τοῦ κέγχρου
δὶς τῆς ἡμέρης, καὶ ἐπιπινέτω οἶνον λευκὸν ὑδαρέα. καὶ ἢν ὑπερφύγῃ τεσσερεσκαίδεκα ἡμέρας,
ὑγιὴς γίνεται.

46. Ἄλλη πλευρῖτις· πυρετὸς ἴσχει καὶ βρυγ-
μὸς καὶ βὴξ ξηρή, καὶ ἐκβήσσεται χλωρά, ἔστι

---

[1] Θ: καὶ πινέτω Μ.   [2] Potter: ηδυνατα Θ: ἡδύτατα Μ.
[3] ἢ om. Μ.

meal; let him drink a little dilute strong white wine; make the wine so dilute as to have no odour. When the fever remits, for two days have the patient drink the millet as gruel twice daily, and eat seasoned beets; after that, make boiled puppy or fowl, and have him drink the sauce and eat a little of the meat. From then on, in particular for as long as he is subject to the disease, let him breakfast on millet, and towards evening eat very small portions of very mild foods.

45. Another pleurisy: fever is present, coughing, chills, and pain in the side and sometimes around the collar-bone; the patient expectorates somewhat bilious pus which, if he happens to have tears, is also charged with blood. To this patient apply fomentations wherever the pain is severest, and wash him in hot water unless his fever is very great—if the fever is very great, do not wash. Give him honeycomb soaked in water to drink, adding water until the mixture is just slightly sweetish, and let him take as gruel millet-juice twice daily, and after that drink dilute white wine. If he survives for fourteen days, he recovers.

46. Another pleurisy: fever, chattering of the teeth, and a dry cough are present, and the patient coughs up yellow-green or sometimes livid sputum.

δ' ὅτε πελιδνά. καὶ τὸ[1] πλευρὸν ὀδύνη λαμβάνει,
καὶ τὸ μετάφρενον ὑπέρυθρον γίνεται, χλιαίνεται
δὲ τὴν κεφαλὴν καὶ τὰ στήθεα, ποτὲ δὲ τὴν
κοιλίην καὶ τοὺς πόδας καὶ τὰ σκέλεα. καὶ
ἀνακαθήμενος μᾶλλον βήσσει, καὶ ἡ γαστὴρ
ταράσσεται, καὶ τὸ ἀποπάτημα χλωρὸν καὶ
κάκοδμον. οὗτος ἐν εἴκοσιν ἡμέρῃσιν ἀποθνῄσκει·
ἢν δὲ ταύτας ἐκφύγῃ, ὑγιὴς γίνεται.

Τοῦτον, ἔστ' ἂν τεσσερεσκαίδεκα ἡμέραι παρέλθωσι, διδόναι πίνειν τὸ ἀπὸ τοῦ κρίμνου, καὶ
μεταμίσγειν οἶνον λευκόν, οἰνώδεα, ὑδαρέα· ῥυφεῖν δὲ τὸν χυλὸν τῆς πτισάνης ψυχρὸν δὶς τῆς
ἡμέρης· ἀντὶ δὲ τοῦ μέλιτος ὑπὸ τὸν χυλὸν ὑπο-
μίσγειν[2] ῥοιῆς χυλὸν οἰνώδεος, ὅταν ἤδη ὁ χυλὸς
ἑφθὸς ᾖ· καὶ λούειν μὴ πολλῷ. ἐπὴν δὲ τεσσερεσκαίδεκα ἡμέραι παρέλθωσιν, ἔπειτα ἀριστιζέσθω
τὸν κέγχρον· ἐς ἑσπέρην δὲ τοῖσι κρέασι τοῖσιν
ὀρνιθείοισι καὶ τῷ ζωμῷ καὶ σιτίοισιν ὀλίγοισι
χρήσθω. τὴν δὲ νοῦσον ὀλίγοι ἐκφυγγάνουσι.

47. Περιπλευμονίη· πυρετὸς ἴσχει ἡμέρας τεσσερεσκαίδεκα τὸ ἐλάχιστον, τὸ δὲ μακρότατον
δυῶν δεούσας εἴκοσι· καὶ βήσσει ταύτας τὰς ἡμέρας ἰσχυρῶς. καὶ ἀποχρέμπτεται τὸ μὲν πρῶτον
σίαλον παχὺ καὶ καθαρὸν ἑβδόμῃ καὶ ὀγδόῃ, ἐπὴν
δ' ὁ πυρετὸς λάβῃ, ἐνάτῃ καὶ δεκάτῃ ὑπόγλυκυ
καὶ πυῶδες, ἔστ' ἂν αἱ τεσσερεσκαίδεκα ἡμέραι |

___

[1] Μ: τότε Θ.   [2] Θ: μίσγειν Μ.

Pain occupies his side, his back becomes reddish, and he grows warm in his head and chest, sometimes also in his cavity, feet and legs. On sitting up, he coughs more; his belly is set in motion, and the faeces are yellow-green and ill-smelling. This patient dies in twenty days; if he survives that many, he recovers.

For fourteen days give this patient water made from groats to drink, adding dilute strong white wine; as gruel let him drink cold barley-water twice daily; in place of honey, add juice of the vinous pomegranate to the barley-water after it has been boiled; wash the patient in a small amount of water. When the fourteen days have passed, let him breakfast on millet and, towards evening, eat meats of fowl with their sauces, and a few cereals. Few escape this disease.

47. Pneumonia: fever is present for between fourteen and eighteen days, and during this period the patient coughs violently. First, he expectorates thick clear sputum on the seventh and eighth days after the fever has set in, and then on the ninth and tenth days sweetish and purulent sputum, which continues until the fourteen days have

66 παρέλθωσι. καὶ ἢν μὲν ἐν τῇ πεντεκαιδεκάτῃ
ἡμέρῃ ξηρανθῇ ὁ πλεύμων καὶ ἐκβήξῃ, ὑγιάζεται·
εἰ δὲ μή, δυῶν δεούσαις εἴκοσι προσέχειν· καὶ ἢν
μὲν ἐν ταύτῃσι παύσηται τοῦ βήχματος, ἐκφεύ-
γει. ἢν δὲ μὴ παύηται, εἴρεσθαι εἰ γλυκύτερον τὸ
σίαλον, καὶ ἢν φῇ, ἡ νοῦσος ἐνιαυσίη γίνεται· ὁ
γὰρ πλεύμων ἔμπυος γίνεται.

Τούτῳ χρὴ τὰς μὲν πρώτας ἡμέρας οἶνον διδόναι γλυκύν, λευκόν, ὑδαρέα, κατ' ὀλίγον πυκινὰ
πίνειν· ῥυφάνειν δὲ τῆς πτισάνης τὸν χυλὸν διδόναι, μέλι παραμίσγων, τρὶς τῆς ἡμέρης, ἔστ' ἂν
αἱ ὀκτωκαίδεκα ἡμέραι παρέλθωσι καὶ ὁ πυρετὸς
παύσηται. κινδυνεύει δὲ μάλιστα ἐν τῇσιν ἑπτὰ
ἢ ἐν τῇσι τεσσερεσκαίδεκα· ἐπὴν δὲ τὰς ὀκτωκαίδεκα ἡμέρας ὑπερβάλῃ, οὐκ ἔτι ἀποθνήσκει, ἀλλὰ
πτύει πύον, καὶ τὰ στήθεα πονέει, καὶ βήσσει.
ὅταν οὕτως ἔχῃ, πιπίσκειν τὸ σὺν τῷ ἐλελισφάκῳ
νῆστιν, καὶ ῥυφάνειν ἔτνος, στέαρ συμμίσγων
πλέον, ἢν μὴ θάλπος ᾖ· ἢν δ' ᾖ, μὴ ῥυφανέτω,
ἀλλὰ σιτίοισι χρήσθω ἁλυκοῖσι καὶ λιπαροῖσι καὶ
τοῖσι θαλασσίοισι μᾶλλον ἢ κρέασι. καὶ ἢν μή[1]
σοι δοκέῃ καθαίρεσθαι κατὰ λόγον, ἐγχεῖν καὶ
πυριᾶν· ἢν μὲν παχὺ ᾖ τὸ πύον, πυριᾶν· ἢν δὲ
λεπτόν, ἐγχεῖν. καὶ τῶν σιτίων ἔχεσθαι ὡς
μάλιστα, καὶ τῶν δριμέων ἀπέχεσθαι καὶ κρεῶν
βοείων καὶ οἰείων[2] καὶ χοιρείων.

---

[1] μὴ om. M.   [2] Later mss: ὑείων ΘΜ.

passed. If on the fifteenth day the lung becomes dry and the patient has coughed everything up, he recovers; if not, you must turn your attention to the eighteenth day, and, if he stops expectorating then, he escapes too. If, however, he does not stop, ask him whether his sputum is sweetish; if he says it is, the disease will last for a year; for the lung is suppurating internally.

On the first days administer to this patient dilute sweet white wine frequently a little at a time; as gruel give barley-water with honey three times a day, until eighteen days have passed and the fever has stopped. The patient is in the most danger for seven or fourteen days; when he has got beyond the eighteenth day, there is no longer any chance of death, but he does expectorate pus, suffer pain in the chest, and cough. When the case is such, let him drink, in the fasting state, a potion with salvia and, unless fever heat is present, as gruel a thick soup containing a generous portion of fat; if fever is present, let him not take gruel, but eat salty and rich foods, sea-foods more than meats. If you think the patient is not being cleaned as he should be, administer an infusion or a vapour-bath: if the pus is thick, a vapour-bath, if it is thin, an infusion. Have him take foods in generous amounts, but avoid sharp vegetables, beef, mutton and pork.

Ὅταν ἐκ περιπλευμονίης ἔμπυος γίνηται,
πυρετὸς ἴσχει καὶ βὴξ ξηρὴ καὶ δυσπνοίη· καὶ οἱ
πόδες οἰδέουσι, καὶ οἱ ὄνυχες ἕλκονται τῶν χειρῶν
καὶ τῶν ποδῶν. τοῦτον, ἐπὴν δεκαταῖος γένηται, ἐφ'[1] ἧς ἂν ἄρξηται ἔμπυος γίνεσθαι, λούσας
πολλῷ θερμῷ, τρίψας ἄρου ῥίζαν, ὅσον ἀστράγα-
λον μέγεθος, καὶ ἁλὸς χόνδρον, καὶ μέλι καὶ
ὕδωρ, καὶ ἄλειφα ὀλίγον, ἐξειρύσας τὴν γλῶσσαν,
ἐγχέαι χλιαρόν. ἔπειτα κινῆσαι τῶν ὤμων[2] | ἢν
μέντοι ὑπὸ τούτου τὸ πύον ῥαγῇ. εἰ δὲ μή, ἕτερον ποιῆσαι· σίδια δριμέα ἐκχυλώσας[3] καὶ κυκλά-
μινον—ὅσον ὀξύβαφον τῶν σμικρῶν ἑκατέρου
ἔστω—ἔπειτα ὀπὸν σιλφίου τρίψας ὅσον κύαμον,
διεῖναι, καὶ συμμῖξαι γάλακτος ὅσον ὀξύβαφον
αἴγειον ἢ ὄνειον· τοῦτο χλιαρὸν ἐγχεῖν. ἢν δ' ὑπὸ
τούτων μὴ ῥαγῇ, ῥαφάνου φλοιὸν καὶ ἄνθος χαλκοῦ ὅσον τρεῖς κυάμους τρίψας λεῖον—διπλάσιον
δ' ἔστω τῆς ῥαφάνου—ἐλαίῳ διεῖναι, ὅσον τεταρ-
τημορίῳ[4] κοτύλης· τοῦτο ἐγχεῖν χλιαρόν. καὶ ἢν
ῥαγῇ τὸ πύον, σιτίοισιν ὡς ἁλμυρωτάτοισι καὶ
λιπαρωτάτοισι χρήσθω. καὶ ἢν μὴ ἴῃ τὸ πύον,
κατ' ὀλίγον πυριᾶν κατὰ τὸ στόμα σίου χυλῷ,
οἴνῳ Τορνίῳ, γάλακτι βοείῳ ἢ αἰγείῳ· ἴσον δ'
ἑκάστου συμμίξας· ἔστω δ' ὅσον τρεῖς κοτύλαι

[1] Θ: ἀφ' Μ.   [2] Θ: τὸν ὦμον Μ.   [3] Θ: ἐκχυμώσας Μ
[4] Μ: -μορον Θ.

DISEASES II

When a person suppurates internally after pneumonia, fever is present together with a dry cough and difficulty in breathing; his feet swell, and the nails of both his hands and his feet become curved. On the tenth day after the internal suppuration has begun, wash the patient in copious hot water and, grinding together cuckoo-pint root to the amount of a vertebra, a lump of salt, honey, water, and a little anointing oil, draw out the tongue and infuse this warm; then shake the patient by his shoulders. If, with this, the pus breaks out, fine; if not, make the following alternative: squeeze juice from bitter pomegranate-peels, from cyclamen—let there be a small oxybaphon of each—and from silphium, to the amount of a bean; combine these, add an oxybaphon of goat's or ass's milk, and infuse this warm. If with these measures there is still no break, grate radish skin fine, grind flower of copper equal in amount to three beans—let the amount of radish be twice that—soak these in one quarter cotyle of oil, and infuse warm. If the pus breaks out, let the patient take very salty and rich foods; but if the pus does not move, administer gently through the mouth a vapour-bath consisting of water-parsnip juice, Toronian[1] wine, and goat's or cow's milk: mix together an equal amount of each of these so that there will be three cotylai altogether;

[1] Littré concludes that Τορνίῳ must refer to some locality unknown to us. I suggest Torone, a city in the Chalcidice not far from Mende; Mendean wine is prescribed in *Internal Affections* 13, 16–18 and 24.

ἔπειτα ἐμβάλλειν ἰπνοῦ ὄστρακα διαφήνας· τοῦτο
ἑλκέτω διὰ τοῦ αὐλοῦ φυλασσόμενος ὅπως μὴ
κατακαίηται. ἐπὴν δὲ καθαρώτερον πτύῃ, ἐγχεῖν
αὐτῷ κνίδης σπέρμα, λιβανωτόν, ὀρίγανον, ἐν
οἴνῳ λευκῷ καὶ μέλιτι καὶ ἐλαίῳ ὀλίγῳ[1] ἐγχεῖν
δὲ διὰ τρίτης ἡμέρης, μετὰ δέ, βούτυρον, ῥητίνην
ἐν μέλιτι διατήκων· καὶ σιτίοισι μηκέτι χρῆσθαι
ἁλμυροῖσι μηδὲ λιπαροῖσι. πινέτω δὲ νῆστις τὰς
ἐν μέσῳ ἡμέρας τῶν ἐγχύτων,[2] ἐλελίσφακον,
πήγανον, θύμβραν, ὀρίγανον, ἴσον ἐν οἴνῳ ἀκρήτῳ, ὅσον ὀξύβαφον συμπάντων ἐπιπάσσων.

Ἢν δὲ μὴ ῥαγῇ ὑπὸ τῶν ἐγχύτων· πολλάκις
γὰρ ἐκρήγνυται ἐς τὴν κοιλίην, καὶ αὐτίκα δοκέει
ῥᾷον[3] εἶναι, ὅταν ἐκ στενοῦ ἐς εὐρυχωρίην ἔλθῃ.
70 ὅταν ὁ χρόνος | πλείων γίνηται, ὅ τε πυρετὸς
ἰσχυρότερος[4] καὶ ἡ βὴξ ἐπιλαμβάνει, καὶ τὸ πλευρὸν ὀδυνᾶται, καὶ ἐπὶ μὲν τὸ ὑγιὲς οὐκ ἀνέχεται
ἀνακείμενος,[5] ἐπὶ δὲ τὸ ἀλγέον· καὶ οἱ πόδες
οἰδέουσι καὶ τὰ κοῖλα τῶν ὀφθαλμῶν. τοῦτον
ὅταν ἡμέρη πέμπτη καὶ δεκάτη γένηται ἀπὸ τῆς
ἐκρήξιος, λούσας πολλῷ θερμῷ, καθίσας ἐπὶ ἐφέ-
δρου, ὅ τι μὴ ὑποκινήσει, ἕτερος μὲν τὰς χεῖρας
ἐχέτω, σὺ δὲ τῶν ὤμων σείων, ἀκοάζεσθαι ἐς
ὁπότερον ἂν τῶν πλευρέων ψοφέῃ· βούλεσθαι δ᾽
ἐς τὸ[6] ἀριστερὸν ταμεῖν· ἧσσον γὰρ θανατῶδες.
ἢν δέ σοι ὑπὸ τοῦ πάχεος καὶ τοῦ πλήθεος μὴ

then put glowing oven sherds into the mixture, and let the patient draw in the vapour through a reed, taking care not to burn himself. When the expectorations become cleaner, infuse stinging-nettle seeds, frankincense and marjoram in white wine, honey and a little oil; infuse this every other day, and afterwards give the patient butter and resin melted in honey; let him no longer take salty or rich foods. On the days between infusions, let him drink, in the fasting state, salvia, rue, savory and marjoram—an equal amount of each—in unmixed wine, sprinkling in an oxybaphon of them all together.

If no external break occurs with these infusions—for often the pus breaks out into the cavity so that the patient at first seems to be better, since his pus has moved from a narrow space into an open one—as time passes more violent fever and coughing come on, the patient has pains in the side, and he cannot tolerate reclining on his healthy side, but only on the painful one; his feet swell up, and also the hollows of his eyes. When the fifteenth day after the pus has broken out into the cavity arrives, wash this patient in copious hot water, and seat him on a chair that does not move; have someone else hold his arms, and you shake him by the shoulders, listening on which of his sides there is a sound; prefer to incise on the left side, for it is less dangerous. If, because of the thickness and

---

[1] ὀλίγῳ om. Θ. [2] M: τῷ ἐγχύτῳ Θ. [3] Θ: ῥάων M.
[4] Θ: ἰσχυρὸς M. [5] Θ: κατα- M. [6] M: τὸν Θ.

ψοφέῃ—ποιέει γὰρ τοῦτο ἐνίοτε—ὁπότερον ἂν
ἀποιδέῃ τῶν πλευρέων καὶ ὀδυνᾶται μᾶλλον,
τοῦτο τάμνειν ὡς κατωτάτω ὄπισθεν τοῦ οἰδήμα-
τος μᾶλλον ἢ ἔμπροσθεν, ὅπως σοι ἡ ἔξοδος τῷ
πύῳ εὔροος ᾖ. τάμνειν δὲ μεταξὺ τῶν πλευρέων
στηθοειδέϊ μαχαιρίδι τὸ πρῶτον δέρμα· ἔπειτα
ὀξυβελέϊ, ἀποδήσας ῥάκει, τὸ ἄκρον τῆς μαχαιρί-
δος λιπὼν ὅσον τὸν ὄνυχα τοῦ δακτύλου τοῦ με-
γάλου, καθεῖναι ἔσω. ἔπειτα ἀφεὶς τὸ πύον ὅσον
ἄν σοι δοκέῃ, μοτοῦν ὠμολίνῳ μοτῷ, λίνον ἐκδή-
σας· ἀφεῖναι δὲ τὸ πύον ἅπαξ τῆς ἡμέρης· ἐπὴν
δὲ γένηται δεκαταῖος, ἀφεὶς ἅπαν τὸ πύον,
ὀθονίῳ μοτοῦν. ἔπειτα ἐγχεῖν οἶνον καὶ ἔλαιον
χλιαίνων αὐλίσκῳ, ὡς μήτε ὁ πλεύμων ἐξαπίνης
ἑωθὼς βρέχεσθαι τῷ πύῳ ἀποξηρανθῇ· ἐξιέναι δὲ
τὸ ἔγχυμα τὸ μὲν ἕωθεν ἐς ἑσπέρην, τὸ δ'
ἑσπερινὸν ἕωθεν. ἐπὴν δὲ τὸ πύον λεπτὸν οἷον
ὕδωρ ᾖ, καὶ γλίσχρον τῷ δακτύλῳ ψαυόμενον, καὶ
ὀλίγον, ἐντιθέναι μοτὸν κασσιτέρινον κοῖλον.
ἐπὴν δὲ παντάπασι ξηρανθῇ ἡ κοιλίη, ἀποτάμ-
νων τοῦ μοτοῦ κατὰ σμικρόν, συμφύειν τὸ ἕλκος,
72 ἔστ' ἂν ἐξέλῃς | τὸν μοτόν. σημεῖον δὲ ἢν μέλλῃ
ἐκφεύξεσθαι· ἢν μὲν τὸ πύον ᾖ[1] λευκὸν καὶ καθαρὸν[2] καὶ ἶνες αἵματος ἐνέωσιν, ὡς τὰ πολλὰ
ὑγιὴς γίνεται· ἢν δὲ οἷον λεκιθοειδὲς ἀπορρυῇ τῇ
πρώτῃ, ἢ τῇ ὑστεραίῃ ἀπορρυῇ παχύ, ὑπόχλω-

---

[1] τὸ πύον ᾖ M: ᾖ τὸ πύον ᾖ Θ.   [2] καὶ καθαρὸν om. Θ.

abundance of pus, there is no sound for you to hear—for sometimes this happens—on whichever side there is swelling and more pain, make an incision as low down as possible, behind the swelling rather than in front of it, in order that the exit you make for the pus will allow freedom of flow. First cut the skin between the ribs with a bellied scalpel; then wrap a lancet with a piece of cloth, leaving the point of the blade exposed a length equal to the nail of your thumb, and insert it. When you have removed as much pus as you think appropriate, plug the wound with a tent of raw linen, and tie it with a cord; draw off pus once a day; on the tenth day, draw all the pus, and plug the wound with linen. Then make an infusion of warm wine and oil with a tube, in order that the lung, accustomed to being soaked in pus, will not be suddenly dried out; discharge the morning infusion towards evening, and the evening one in the morning. When the pus is thin like water, sticky when touched with a finger, and small in amount, insert a hollow tin drainage tube. When the cavity is completely dried out, cut off the tube little by little, and let the ulcer unite before you remove the tube. A sign whether the patient is going to escape: if the pus is white and clean, and contains streaks of blood, he generally recovers; but if it flows out on the first day yolk-coloured, or on the following day thick, slightly

ρον, ὄζον, ἀποθνήσκουσιν, ἐπειδὰν ἐκρυῇ τὸ πύον.

48. Ὅταν πλευμᾷ, τὸ σίαλον παχύ, ὑπόχλω-
ρον, γλυκὺ βήσσεται, καὶ βρυγμός, καὶ ὀδύνη ἐς
τὸ στέρνον καὶ ἐς τὸ μετάφρενον. καὶ συρίζει ἐν
τῇ φάρυγγι λεπτόν, καὶ ἡ φάρυγξ σκληρὴ[1] γίνεται, καὶ τὰ κύλα[2] ἐρυθρά, καὶ ἡ φωνὴ βαρέη· καὶ
οἱ πόδες οἰδίσκονται, καὶ οἱ ὄνυχες ἕλκονται· καὶ
καταλεπτύνονται καὶ τὰ ἄνω μινύθει. καὶ μυ-
σάσσεται τὸ σίαλον, ἐπὴν ἀποχρεμψάμενος ἔχῃ
ἐν τῷ στόματι· καὶ βήσσει τοὺς ὄρθρους καὶ μεσο-
νύκτιος[3] μάλιστα· βήσσει δὲ καὶ τὸν ἄλλον χρόνον. καὶ λαμβάνει μᾶλλον γυναῖκα νεωτέρην ἢ
πρεσβυτέρην. τούτῳ ἢν μὲν αἱ τρίχες ἤδη ἐκ τῆς
κεφαλῆς ῥέωσι καὶ ψιλῶται ἤδη ἡ κεφαλὴ ὡς ἐκ
νούσου, καὶ πτύοντι ἐπ' ἄνθρακας βαρὺ ὄζῃ τὸ
σίαλον, φάναι αὐτὸν ἀποθανεῖσθαι ἐντὸς ὀλίγου
χρόνου, τὸ δὲ κτεῖνον ἔσεσθαι διάρροιαν. ἐπὴν
γὰρ ἤδη τὸ πύον τὸ περὶ τὴν καρδίην σήπηται,
τοῦτο ὄζει κνίσης ἐπὶ τοῖσιν ἄνθραξι, καὶ συν-
θερμαινόμενος ὁ ἐγκέφαλος ῥεῖ ἅλμην, ἢ κινεῖ
τὴν κοιλίην· σημεῖον δὲ τοῦτο[4] ῥέουσιν[5] ἐκ τῆς
κεφαλῆς τρίχες.

Τοῦτον μὴ ἰᾶσθαι ὅταν οὕτως ἔχῃ· ἢν δὲ κατ'
ἀρχὰς ἐπιτύχῃς τῇ νούσῳ, φάκιον δὸς πιεῖν. εἶτα
διαλιπὼν μίαν ἡμέρην ἐλλέβορον δοῦναι κεκριμέ-

---

[1] Θ: ξηρὴ Μ.   [2] Θ: κοῖλα Μ.   [3] Θ: -τίου Μ.   [4] Θ: -του Μ.
[5] Θ adds αἱ.

yellow-green, and stinking, when it has flowed out
the patient dies.

48. When there is a disease of the lung,[1] thick
sweet yellow-green sputum is coughed up, the teeth
chatter, and pain occupies the chest and back; the
throat whistles quietly and becomes stiff, the areas
under the eyes become red, and the voice is deep;
the feet swell up, and the nails become curved;
these patients become very thin, and the upper
parts of their bodies are wasted. The patient is dis-
gusted by the sputum when it is in his mouth after
being coughed up; he coughs most early in the
morning and in the middle of the night, although
he coughs at other times, too. This disease is more
frequent in younger women than in older ones. If
hair is falling out of the head, which is already on
the point of becoming bald from the disease, and if,
when the patient spits on to coals, his sputum has a
heavy odour, tell him that he is about to die before
long, and that what kills him will be diarrhoea; for
when the pus about the cardia is in a state of putrefaction, if it is thrown onto coals it smells like burnt
fat; and then the brain, being heated, pours forth a
salty fluid that sets the cavity in motion; the fact
that hair falls out of the head shows this.

When the case is such, do not treat this patient.
If, however, you happen to be present at the onset
of the disease, give a decoction of lentils to drink;
then, leaving an interval of one day, give hellebore

[1] Consumption.

νον ὅπως τὴν κάτω κοιλίην μὴ κινήσῃ· καὶ ἐπὴν
ἐς τὸ στόμα¹ τῆς νυκτὸς φοιτᾷ αὐτῷ ἅλμη, πρὸς
τὰς ῥῖνας αὐτῷ προστίθει φάρμακα πυκνότερα· ἢν
δὲ μὴ ῥέῃ, προστίθει μέν, διὰ πλέονος δὲ χρόνου,
καὶ τοῦ | μηνὸς ἅπαξ προσπιπίσκοντα ἐλλέβορον,
ὅσον τοῖσι δυσὶ δακτύλοισιν ἆραι, ἐν οἴνῳ γλυκεῖ
κεκρημένῳ· φάκιον δὲ αὐτίκα δοῦναι ἐπιπίνειν·
φάρμακα δὲ ὡς ἐλάχιστα πινέτω. ἢν μὴ οἱ πυρετοὶ ὀξύτεροι ἐπιλαμβάνωσιν· ἢν δὲ λαμβάνωσι,²
τὴν ῥίζαν τὴν λευκὴν [καὶ]³ τοῦ ἐλλεβόρου
λείχειν ἐν μέλιτι δίδου⁴ οὕτω γὰρ ἥκιστα τὴν
κοιλίην κινήσει. ἢν δὲ στρόφος ἐγγίνηται ἐν τῇ
κάτω κοιλίῃ, πρῶτον μὲν κλύσαι κείνῳ ἐς ὃ ὁ κόκ-
κος συμμίσγεται· ἢν δὲ μηδ' οὕτω παύσηται, γάλακτι ὀνείῳ ἑφθῷ κάθηρον· φάρμακον δὲ μὴ δίδου
κατωτερικόν. ἢν δὲ πρὸ τοῦ φακίου⁵ προπίνων
τὸν ἐλλέβορον χολὴν ἐμέῃ, αὐτῷ τῷ φακίῳ ἐμείτω. σιτίοισι δὲ χρήσθω, ἢν μὲν οἱ πυρετοὶ ὀξέες
ἔχωσι, κρέασι μηλείοισι ἑφθοῖσι καὶ ὀρνιθείοισι
καὶ κολοκύντῃ καὶ τεύτλοισι· ζωμὸν δὲ μὴ ῥυφείτω, μηδ' ἐμβάπτεσθαι⁶· ἰχθύσι δὲ χρήσθω σκορ-
πίοισι καὶ σελάχεσιν ἑφθοῖσι. θερμὸν δὲ μηδὲν
ἐσθιέτω μηδὲ λούσθω, ἢν ὁ πυρετὸς ἔχῃ πολύς,
μηδὲ λαχάνοισι χρήσθω δριμέσιν, ὅτι μὴ θύμβρῃ ἢ
ὀριγάνῳ· οἶνον δὲ λευκὸν πινέτω. ἢν δ' ἄπυρος

¹ Cornarius: σῶμα ΘΜ.   ² ἢν δὲ λαμβάνωσι om. Μ.
³ Del. I.   ⁴ Θ: δ' οὖ Μ.   ⁵ Θ: φαρμακίου Μ.   ⁶ Θ: βάπτε-
σθαι Μ.

that has been diluted with water to prevent it from setting the lower cavity in motion. When salty fluid runs into the patient's mouth during the night, apply frequent medications to the nostrils; if no such flux takes place, apply the same medications, but over a longer time. Once a month have the patient first drink hellebore—as much as you pick up with two fingers, in sweet wine diluted with water—and immediately afterwards give him a decoction of lentils to drink. Let him drink as few medications as possible. If sharper fevers do not supervene, fine; but if they do, give the patient white hellebore root in honey to take, for in this form the hellebore will be least likely to set the cavity in motion. If colic arises in the lower cavity, first apply an enema containing Cnidian berry; if the colic does not stop with this treatment, clean the cavity out with boiled ass's milk; do not give any medication to clean downwards. If the patient drinks hellebore before the decoction of lentils, and he vomits bile, from then on use the decoction alone as emetic. If sharp fevers are present, let the patient eat boiled mutton, fowl, gourd and beets, but not drink the sauce or dip anything into it; of fish let him have boiled scorpion fish and selachians; if a very great fever is present, though, let him neither eat anything hot, nor bathe, nor take sharp vegetables other than savory or marjoram, and have him drink white wine. If the patient is

ᾖ, θέρμαι δὲ[1] λαμβάνωσιν ἄλλοτε καὶ ἄλλοτε,
ἐσθιέτω ἰχθῦς ὡς ἀρίστους καὶ πιοτάτους, καὶ
λιπαρὰ καὶ ἁλμυρὰ καὶ γλυκέα ὡς μάλιστα. καὶ
περιπάτοισι χρήσθω μήτε ἐν ἀνέμῳ μήτε ἐν ἡλίῳ·
καὶ ἐμείτω ἀπὸ τῶν σιτίων, ὅταν οἱ δοκέῃ καιρὸς
εἶναι· καὶ λοῦσθαι χλιαρῷ πλὴν τῆς κεφαλῆς.
σιτίων δὲ ἄρτος ἀμείνων, ὅσοι μὴ μαζοφάγοι εἰσί·
τούτοισι δὲ ἀμφότερα συμμίσγειν.

49. Ἄλλη φθόη· βὴξ ἔχει, καὶ τὸ πτύσμα
πολλὸν καὶ ὑγρόν, καὶ ἐνίοτε ῥηϊδίως ἀναβήσσε-
ται οἷον χάλαζα τὸ πύον, καὶ διατριβόμενον ἐν
τοῖσι δακτύλοισι σκληρὸν καὶ κάκο|δμον γίνεται.
ἡ δὲ φωνὴ καθαρὴ καὶ ἀνώδυνος, καὶ οἱ πυρετοὶ
οὐ λαμβάνουσι, θέρμη δ' ἐνίοτε· ἄλλως τε καὶ
ἀσθενής. τοῦτον χρὴ ἑλλέβορον πιπίσκειν καὶ
φάκιον, καὶ εὐωχέειν ὡς μάλιστα, ἀπεχόμενον
δριμέων καὶ κρεῶν βοείων καὶ χοιρείων καὶ
οἰείων[2] καὶ γυμνάζεσθαι ὀλίγα καὶ περιπατεῖν,
καὶ ἀπὸ σίτων ἐμέτοισι χρῆσθαι, καὶ λαγνείης
ἀπέχεσθαι. αὕτη ἡ νοῦσος γίνεται ἑπτὰ ἔτεα
ἢ ἐννέα· οὗτος ἢν ἐξ ἀρχῆς θεραπευθῇ, ὑγιὴς
γίνεται.

50. Ἢν ἀφθήσῃ ἡ σῦριγξ τοῦ πλεύμονος, πυρετὸς ἴσχει ἰσχυρὸς[3] καὶ ὀδύνη μέσον τὸ στῆθος,
καὶ τοῦ σώματος κνησμός· καὶ ἡ φωνὴ βραγχώ-
δης, καὶ τὸ σίαλον ὑγρὸν καὶ λεπτὸν πτύει, ἐνί-

---

[1] Μ: τε Θ.   [2] Later mss: ὑείων ΘΜ.   [3] Θ: βληχρὸς Μ.

without true fever, but fever heat is present now
and then, let him eat the best and richest of fishes,
and generous amounts of rich, salty and sweet
foods, and take walks out of the wind and sun;
induce vomiting by means of foods, when you think
the time is opportune, and wash the patient in
warm water, except for his head. Bread is better
than other cereals for persons that do not eat
barley-cakes; for those that do, mix both together.

49. Another consumption: there is coughing, the
sputum is copious and moist, and sometimes the
patient without difficulty coughs up pus that
resembles hail stones which, on being rubbed
between the fingers, are hard and evil-smelling.
The voice is clear, the patient is free of pain, and
there are no fevers, although sometimes fever heat;
the patient is especially weak. You must make this
patient drink hellebore and a decoction of lentils,
and feed him as well as possible, while avoiding
sharp vegetables, beef, pork and mutton; have him
do a few exercises, take walks, vomit after meals,
and refrain from venery. This disease lasts for
seven or nine years; if the patient is treated from
the beginning, he recovers.

50. If the pipe of the lung becomes aphthous,
there is violent fever, pain in the middle of the
chest, and itching of the body; the patient's voice is
hoarse, and he expectorates thin moist sputum;

οτε δὲ παχὺ καὶ οἷον πτισάνης χυλόν. καὶ ἐν τῷ
στόματι ὀδμή οἱ ἐγγίνεται βαρέη οἷον ἀπὸ ἰχθύων
ὠμῶν· καὶ ἄλλοτε καὶ ἄλλοτε ἐν τῷ σιάλῳ ἐμφαί-
νεται σκληρά, οἷον μύκης ἀφ' ἕλκεος. καὶ τὰ
ἄνω λεπτύνεται, μάλιστα δὲ ἅπας· καὶ οἱ κύκλοι
τοῦ προσώπου ἐρυθριῶσι, καὶ οἱ ὄνυχες τῷ χρόνῳ
ἕλκονται καὶ ξηροὶ καὶ χλωροὶ γίνονται. τελευτᾷ
δὲ αὐτίκα, ἢν μὴ θεραπευθῇ, αἷμα πτύων καὶ
πύον· ἔπειτα καὶ πυρετοὶ ἰσχυροὶ ἐπιγινόμενοι
κατ' οὖν ἔκτειναν· ἢν δὲ θεραπευθῇ, πλεῖστοι ἐκφυγγάνουσιν ἐκ ταύτης τῆς φθίσιος.

Θεραπεύειν δὲ χρή, φάκια πιπίσκοντα ἐμέειν·
ἢν δέ σοι καιρὸς δοκέῃ εἶναι ἐλλέβορον πίνειν, ἢν
μὲν δυνατὸς ᾖ ὤνθρωπος, αὐτόθεν· ἢν δὲ μή,
παραμίσγειν τῷ φακίῳ ἥμισυ πόσιος, διαλείπων
ἐν πέμπτῃ ἢ ἐν ἕκτῃ πόσει. τὴν δὲ κάτω κοιλίην μὴ κινέειν φαρμάκῳ, ἢν μὲν[1] οἱ πυρετοὶ
λαμβάνωσιν ἰσχυροί· ἢν δὲ μὴ λαμβάνωσι, γάλακτι ὄνου ὑποκαθαίρειν· ἢν δ' ἀσθενὴς ᾖ ὥστε
πίνειν, ὑποκλύσαι.[2] ἧττον δὲ κεφαλήν· καὶ ἢν μὲν
τὸ σίαλον ἐς τὸ στόμα ἴῃ πολλὸν καὶ ἁλμυρόν,
πρὸς τὰς ῥῖνας προσθεῖναι ὅ τι χολὴν μὴ ἄξει· ἢν
δὲ μὴ ἴῃ[3] τὸ ῥεῦμα ἐς τὸ στόμα, μὴ προστιθέναι |
πρὸς τὴν κεφαλήν. ἐπὴν δὲ τὸ σίαλον δυσῶδες
ᾖ, τὰς μεταξὺ τῶν φακίων ἐγχεῖν ἐς τὸν πλεύμο-

78

---

[1] Potter: μὴ ΘΜ.   [2] Θ has lost a leaf containing -σαι ...
(51) γαλακτο-.   [3] Cornarius: ᾖ Μ.

sometimes, though, the sputum is thick and like barley-water. A heavy smell arises in the patient's mouth, like that of raw fish, and from time to time there appear in his sputum hard pieces like the fleshy excrescences of an ulcer. The patient becomes lean in his upper parts, and his whole body is very wasted. His cheeks blush, and after a while his nails become curved, dry and yellow-green. This patient dies at once unless he is treated: he expectorates blood and pus, and then violent fevers come on and kill him. If it is treated, most escape from this consumption.

You must treat by having the patient drink a decoction of lentils, and then vomit. If you think it is the right time to give hellebore, if the person is able, let him take it straight off, but if he is not, mix half a draught of hellebore with lentils, and stop the administration with the fifth or sixth draught. If violent fevers are present, do not give a medication to set the lower cavity in motion, but if they are not, clean downwards with ass's milk; if the patient is too weak to drink this, employ an enema. Give a gentle medication for the head; if copious salty sputum runs into the mouth, apply a medication to the nostrils, but one that will not draw bile; if no flux goes into the mouth, do not make any application to the head. When the sputum is ill-smelling, on the days between the lentil

να φάρμακον· μίαν δὲ διαλιπὼν ἐπὴν ἐγχῇς[1] ἡμέρην, θυμιᾶν. σιτίοισι δὲ χρῆσθαι κρέασι μηλείοισι
καὶ ὀρνιθείοισι, καὶ ἰχθύσι σελάχεσι καὶ σκορπίοι-
σιν ἐφθοῖσι· διὰ τετάρτης ἡμέρης τάριχον ἐσθιέτω
ὡς ἄριστον καὶ πιότατον, καὶ ἀριστάτω μὲν μᾶζαν, δειπνείτω δὲ συμμίσγων καὶ ἄρτον· καὶ μήτε
ῥυφανέτω μηδέν, μήτε κυκεῶνα πινέτω, ἢν ἐσθίειν δυνατὸς ᾖ. τὰ δ᾽ ὄψα ἡδύνειν σησάμῳ ἀντὶ
τοῦ τυροῦ, καὶ κοριάννῳ καὶ ἀνήθῳ· σιλφίῳ δὲ μηδὲν χρῆσθαι μηδὲ λαχάνῳ δριμεῖ, ὅ τι μὴ ὀριγάνῳ
ἢ θύμῳ ἢ πηγάνῳ. περιπάτοισι δὲ χρήσθω καὶ
πρὸ τοῦ σιτίου καὶ μετὰ τὸ σιτίον, φυλασσόμενος
τὸν ἄνεμον καὶ τὸν ἥλιον· θωρηξίων ἀπεχέσθω
καὶ ἀφροδισίων· λούσθω δὲ χλιαρῷ, πλὴν τῆς
κεφαλῆς, ταύτην δὲ ὡς διὰ πλείστου χρόνου.

51. Νωτιὰς φθίσις ἀπὸ τοῦ μυελοῦ γίνεται·
λαμβάνει δὲ μάλιστα νεογάμους καὶ φιλολάγνους.
γίνονται δὲ ἄπυροι, καὶ ἐσθίειν ἀγαθοί, καὶ τή-
κονται· καὶ ἢν ἐρωτᾷς αὐτόν, φήσει οἱ ἄνωθεν
ἀπὸ τῆς κεφαλῆς κατὰ τὴν ῥάχιν ὁδοιπορέειν
οἷον μύρμηκας. καὶ ἐπὴν οὐρέῃ ἢ ἀποπατέῃ,
προέρχεταί οἱ θορὸς πολὺς καὶ ὑγρός· καὶ γενεὴ
οὐκ ἐγγίνεται, καὶ ὀνειριάει, καὶ ἢν συγκοιμηθῇ
γυναικί, καὶ ἢν μή. καὶ ὅταν ὁδοιπορήσῃ ἢ δρά-
μῃ, ἄλλως τε καὶ πρὸς αἶπος, ἆσθμά μιν καὶ
ἀσθενείη ἐπιλαμβάνει, καὶ τῆς κεφαλῆς βάρος,

[1] Potter: ἔχῃς M.

decoctions infuse a medication into the lung; then,
leaving one day from when you infused, apply a
fumigation. As food give mutton and fowl, and
boiled selachians and scorpion fish; every third day
have the patient eat the best and fattest salt-fish,
breakfast on barley-cakes, and at dinner take bread
as well; let him not drink any gruel or cyceon, if he
is able to eat food. Let him season his main dishes
with sesame instead of cheese, and with coriander
and dill; let him not have silphium or any sharp
vegetables except marjoram, thyme or rue. Have
the patient take walks before and after his meals,
avoiding both wind and sun, refrain from drunkenness and venery, and bathe in warm water, except
for his head; it he should wash only at very great
intervals.

51. Consumption of the back arises from the
marrow; most frequently it occurs in newly-weds
and those fond of venery. They are without fever
and eat well, but still they melt away. If you ask
the patient, he will say that starting from his head
he feels something crawling down his spine, like
ants. When he passes urine or goes to stool, copious moist semen comes forth; he begets no off-
spring, and he has nocturnal emissions whether he
sleeps with a woman or not. When he walks or
runs, especially against a grade, panting and weakness come over him, his head feels heavy, and

καὶ τὰ ὦτα ἠχέει. τοῦτον χρόνῳ ὅταν ἐπιλάβωσι
πυρετοὶ ἰσχυροί, ἀπ' οὖν ὤλετο ὑπὸ λιπυρίου.

Ὅταν οὕτως ἔχῃ, ἢν ἐξ ἀρχῆς μεταχειρίσῃ,
πυριάσας αὐτὸν ὅλον, φάρμακον δοῦναι πίνειν |
80 ἄνω, καὶ μετὰ τοῦτο τὴν κεφαλὴν καθῆραι, μετὰ
δὲ πῖσαι κάτω (ἐγχειρέειν δὲ βούλεσθαι μάλιστα
τοῦ ἦρος) καὶ μεταπῖσαι ὀρὸν ἢ γάλα ὄνειον·
βόειον δὲ γάλα διδόναι πιέειν τεσσεράκοντα ἡμέ-
ρας· ἐς ἑσπέρην δὲ ἕως ἂν γαλακτοποτέῃ[1] χόνδρον διδόναι ῥυφεῖν, σιτίων δὲ ἀπεχέσθω. ἐπὴν
δὲ παύσηται γαλακτοποτέων, σιτίοισι διακομί-
ζειν αὐτὸν μαλθακοῖσιν ἐξ ὀλίγου ἀρχόμενος, καὶ
παχῦναι ὡς μάλιστα. καὶ ἐνιαυτοῦ θωρηξίων
ἀπεχέσθω καὶ ἀφροδισίων καὶ ταλαιπωρίων ὅ τι
μὴ περιπάτοισι, φυλασσόμενος τὰ ψύχεα καὶ τὸν
ἥλιον. λούσθω δὲ χλιαρῷ.

52. Πλεύμονος· τὸ σίαλον παχὺ καὶ λιγνυῶ-
δες βήσσεται, καὶ ἡ χροιὴ μέλαινα καὶ ὑποιδαλέη,
καὶ ὀδύναι λεπταὶ ὑπὸ τὸ στῆθος καὶ ὑπὸ τὰς
ὠμοπλάτας, καὶ δυσελκέες γίνονται. ἧσσον δ' ἐπι-
κίνδυνος τοῦ ἑτέρου οὗτος, καὶ ἐκφυγγάνουσι πλέονες.

Τοῦτον χρὴ ἐλλέβορον πιπίσκειν καὶ αὐτὸν καὶ
τοῖσι φακίοισι μίσγοντα, καὶ ἐγχεῖν ἐς τὸν πλεύ-
μονα· καὶ θυμιᾶν, καὶ εὐωχέειν ἀπεχόμενον κρεῶν
βοείων καὶ οἰείων[2] καὶ χοιρείων καὶ λαχάνων δρι-

---

[1] Θ resumes with -ποτέῃ.   [2] Later mss: ὑείων ΘΜ.

his ears ring. When, in time, violent fevers befall this patient, he perishes from one of the remittent variety.

When the case is such, if you deal with it from the beginning, apply a general vapour-bath, and give a medication that acts upwards to drink; after that clean out the head, and next have the patient drink a medication that acts downwards—prefer to take on such a case in the spring—and then after the cleanings give whey or ass's milk to drink; give cow's milk for forty days. As long as the patient is drinking milk, towards evening give him spelt as gruel, and have him abstain from foods. When he stops drinking milk, restore him with mild foods, starting off with small amounts, and make him as fat as possible. For one year let the patient refrain from drunkenness, venery, and exertions other than walks, during which he must avoid cold and sun; have him bathe in warm water.

52. Disease of the lung: thick dark-coloured sputum is coughed up, the skin is dark and somewhat swollen, mild pains occupy the chest and the region under the shoulder-blades, and patients heal poorly. This patient is in less danger than the one above, and most survive.

Have him drink hellebore, both by itself and mixed with a decoction of lentils, make an infusion into the lung, and apply a fumigation. Feed the patient well, but have him avoid beef, mutton, pork, and sharp vegetables other than marjoram or

μέων, ὅ τι μὴ ὀριγάνῳ ἢ θύμβρῃ. καὶ περιπάτοισι χρήσθω· ἐξ ἠοῦς δὲ πρὸς αἶπος ὁδοιπορέειν
νῆστιν· ἔπειτα πίνειν τῶν φύλλων ἐπ᾽ οἴνῳ ἐπι-
πάσαντα κεκρημένῳ· τὸ δὲ λοιπὸν σιτίοισι
χρήσθω τοῖς εἰρημένοισιν.

53. Ἢν τρωθῇ ἡ ἀρτηρίη, βὴξ ἔχει καὶ αἷμα
βήσσεται· καὶ λανθάνει ἡ φάρυγξ πιμπλαμένη
τοῦ αἵματος, καὶ ἐκβάλλει θρόμβους, καὶ ὀδύνη
γίνεται ἐκ τοῦ στήθεος ἐς τὸ μετάφρενον ὀξέη,
καὶ τὸ σίαλον γλίσχρον καὶ πολύ, καὶ ἡ φάρυγξ
82   ξηρή, καὶ | πυρετὸς καὶ ῥῖγος ἐπιλαμβάνει, καὶ
κέρχνεται ἡ φάρυγξ οἶον ὑπὸ λιπαροῦ. καὶ ἐς
τὰς¹ μὲν πεντεκαίδεκα ἡμέρας τοιαῦτα πάσχει.
μετὰ δὲ πύον πτύει, καὶ οἶον² ἕλκεος κροτώνας·³
καὶ αὖτις βήξ, καὶ ἐρράγη οὖν τὸ αἷμα, καὶ μετὰ
τὸ πύον παχύτερον πτύει· καὶ ὁ πυρετὸς ἰσχυρότερος γίνεται, καὶ τελευτᾷ ἐς πλεύμονα· καὶ
καλέεται ῥηγματίας πλεύμος.⁴

Ἢν δὲ μετὰ τὸ πρῶτον αἷμα μὴ πτύσῃ πύον,
παυσάμενον χρὴ ταλαιπωρίης καὶ γυμνασίων καὶ
ἐπ᾽ ὄχημα μὴ⁵ ἀναβαίνειν, σιτίων ἀπεχόμενον
ἁλμυρῶν καὶ λιπαρῶν καὶ πιόνων καὶ λαχάνων
δριμέων. καὶ ἐπὴν αὐτὸς ἑωυτοῦ δοκέῃ ἄριστα
τοῦ σώματος ἔχειν, καῦσαι τὰ στήθεα ἐπὶ⁶ τὸ
μετάφρενον ἐν μοίρῃ ἑκάτερον· καὶ ἐπὴν τὰ ἕλκεα

¹ Potter: ἔσται Θ: ἔς τε Μ.   ² Θ: οἷα Μ.   ³ κρότωνας
Vander Linden: κρότωνες ΘΜ.   ⁴ Θ: πλεύμονος Μ.   ⁵ μὴ
om. Μ.   ⁶ Θ: κατὰ Μ.

savory. Let him take walks: have him begin at
dawn, in the fasting state, and walk against a
grade; then let him sprinkle herbs over wine mixed
with water, and drink it; from then on, have him
eat the foods mentioned.

53. If the bronchial tube is injured, the patient
coughs up blood, his throat becomes full of blood
without his noticing it, and he casts up clots.
Sharp pain extends from his chest to his back, his
sputum is sticky and copious, his throat is dry,
fever and chills come on, and his throat makes a
rough noise as if there were fat in it. The patient
suffers these things for fifteen days. After that he
expectorates pus and material like the fragments of
an ulcer; once again there is coughing, blood breaks
out, and after that the pus coughed up is thicker;
the fever becomes more violent, and finally the
disease enters the lung; this is called "pneumonia
with tears".

If after the first blood the patient does not expectorate pus, he must cease from exertions, exercises
and riding in a wagon, and refrain from foods that
are salty, rich or fat, and from sharp vegetables.
When his body seems to be spontaneously in
optimal condition, cauterize the chest at the
back on each side in equal proportions. When he

ΠΕΡΙ ΝΟΥΣΩΝ Β

ὑγιὴς γένηται, ἐνιαυτὸν ἀπεχέσθω θωρηξίων, καὶ
μὴ ὑπερπιμπλάσθω,[1] μηδὲ τῇσι χερσὶ ταλαιπωρέειν, μηδ' ἐπ' ὄχημα ἀναβαίνειν· ἀλλὰ παχύνειν
αὐτὸν ὡς μάλιστα τὸ σῶμα.

54. Ἐπὴν ἀρτηρία[2] σπασθῇ τοῦ πλεύμονος,
τὸ πτύσμα λευκὸν πτύει, ἐνίοτε δὲ αἱματώδεα·
ἀφρονέει καὶ πυρετὸς ἴσχει, καὶ ὀδύνη τὸ στῆθος
καὶ τὸ μετάφρενον καὶ τὸ πλευρόν· καὶ ἢν στρα-
φῇ, βήσσεται καὶ πτάρνυται. τοῦτον ᾗ ἂν ἡ[3]
ὀδύνη ἔχῃ, χλιάσματα προστιθέναι, καὶ διδόναι
προρυφάνειν· κενταύριον καὶ δαῦκον καὶ ἐλελισφάκου φύλλα τρίβων,[4] μέλι καὶ ὄξος ἐπιχέων
καὶ ὕδωρ, διδόναι καταρρυφάνειν· καὶ πτισάνης
χυλὸν προρυφανέτω, καὶ ἐπιπινέτω οἶνον ὑδαρέα.
ἐπὴν δὲ τῆς ὀδύνης παύσηται, ἐλελίσφακον κόψας καὶ[5] σήσας, καὶ ὑπερικὸν καὶ ἐρύσιμον λεῖον
καὶ ἄλφιτον, ἴσον ἑκάστου, ταῦτ' ἐπιβαλὼν ἐπ'
οἶνον | κεκρημένον, διδόναι πίνειν νῆστι· καὶ ἢν
μὴ νῆστις ᾖ, διδόναι δὲ ῥυφάνειν ἔτνος ἄναλτον.
ἢν δὲ θάλπος ᾖ, σιτίοισι διαχρῆσθαι ὡς μαλθακωτάτοισιν, ἀνάλτοισι καὶ ἀκνίσοις, ἐπὴν ἤδη ἐπιεικῶς ἔχῃ τὸ σῶμα καὶ τὸ στῆθος καὶ τὸν νῶτον.

Ἢν δ' ἀμφότερα σπασθῶσι, βὴξ ἴσχει, καὶ τὸ
σίαλον βλέπεται παχὺ λευκόν, καὶ ὀδύνη ὀξέη

84

[1] Θ: -σθαι Μ.   [2] Potter: ἄρθρα ΘΜ: ἄορτρα Vander Linden
(cf. Galen XIX.82).   [3] ἡ om. Μ.   [4] Μ adds καί.
[5] καὶ om. Μ.
290

recovers from the burns, have him refrain from drunkenness, overfullness, exertions with the arms, and riding in a wagon for a year; make his body as fat as possible.

54. When the bronchial tube of a lung is torn, the patient expectorates white or sometimes bloody sputum; he behaves irrationally, and there is fever, and pain in his chest, back and side; if he turns himself, he coughs and sneezes. Apply fomentations to this patient wherever there is pain, after first giving him the following gruel to drink: grind centaury, dauke, and salvia leaves, add honey, vinegar and water, and give this to the patient to swallow; you can also have him first drink barley-water gruel, and afterwards dilute wine. When the pain goes away, pound and sieve salvia together with hypericum, fine hedge-mustard, and meal—an equal amount of each—sprinkle this over wine mixed with water, and give it to the fasting patient to drink; if he is not in the fasting state, give him thick unsalted soup to drink. If the weather is warm, employ very soft unsalted foods without savoury odours, once the body, chest and back are in a relatively good state.

If the bronchial tubes are torn on both sides, there is coughing, the sputum looks thick and white, and sharp pain occupies the chest, the area

ἴσχει ἐς τὸ στῆθος καὶ ὑπὸ τὰς ὠμοπλάτας καὶ τὸ
πλευρόν. καῦμα ἔχει, καὶ καταπίμπλαται φώ-
δων, καὶ ξυσμὴ ἔχει, καὶ οὐκ ἀνέχεται οὔτε καθήμενος οὔτε κείμενος οὔτε ἑστηκώς, ἀλλὰ δυσ-
θετέει.[1] οὗτος τεταρταῖος μάλιστα ἀποθνήσκει· ἢν
δὲ ταύτας ὑπερφύγῃ, ἐλπίδες μὲν οὖν[2] πολλαί·
κινδυνεύει δὲ καὶ ἐν τῇσιν ἑπτά· ἢν δὲ καὶ ταύτας διαφύγῃ, ὑγιάζεται. τοῦτον, ὅταν οὕτως
ἔχῃ, λούειν πολλῷ θερμῷ δὶς τῆς ἡμέρης, καὶ
ὅταν ἡ ὀδύνη ἔχῃ, χλιάσματα προστιθέναι· καὶ
πίνειν[3] διδόναι μέλι καὶ ὄξος, ῥυφάνειν καὶ χυλὸν
πτισάνης, καὶ ἐπιπίνειν οἶνον λευκὸν οἰνώδεα.
ἢν δὲ πρὸς τὸ λουτρὸν καὶ τὰ χλιάσματα πονέῃ
καὶ μὴ ἀνέχηται, προσφέρειν αὐτῷ ψυχρὰ[4] ῥάκεα
ἡμιτυβίου, βάπτων ἐς ὕδωρ ἐπὶ τὰ στήθεα ἐπι-
τιθέναι καὶ ἐπὶ τὸν νῶτον. καὶ πίνειν διδόναι
κηρίον ἐν ὕδατι ἀποβρέχων ὡς ψυχρότατον, καὶ
τὸν χυλὸν ψυχρὸν καὶ ὕδωρ ἐπιπίνειν, καὶ κεῖσθαι
πρὸς τὸ ψῦχος. ταῦτα ποιέειν· ἡ δὲ νοῦσος
θανατώδης.

55. Ἢν ἐρυσίπελας ἐν πλεύμονι γένηται, βὴξ
ἔχει, καὶ τὸ σίαλον ἀποπτύει πολὺ καὶ ὑγρόν, οἷον
ἀπὸ βρόγχου· ἔστι δὲ οὐχ αἱματῶδες. καὶ ὀδύνη
ἔχει τὸ μετάφρενον καὶ τοὺς κενεῶνας καὶ τὰς
λαπάρας, καὶ τὰ σπλάγχνα μύζει καὶ ἐμέει λάπ-

---

[1] Potter (cf. ch. 67): δυσθενέει ΘΜ.   [2] Potter (cf. ch. 58):
οὐ ΘΜ.   [3] M adds δέ.   [4] ψυχρὰ om. M.

under the shoulder-blade, and the side. Burning heat comes on, the patient is covered with blisters, he itches, and he tolerates neither sitting, nor lying, nor standing, but is greatly distressed. He usually dies on the fourth day; if he survives for that many, there is good hope, although he is still in danger for seven days; if he escapes these too, he recovers. When the case is such, wash the patient in copious hot water twice a day, and when pain is present apply fomentations; give honey and vinegar to drink, barley-water as gruel, and afterwards dilute white wine. If, with the bath and the fomentations, there is such pain that the person cannot stand it, soak linen cloths in cold water, and apply them to the chest and back. To drink give honeycomb well-soaked in the coldest water, and afterwards cold barley-juice and water; let the patient lie exposed to the cold. Do these things; the disease is often mortal.

55. If erysipelas occurs in the lung, there is coughing, and the patient expectorates copious moist sputum like that produced in a sore throat; it is not bloody. Pain occupies his back, flanks and sides, his inward parts rumble, he vomits up scum

πην καὶ οἷον ὄξος, καὶ τοὺς ὀδόντας αἱμωδιᾳ·
86 καὶ πυρετὸς καὶ | ῥῖγος καὶ δίψα λαμβάνει. καὶ
ὅταν τι φάγῃ, ἐπὶ τοῖσι σπλάγχνοισι μύζει, καὶ
ἐρεύγεται ὀξύ, καὶ ἡ κοιλίη τρύζει, καὶ ναρκᾳ τὸ
σῶμα. καὶ ὅταν ἐμέσῃ, δοκέει ῥάων εἶναι· ὅταν δὲ
μὴ ἐμέσῃ, ἀπιούσης τῆς ἡμέρης, στρόφος καὶ
ὀδύνη ἐγγίνεται ἐν τῇ γαστρί, καὶ ὁ[1] ἀπόπατος
ὑγρὸς γενόμενος διεχώρησεν. ἡ δὲ νοῦσος μάλιστα γίνεται ἐκ θωρηξίων καὶ ἐκ κρεηφαγίων καὶ
ἐξ ὕδατος μεταβολῆς· ἴσχει δὲ καὶ ἄλλως.

Τοῦτον φάρμακον πιπίσκειν κάτω, καὶ μεταπιπίσκειν γάλα ὄνου, ἢν μὴ σπληνώδης ᾖ φύσει·
ἢν δὲ σπληνώδης ᾖ, μὴ καθαίρειν μήτε χυλῷ[2]
μήτε γάλακτι μήτε ὀρῷ, ἀλλ' ὅ τι ὀλίγον ἐσελθὸν
πολὺ ἐξάξει. ὑποκλύζειν δὲ τὰς κοιλίας, καὶ
βαλάνους προστιθέναι, ἢν μὴ ἡ κοιλίη ὑπάγῃ,[3]
ἐν πάσῃσι τῇσι νούσοισι. καὶ ψυχρολουτέειν ἐν
ταύτῃ τῇ νούσῳ, καὶ γυμνάζεσθαι, ὅταν οἱ πυρετοὶ
ἀνῶσι καὶ δοκέῃ ἐπιεικῶς ἔχειν τοῦ σώματος.
καὶ τοῦ ἦρος καὶ τοῦ μετοπώρου ἔμετόν οἱ ποιέ-
ειν·[4] σκορόδων δὲ κεφαλὰς καὶ ὀριγάνου δραχμίδα
ὅσην τρισὶ[5] δακτύλοισι περιλαβεῖν, ἑψεῖν ἐπι-
χέαντα δύο κοτύλας οἴνου γλυκέος καὶ κοτύλην
ὄξους ὡς ὀξυτάτου καὶ μέλιτος ὅσον τεταρτη-

---

[1] ὁ om. M.   [2] Θ: -οῖσι M.   [3] Θ: ὑποχωρέῃ M.   [4] οἱ ποιέειν Θ: ἐμποιέειν M.   [5] Littré, following Mack's incorrect
collation of Θ: τοῖσι ΘM.

and material like vinegar, and his teeth are set on
edge; fever, chills and thirst are present. When the
patient eats anything, he rumbles in his inward
parts and suffers from oxyrygmia, his cavity sends
up fluid, and his body becomes numb. When he has
vomited, he seems to be better, but when he does
not vomit, as the day wanes he suffers colic and
pain in his belly, and he passes watery stools. In
most cases, this disease arises from drunkenness,
from eating meat, or from a change of water; it can
also occur in other circumstances.

Have this patient drink a medication that acts
downwards, and afterwards ass's milk, unless he
has a splenic diathesis; if he has a splenic diathesis,
do not clean him with juices, milk or whey, but
with a medication that, entering in a small amount,
will draw much out. (Apply enemas and suppositories in all diseases, if the cavity does not come
down.) Also, bathe the patient in cold water, in this
disease, and have him do exercises, when he is
without fever and his body seems to be in fairly
good condition. In spring and fall induce vomiting:
boil garlic heads and a pinch (the amount you take
with three fingers) of marjoram; pour in two cotylai
of sweet wine, one of very acid vinegar, and a

μόριον, ἑψεῖν δ' ἔστ' ἂν[1] ἡ τρίτη μοῖρα λειφθῇ, καὶ
ἔπειτα γυμνάσας τὸν ἄνθρωπον καὶ λούσας ὕδατι
χλιαρῷ πῖσαι θερμόν·[2] καὶ ἐπιπιπίσκειν[3] φάκιον,
μέλι καὶ ὄξος συμμίσγων, ἔστ' ἂν ἐμπλησθῇ.
ἔπειτα ἐμείτω, καὶ τὴν ἡμέρην ταύτην πιὼν
ἄλφιτον καὶ ὕδωρ ἐκνηστευέτω· ἐς ἑσπέρην δὲ
τεῦτλα φαγέτω καὶ μάζης μικρόν, καὶ πινέτω
οἶνον ὑδαρέα. ἀνὰ δὲ τὸν ἄλλον χρόνον ἐμείτω
τοῖσι φακίοισι καὶ ἀπὸ σιτίων.

Καὶ ἢν ἀφίστηται ἡ ὀδύνη ὑπὸ τὰς ὠμοπλάτας, σικύην προσβάλλειν, καὶ τὰς φλέβας
ἀποτύψαι τὰς ἐν τῇσι χερσί· σιτίοισι δὲ χρήσθω[4]
ἀνάλτοισι καὶ μὴ λιπαροῖσι μηδὲ πίοσι· δριμέα
δὲ καὶ ὀξέα ἐσθιέτω καὶ ψυχρὰ πάντα, καὶ περιπάτοισι χρήσθω. ταῦτα ποιέων ἄριστ' ἂν διαιτῶτο,
καὶ διὰ πλείστου χρόνου ἡ νοῦ|σος γίνοιτο· ἔστι
δὲ οὐ θανατώδης, ἀλλ' ἀπογηράσκοντας ἀπολείπει.
εἰ δὲ βούλοιο νεώτερον ὄντα θᾶσσον ἀπαλλάξαι
τῆς νούσου, καθήρας αὐτόν, καῦσαι τὰ στήθεα καὶ
τὸ μετάφρενον.

56. Νωτιάς· ῥῖγος καὶ πυρετὸς καὶ βὴξ καὶ
δύσπνοια λαμβάνει, καὶ τὸ σίαλον πτύει χλωρόν,
ἔστι δ' ὅτε ὕφαιμον· καὶ πονέει μάλιστα τὸ μετάφρενον καὶ τοὺς βουβῶνας, καὶ ἡμέρῃ τρίτῃ ἢ
τετάρτῃ οὐρέει αἱματῶδες. καὶ ἀποθνήσκει

---

[1] δ' ἔστ' ἂν Θ: δὲ ὅταν Μ.   [2] Μ: -μῷ Θ.   [3] Jouanna
(p. 196): ἐπιπίσκειν Θ: πιπίσκειν Μ.   [4] Θ: -σθαι Μ.

quarter cotyle of honey; boil until one third is left; then have the person do exercises, wash him in warm water, and have him drink the potion warm; afterwards have him drink a decoction of lentils, to which have been added honey and vinegar, until he is full; then let him vomit, and during that day eat nothing, but drink meal and water. Towards evening, let him eat beets and a little barley-cake, and drink dilute wine. From then on, have him vomit with lentil decoctions and by means of foods.

If the pain withdraws beneath the shoulder-blades, apply a cupping instrument, and incise the vessels of the arms. Have the patient eat foods that are not salty, rich or fat, and eat everything that is sharp, acid and cold; let him take walks. In doing these things, he will be following the best regimen, and the disease will stretch out over a long period of time; it is not mortal, but leaves people only when they grow old. Should you wish to relieve someone younger of the disease more quickly, clean him out, and cauterize his chest and back.

56. Disease of the back: there are chills, fever, coughing, and difficulty in breathing; the patient expectorates yellow-green sputum sometimes charged with blood; he suffers pains mainly in his back and groins, and on the third or fourth day he passes bloody urine. He dies on the seventh day;

ἑβδομαῖος· ἐπὴν δὲ τὰς τεσσερεσκαίδεκα ὑπερφύγῃ, ὑγιὴς γίνεται· ἐκφυγγάνει δ' οὐ μάλα.

Τούτῳ διδόναι μελίκρητον ἀναζέσας ἐν καινῇ
χύτρῃ, ψύχων, σελίνου φλοιὸν ἀποτέγγων ἢ μα-
ράθου· τοῦτο διδόναι πίνειν, καὶ πτισάνης χυλὸν
δὶς τῆς ἡμέρης, καὶ ἐπιπίνειν οἶνον λευκὸν ὑδαρέα. ᾗ δ' ἂν ἡ[1] ὀδύνη προσίστηται, καὶ[2] χλιαίνειν
θερμῷ καὶ λούειν, ἢν μὴ ὁ πυρετὸς πολὺς ἔχῃ.
ἐπὴν δὲ[3] τεσσερεσκαίδεκα ἡμέραι παρέλθωσιν,
ἀριστίζεσθαι μὲν τὸν κέγχρον, ἐς ἑσπέρην δὲ
κρέας[4] σκυλακίου ἢ ὀρνίθεια ἑφθὰ ἐσθίειν, καὶ τοῦ
ζωμοῦ ῥυφάνειν· σιτίοισι δὲ ὡς ἐλαχίστοισι χρῆσθαι τὰς πρώτας ἡμέρας.

57. Ἐπὴν φῦμα ᾖ[5] ἐν τῷ πλεύμονι, βὴξ ἔχει
καὶ ὀρθοπνοίη καὶ ὀδύνη ἐς τὸ στῆθος ὀξέη καὶ ἐς
τὰ πλευρά, καὶ ἐς τὰς[6] μὲν τεσσερεσκαίδεκα ἡμέ-
ρας·[7] τοῖσι γὰρ πλείστοισι τοσαύτας ἡμέρας μάλιστα φλεγμαίνει τὸ φῦμα. καὶ τὴν κεφαλὴν δὲ
ἀλγέει καὶ τὰ βλέφαρα, καὶ ὁρᾶν οὐ δύναται, καὶ
τὸ σῶμα ὑπόπυρρον γίνεται καὶ φλεβῶν ἐμπίμπλαται.

Τοῦτον[8] λούειν πολλῷ θερμῷ, καὶ μελίκρητον
διδόναι πίνειν ὑδαρές, καὶ τῆς πτισάνης τὸν
χυλὸν ῥυφάνειν, καὶ οἶνον ὑδαρέα ἐπιπίνειν. ἢν

---

[1] δ' ἂν ἡ Θ: ἂν Μ.   [2] καὶ om. Μ.   [3] Μ adds αἱ.
[4] Θ: κρέα Μ.   [5] ᾖ Θ: φυῇ Μ.   [6] Potter: ἔσται Θ: ἔστε Μ.
[7] Μ adds πάσχει.   [8] Μ: -το Θ.

when he survives for fourteen days, he recovers. Escape is not common.

Give the patient melicrat: boil it up in a new pot, cool, and soak celery or fennel bark in it; give this to drink, and also barley-water twice daily; afterwards have the patient drink dilute white wine. Wherever there is pain apply fomentations, and wash with hot water, unless great fever is present. After fourteen days, let the patient breakfast on millet, and towards evening eat boiled meat of puppy or fowl, and drink the sauce; on the first days let him have as little food as possible.

57. When a tubercle forms in the lung, coughing, orthopnoea, and sharp pains in the chest and sides are present for fourteen days; for in the majority of patients the tubercle is most swollen for that many days. The patient has pain in his head and eyelids, he cannot see, and his body becomes reddish and covered with vessels.

Wash this patient in copious hot water, and give him dilute melicrat to drink, barley-water gruel, and afterwards dilute wine. If the pain presses,

δ' ἡ ὀδύνη πιέζῃ, χλιαίνειν· ἐπὴν δὲ παύσηται,
σιτίοισιν ὡς μαλθακωτάτοισι χρῆσθαι. ἢν δ' ἀπηλ-
λαγμένον τῆς νούσου δυσπνοίη λαμβάνῃ, | ἐπὴν
πρὸς ὀρθὸν χωρίον ἴῃ ἢ σπεύσῃ τι ἄλλως, φάρμακον διδόναι, ὑφ' οὗ ἡ κοιλίη ἡ κάτω μὴ κινήσε-
ται. καὶ ἢν ἅμα τῷ πτύσματι[1] πύον ἕπηται, ἢν
μέντοι[2] τὸ πύον ᾖ λευκὸν καὶ ἶνες ἐν αὐτῷ ὕφαι-
μοι ἔωσιν, ἐκφυγγάνει· ἢν δὲ πελιδνὸν καὶ χλωρὸν καὶ κάκοδμον, ἀποθνήσκει. καθαίρονται δ' ἐν
τεσσεράκοντα ἡμέρῃσιν ἀφ' ἧς ἂν ῥαγῇ, πολλοῖσι
δὲ καὶ ἐνιαυσίη γίνεται ἡ νοῦσος· ποιέειν δὲ χρὴ
τοῦτον ἅπερ τὸν ἔμπυον. ἢν δὲ μὴ ῥαγῇ, ἐνίοισι
γὰρ τῷ χρόνῳ ἀφίσταται ἐς[3] τὸ πλευρὸν καὶ ἐξοι-
δίσκεται· τοῦτον χρή, ἢν τοιοῦτο γένηται, τάμνειν
ἢ καῦσαι.

58. Ἢν πλησθῇ ὁ πλεύμων, βὴξ ἴσχει καὶ ὀρθοπνοίη καὶ ἆσθμα, καὶ τὴν γλῶσσαν ἐκβάλλει,
καὶ πίμπραται[4] καὶ ὀδύνη ὀξέη ἴσχει ἐς τὸ στῆθος
καὶ ὑπὸ τὰς ὠμοπλάτας, καὶ καταπίμπλαται
φῴδων καὶ ψυγμὸς[5] ἔχει· καὶ οὐκ ἀνέχεται οὔτε
καθήμενος οὔτ' ἀνακείμενος οὔθ' ἑστηκώς, ἀλλὰ
δυσθετεῖ.[6] οὗτος τεταρταῖος μάλιστα ἀποθνήσκει·
ἢν δὲ ταύτας ὑπερφύγῃ, ἐλπίδες ὡς τὰ πολλὰ·
κινδυνεύει δὲ καὶ ἐν τῇσιν ἑπτά· ἢν δὲ καὶ ταύτας
ὑπερφύγῃ, ὑγιάζεται.

---

[1] Potter: ἐμέσματι ΘΜ.   [2] Θ: μὲν Μ.   [3] Ermerins: ὡς ΘΜ.
[4] Θ: πίμπλαται Μ.   [5] Θ: ξυσμὸς Μ.   [6] Potter: δυσθενεῖ ΘΜ.

foment him; when it stops, give very soft foods. If a patient that has been relieved of the disease experiences difficulty in breathing when he walks against rising ground, or exerts himself in any other way, give him a medication that does not set the lower cavity in motion. If together with the sputum there follows pus, and if the pus is white and contains bloody streaks, the patient escapes; but if the pus is livid, yellow-green and evil-smelling, he dies. Patients are cleaned in forty days from when the pus breaks out, and in many cases the disease lasts for a year; you must handle this patient the same as one with internal suppuration. If the pus does not break out—for in some patients after a time the pus migrates to the side and a swelling arises—you must, if this happens, incise or cauterize.

58. If the lung fills up, there are coughing, orthopnoea and panting, the patient protrudes his tongue, and he burns with fever; sharp pains occupy his chest and the region under his shoulder-blades, he is covered with blisters, and he becomes cold; he can tolerate neither sitting, nor lying, nor standing, but is greatly distresed. The patient usually dies on the fourth day; if he survives for that many, there is good hope, although he does remain in danger for seven days; if he escapes these, too, he recovers.

Τοῦτον ὅταν οὕτως ἔχῃ, λούειν[1] θερμῷ δὶς τῆς
ἡμέρης, καὶ ὅταν ἡ[2] ὀδύνη ἔχῃ, χλιάσματα προστιθέναι, καὶ πίνειν διδόναι μέλι καὶ ὄξος ἐφθόν,
καὶ ῥυφάνειν χυλὸν πτισάνης καὶ ἐπιπίνειν οἶνον.
ἢν δὲ πρὸς τὸ λουτρὸν καὶ τὰ χλιάσματα πονέῃ
καὶ μὴ ἀνέχηται, προσφέρειν αὐτῷ ψύγματα,[3]
καὶ πίνειν διδόναι κηρίον ἐν ὕδατι ἀποβρέχων ὡς
ψυχρότατον, καὶ κεῖσθαι πρὸς τὸ ψῦχος. ταῦτα
ποιέειν· ἡ δὲ νοῦσος χαλεπὴ καὶ θανατώδης.

92   59. Ἢν ὁ πλεύμων πρὸς τὸ πλευρὸν προσπέσῃ, βὴξ ἴσχει καὶ ὀρθοπνοίη, καὶ τὸ[4] σίαλον
βήσσεται λευκόν, καὶ ὀδύνη τὸ στῆθος καὶ τὸ μετάφρενον ἴσχει, καὶ ὠθέει προσκείμενος. καὶ δοκέει
τι ἐγκεῖσθαι βαρὺ ἐν τοῖσι στήθεσι, καὶ κεντέουσιν
ὀδύναι ὀξέαι, καὶ τρίζει [τὸ δέρμα][5] οἷον μάσθλης,
καὶ τὴν πνοὴν ἐπέχει. καὶ ἐπὶ μὲν τὸ πονέον
ἀνέχεται ἀνακείμενος,[6] ἐπὶ δὲ τὸ ὑγιὲς οὔ, ἀλλὰ
δοκέει τι αὐτῷ οἷον ἐκκρέμασθαι βαρὺ ἐκ τοῦ
πλευροῦ, καὶ διαπνεῖν δοκέει ἐκ[7] τοῦ στήθεος.

Τοῦτον λούειν πολλῷ θερμῷ δὶς τῆς ἡμέρης,
καὶ μελίκρητον πιπίσκειν· καὶ ἐκ τοῦ λουτροῦ,
οἶνον λευκὸν κεραννὺς καὶ μέλι ὀλίγον, καὶ δαύκου
καρπὸν τρίψας καὶ τῆς κενταυρίας, διεὶς τούτοισι,
διδόναι χλιαρὸν καταρρυφάνειν. καὶ προστιθέναι

---

[1] M adds πολλῷ καὶ.   [2] ἡ om. M.   [3] Θ: ψύγμα M.
[4] τὸ om. M.   [5] τὸ δέρμα (Θ) del. Littré: τὸ αἷμα M.
[6] Θ: κατα- M.   [7] Θ: διὰ M.

DISEASES II

When the case is such, wash this patient in hot water twice a day, and when pain is present apply fomentations; give him boiled honey and vinegar to drink, and have him take barley-water gruel, and after that wine. If, with the bath and the fomentations, the pain continues and the patient cannot stand it, apply cold compresses to the body, and give honeycomb steeped in very cold water to drink; also, have him lie exposed to the cold. Do these things; the disease is severe and often mortal.

59. If a lung falls against the side, there are expectoration and orthopnoea, white sputum is coughed up, pain occupies the chest and back, and the lung, lying against the side, exerts pressure. There seems to be something heavy lying inside the chest, sharp pains stab, a sound like leather is heard, and the breath is hindered. This patient will tolerate lying on his diseased side, but not on the healthy one, since then something heavy seems to hang down from the diseased side, and he seems to be breathing out of his chest.

Wash this patient in copious hot water twice a day, and have him drink melicrat; after his bath, mix white wine and a little honey, grind dauke and centaury seed, dissolve them in water, and have the patient drink this warm as a gruel. Pour warm

πρὸς τὸ πλευρὸν ἐς ἀσκίον ἢ ἐς βοείην κύστιν
ὕδωρ χλιαρὸν ἐγχέων, καὶ ταινίῃ συνδεῖν τὰ
στήθεα. καὶ κεῖσθαι ἐπὶ τὸ ὑγιές. καὶ τὸν χυλὸν
διδόναι τῆς πτισάνης χλιαρόν, καὶ ἐπιπίνειν
οἶνον ὑδαρέα.

Ἢν δὲ ἐκ τρώματος τοῦτο γένηται ἢ τμηθέντι
ἐμπύῳ, γίνεται γάρ, τούτῳ κύστιν[1] πρὸς σύριγγα
προσδήσας, ἐμπιμπλάναι τῆς φύσης καὶ ἐσιέναι
ἔσω· καὶ μοτὸν στερεὸν κασσιτέρινον ἐντιθέναι,
καὶ ἀπωθέειν πρόσω. οὕτω διαιτῶν τυγχάνοις ἂν
μάλιστα.

60. Ἐπὴν ἐν πλευρῷ φῦμα φυῇ,[2] βὴξ ἔχει
σκληρὴ καὶ ὀδύνη καὶ πυρετός· καὶ ἔγκειται βαρὺ
ἐν τῷ πλευρῷ, καὶ ὀδύνη ὀξέη ἐς τὸ αὐτὸ αἰεὶ χωρίον λαμβάνει. καὶ δίψα ἰσχυρή, καὶ ἀπερεύγεται
τὸ πῶμα θερμόν· καὶ ἐπὶ μὲν τὸ ἀλγέον οὐκ ἀνέχεται κατακείμενος, ἐπὶ δὲ τὸ ὑγιές· ἀλλ' ἐπὴν
κατακλίνῃ, δοκέει οἷόν | περ λίθος ἐκκρέμασθαι.
καὶ ἐξοιδέει, καὶ ἐξερύθει, καὶ οἱ πόδες οἰδέουσι.

Τοῦτον τάμνειν ἢ καίειν ἔπειτ' ἀφιέναι τὸ
πύον, ἔστ' ἂν γένηται δεκαταῖος, καὶ μοτοῦν
ὠμολίνῳ. ἐπὴν δὲ γένηται δεκαταῖος, ἐξιεὶς τὸ
πύον πᾶν, ἐσιέναι οἶνον καὶ ἔλαιον χλιήνας, ὡς
μὴ ἐξαπίνης ξηρανθῇ,[3] καὶ μοτοῦν ὀθονίῳ· ἐξιεὶς
δὲ τὸ ἐγκεχυμένον, ἐγχεῖν ἕτερον· ποιεῖν δὲ
ταῦτα πέντε ἡμέρας. ἐπὴν δὲ τὸ πύον λεπτὸν

[1] Μ: τοῦτο κύστι Θ.   [2] Θ: ᾖ Μ.   [3] Θ: ἀποξ- Μ.

water into a leather skin or cow's bladder, and apply this to the side; bind the chest with a bandage. Have the patient lie on his healthy side, give him warm barley-water to drink, and after that dilute wine.

If this condition has arisen as the result of a wound or from being incised for internal suppuration—for this happens—attach a pipe to a bladder, fill it with air, and place it in the opening; also introduce a solid tin tube, and force it forward.[1] By prescribing this regimen, you will be most successful.

60. When a tubercle forms in the side, harsh coughing, pain and fever are present; a heaviness lies in the side, and sharp pain presses continually in one place; there is a violent thirst, and the patient regurgitates what he drinks hot. He will not tolerate lying on his painful side, but prefers the healthy one; when he lies down, something like a stone seems to hang down from his side. The chest swells and becomes red, and the feet swell up.

Incise or cauterize this patient; then draw off pus until the tenth day, and plug the wound with a tent of raw linen. On the tenth day, draw out all the pus that remains, inject warm wine and oil to prevent the lung from suddenly becoming dry, and plug it with a tent of linen; draw out what was infused, and infuse anew; do this for five days.

[1] J. B. Gardeil (*Oeuvres d'Hippocrate*, Paris, 1855, II. 161 n. 1) notes that the purpose of this procedure is to prevent pleural adhesions.

ἀπορρυῇ οἷον πτισάνης χυλὸς καὶ ὀλίγον, καὶ
κολλῶδες[1] ἐν τῇ χειρὶ ψαυόμενον ᾖ, κασσιτέρινον
μοτὸν ἐντιθέναι, καὶ ἐπὴν παντάπασι ξηρανθῇ,
ἀποταμών τε τοῦ μοτοῦ ὀλίγον, αἰεὶ ξυμφύειν τὸ
ἕλκος πρὸς τὸν μοτόν.

61. Ἢν ὕδερος ἐν τῷ πλεύμονι γένηται, πυρετὸς καὶ βὴξ ἴσχει, καὶ ἀναπνέει ἀθρόον· καὶ οἱ
πόδες οἰδέουσι, καὶ οἱ ὄνυχες ἕλκονται πάντες,
καὶ πάσχει οἷά περ ἔμπυος γενόμενος, βληχρότε-
ρον δὲ καὶ πολυχρονιώτερον. καὶ ἢν ἐγχέῃς ἢ
θυμιᾷς ἢ πυριᾷς, οὐχ ὁμαρτέει πύον· τούτῳ ἂν
γνοίης ὅτι οὐ πύον, ἀλλὰ ὕδωρ ἐστί. καὶ ἢν
πολλὸν χρόνον προσέχων τὸ οὖς ἀκουάζῃ πρὸς
τὰ πλευρά, ζέει[2] ἔσωθεν οἷον ὄξος. καὶ ἕως μέν
τινος ταῦτα πάσχει· ἔπειτα δὲ ῥήγνυται ἐς τὴν
κοιλίην· καὶ αὐτίκα μὲν δοκέει ὑγιὴς εἶναι καὶ
τῆς νούσου ἀπηλλάχθαι. τῷ δὲ χρόνῳ ἡ κοιλίη
ἐμπίμπραται, καὶ τά τε αὐτὰ κεῖνα πάσχει καὶ
μᾶλλον· ἔνιοι δὲ καὶ οἰδίσκονται τὴν γαστέρα καὶ
τὴν ὄσχην καὶ τὸ πρόσωπον. καὶ ἔνιοι δοκέουσιν
εἶναι ἀπὸ τῆς κοιλίης τῆς κάτω, ὁρῶντες τὴν
γαστέρα μεγάλην καὶ τοὺς πόδας οἰδέοντας· οἰδίσκεται δὲ ταῦτα ἢν ὑπερβάλλῃς τὸν καιρὸν τῆς
τομῆς.

Τοῦτον χρή, ἢν μὲν ἀποιδήσῃ ἔξω, ταμόντα
διὰ τῶν πλευρῶν ἰᾶ|σθαι· ἢν δὲ μὴ ἀποιδέῃ, λού-

96

[1] καὶ κολλῶδες om. M. [2] Littré (*ebullit* Cornarius): ὄζει ΘΜ.
306

DISEASES II

When the pus flows out thin, like barley-water, and in a small amount, and it is viscous to the touch, insert a tin drainage tube; when the pus has dried up completely, cut off the tube a little at a time, and always unite the wound against the tube.

61. If dropsy arises in the lung, there are fever and coughing, and the patient respires rapidly; his feet swell, all his nails become curved, and he suffers the same things as a person that is suppurating internally, only more mildly and over a longer time. If you administer an infusion, fumigation, or vapour-bath, no pus appears; this is how you can tell that there is not pus, but water. If you apply your ear for a long time and listen to the sides, it seethes inside like vinegar. The patient suffers in the way described for a definite time; then there is a break into the cavity, and he at once appears to have recovered and to be free of the disease. With time, however, his cavity fills up, and he suffers the same things again, and more so; some patients also swell up in the belly, scrotum and face. Some people think that this disease originates from the lower cavity, when they see the belly large and the feet swollen, but, in fact, these parts only swell up if you let the proper time for incision go by.

If the patient swells towards the outside, you must treat him by incising between his ribs; if he does not swell, wash him in copious hot water, sit

307

σαντα πολλῷ θερμῷ καθίσαι ὥσπερ τοὺς ἐμπύους,
καὶ ὅπῃ ἂν ψοφέῃ, ταύτῃ τάμνειν· βούλεσθαι δὲ ὡς
κατωτάτω, ὅπως τοι εὔροον ᾖ. ἐπὴν δὲ τάμῃς,
μοτοῦν ὠμολίνῳ, παχὺν καὶ ἔπακρον ποιήσας
τὸν μοτόν· καὶ ἀφιέναι τοῦ ὕδατος[1] φειδόμενος
ὡς ἐλάχιστον.[2] καὶ ἢν μέν σοι περὶ[3] τῷ μοτῷ πύον
περιγένηται πεμπταίῳ ἐόντι ἢ ἑκταίῳ,[4] ὡς τὰ
πολλὰ ἐκφυγγάνει· ἢν δὲ μὴ περιγένηται, ἐπὴν
δ᾽[5] ἐξεράσῃς τὸ ὕδωρ, δίψα ἐπιλαμβάνει καὶ βήξ,
καὶ ἀποθνῄσκει.

62. Ἢν τὸ στῆθος καὶ[6] τὸ μετάφρενον ῥαγῇ,
ὀδύναι ἴσχουσι τὸ στῆθος καὶ τὸ μετάφρενον διαμ-
περές· καὶ θέρμη ἄλλοτε καὶ ἄλλοτε ἐπιλαμβάνει,
καὶ τὸ σίαλον ὕφαιμον βήσσεται, τὸ δ᾽ οἷον θρὶξ
διατρέχει διὰ τοῦ σιάλου αἱματώδης. μάλιστα δὲ
ταῦτα πάσχει, ἢν τῇσι χερσί τι πονήσῃ ἢ ἐπ᾽
ἄμαξαν ἐπιβῇ ἢ ἐφ᾽ ἵππον. τοῦτον καίειν
καὶ ἔμπροσθεν καὶ ἐξόπισθεν μοίρῃ ἴσῃ ἑκάτερον,
καὶ οὕτως ὑγιὴς γίνεται· ἐπισχεῖν δὲ τῶν πόνων
ἐνιαυτόν, καὶ παχῦναι ἐκ τῆς καύσιος.

63. Καυσώδης· πυρετὸς ἴσχει καὶ δίψα ἰσχυρή,
καὶ ἡ γλῶσσα τρηχέη καὶ μέλαινα καὶ χλωρὴ καὶ
ξηρὴ καὶ ἐξέρυθρος ἰσχυρῶς, καὶ οἱ ὀφθαλμοὶ χλω-
ροί· καὶ ἀποπατέει ἐρυθρὸν καὶ χλωρόν, καὶ οὐρέει
τοιοῦτο,[7] καὶ πτύει πολλόν. πολλάκις δὲ καὶ

[1] Μ: αἵματος Θ. [2] Θ: -τα Μ. [3] Θ: ἐν Μ. [4] ἢ ἑκταίῳ
om. Μ. [5] δ᾽ om. Μ. [6] Θ: ἢ Μ. [7] Μ: -τον Θ.

308

him down like those with internal suppuration, and
wherever the sound is heard, incise there; prefer to
make the incision as low down as possible, in order
to assure freedom of flow. After you have incised,
plug the wound with a thick pointed tent of raw
linen; draw off the fluid sparingly in very small
amounts. If on the fifth or sixth day you find pus
around the tent, the patient usually survives; if
not, when you have drawn off all the water, thirst
and coughing come on, and he dies.

62. If the chest and back develop tears, pains
permeate the chest and back through and through;
intermittent fever heat supervenes, and sputum
charged with blood—that is, with a bloody streak
running through it—is coughed up. The patient
suffers these things most if he exerts himself with
his arms or rides a horse or in a wagon. Cauterize
him both anteriorly and posteriorly in equal proportions, and he will recover. After the cautery, the
patient must follow a regimen that fattens, and
avoid exertions for a year.

63. Ardent fever: there are fever and violent
thirst, the tongue is rough, dark, yellow-green, dry,
and severely reddened, and the eyes are yellow-
green. The patient passes stools that are reddish
and yellow-green, urine the same, and he coughs up

μεθίσταται ἐς[1] περιπλευμονίην, καὶ παρακόπτει·
τούτῳ ἂν γνοίης ὅτι περιπλευμονίη γίνεται.
οὗτος ἢν μὲν <μὴ>[2] γένηται περιπλευμονικός, ἢν[3]
τεσσερεσκαίδεκα ἡμέρας ὑπερφύγη, ὑγιὴς γίνε-
ται· ἢν δὲ γένηται, | ἐν ὀκτωκαίδεκα ἡμέρῃσιν,
ἢν μὴ[4] ἀκάθαρτος γενόμενος ἔμπυος γένηται.
τοῦτον χρὴ πίνειν τὸ ἀπὸ τοῦ κρίμνου, καὶ μετα-
πίνειν ὄξος ὡς εὐωδέστατον λευκόν. καὶ ῥυφάνειν τὸν χυλὸν τῆς πτισάνης δὶς τῆς ἡμέρης, ἢν
δ᾽ ἀσθενὴς ᾖ, τρίς· καὶ ἐπιπίνειν οἶνον οἰνώδεα,
λευκόν, ὑδαρέα· καὶ λούειν ὡς ἐλαχίστῳ.[5] ἢν δ᾽
ἔμπυος γένηται, διαιτᾶν ὡς ἔμπυον.

64. Λυγγώδης· πυρετὸς ἴσχει σπερχνός, καὶ
ῥῖγος, καὶ βήξ, καὶ λύγξ, καὶ βήσσει ἅμα τῷ σιάλῳ θρόμβους αἵματος. καὶ ἑβδομαῖος ἀποθνήσκει·
ἢν[6] δέκα ἡμέρας ὑπερφύγη, ῥάων γίνεται,
εἰκοστῇ δ᾽ ἡμέρᾳ ἐμπυΐσκεται. καὶ βήσσει τὰς
πρώτας ἡμέρας πύον ὀλίγον, ἔπειτα ἐπὶ πλέον
καθαίρεται δ᾽ ἐν τεσσεράκοντα ἡμέρῃσι.

Τοῦτον τὰς μὲν πρώτας ἡμέρας πιπίσκειν τὸ
ὄξος καὶ τὸ μέλι ἑφθόν, καὶ μεταμίσγειν ὄξος καὶ
ὕδωρ ὑδαρὲς ποιέων· ῥυφάνειν δὲ χυλὸν πτισάνης
μέλι ὀλίγον παραμίσγων, καὶ οἶνον ἐπιπίνειν λευκὸν οἰνώδεα. ἐπὴν δὲ αἱ[7] δέκα ἡμέραι παρέλθωσιν, ἢν τὸ πῦρ παύσηται καὶ τὸ πτύαλον καθαρὸν

---

[1] Θ: ὡς Μ.   [2] Jouanna (p. 274).   [3] Θ: καὶ Μ.
[4] Μ: μὲν Θ.   [5] Θ: -τα Μ.   [6] Μ adds δὲ.   [7] αἱ om. Μ.

copious sputum. Often the disease changes into pneumonia, and produces derangement—it is by this latter that you can tell it has become pneumonia. If this patient does not become pneumonic, and he survives for fourteen days, he recovers; if he becomes pneumonic, he recovers in eighteen days, unless he becomes unclean and suppurates internally. You must have him drink water made from groats, and afterwards white vinegar of the most fragrant kind. Let him drink barley-water gruel twice a day—if he is weak, three times—and after that dilute strong white wine. Wash him with a very little water. If he suppurates internally, prescribe the regimen for that condition.

64. Disease with hiccups: there are pressing fever, chills, coughing and hiccups, and the patient coughs up clots of blood with his sputum. On the seventh day, he dies; if he survives for ten days, he is better, but on the twentieth day he suppurates internally. On the first days he coughs up little pus, later somewhat more; he is cleaned in forty days.

On the first days have this patient drink vinegar and boiled honey, mixing the vinegar with water so that it is dilute; as gruel have him drink barley-water, to which a little honey has been added, and afterwards strong white wine. After ten days, if the fever heat goes away and the sputum becomes

ᾖ, τὴν πτισάνην ὅλην ῥυφανέτω ἢ τὸν κέγχρον.
ἢν δ᾽ εἰκοσταῖος τὸ πύον πτύσῃ, πινέτω, κόψας
καὶ σήσας τὸν ἐλελίσφακον καὶ πήγανον καὶ θύμ-
βρην καὶ ὀρίγανον καὶ ὑπερικόν, ἴσον ἑκάστου
ξυμμίσγων, ὅσον σκαφίδα σμικρὴν ξυμπάντων,
καὶ ἀλφίτων τὸ αὐτό, ἐπ᾽ οἴνῳ γλυκεῖ κεκρημέ-
νῳ, νῆστι πίνειν. καὶ ῥυφανέτω ἢν χειμὼν ᾖ ἢ
μετόπωρον ἢ ἔαρ· ἢν δὲ θέρος ᾖ, μή· ἀλλ᾽
ἀμυγδάλια τρίβων καὶ σικύου σπέρμα πεφωγμέ-
νον καὶ σήσαμον ἴσον ἑκάστου, σύμπαν δὲ ὅσον
σκαφίδα, ἐπιχέας ὕδατος ὅσον κοτύλην Αἰγι-
ναίαν, ἄλητον ἐπιπάσσων καὶ κηρίον, τοῦτο
ῥυφανέτω μετὰ τὸ πῶμα. σιτίοισι δὲ χρήσθω
λιπαροῖσι καὶ ἁλμυροῖσι καὶ θαλασσίοισι μᾶλλον
ἢ κρέασι· καὶ λούσθω θερμῷ, τὴν δὲ[1] κεφαλὴν
ὡς ἐλάχιστα. ταῦτα ποιέων ἀπαλλάσσεται τῆς
νούσου.

100   65. Λήθαργος· βὴξ ἴσχει, καὶ τὸ σίαλον πτύει
πολὺ καὶ ὑγρόν· καὶ φλυηρεῖ, καὶ ὅταν παύσηται
φλυηρέων, εὕδει· καὶ ἀποπατέει κάκοδμον.
τοῦτον πιπίσκειν τὸ ἀπὸ τοῦ κρίμνου, καὶ μεταπιπίσκειν οἶνον οἰνώδεα λευκόν, καὶ ῥυφάνειν τὸν
χυλὸν τῆς πτισάνης· ξυμμίσγειν δὲ σίδης χυλόν,
καὶ οἶνον ἐπιπίνειν λευκὸν οἰνώδεα· καὶ μὴ λούειν.
οὗτος ἐν ἑπτὰ ἡμέρῃσιν ἀποθνήσκει· ἢν δὲ ταύτας
ὑπεκφύγῃ,[2] ὑγιὴς γίνεται.

¹ δὲ om. M.   ² Θ: ὑπερφύγῃ M.

clean, let him take as gruel whole barley or millet. If on the twentieth day he expectorates pus, have him drink the following: pound and sieve salvia, rue, savory, marjoram and hypericum—an equal amount of each, of all together as much as a small bowl—and mix these together with an equal amount of meal; have him drink this, in the fasting state, in sweet wine mixed with water. Let him take this as gruel in winter, fall or spring, but not in summer; in summer grind small almonds, toasted cucumber seeds, and sesame—an equal amount of each, all together a bowl—add an Aeginetan cotyle of water, sprinkle on flour and honeycomb, and let him take this as gruel after the potion of vinegar and boiled honey. Have the patient eat foods that are rich and salty, sea-foods more than meats; let him wash in hot water, but his head as little as possible. If he does these things, he is relieved of the disease.

65. Lethargy: the patient coughs up copious moist sputum; he talks nonsense, and, when he stops talking nonsense, he falls asleep; he passes ill-smelling stools. Have this patient drink water made from groats, and afterwards strong white wine; let him drink as gruel barley-water to which pomegranate juice has been added, and after that strong white wine. Do not wash him. This patient dies in seven days; if he survives for that many, he recovers.

66. Αὐαντή[1] οὐκ ἀνέχεται ἄσιτος οὐδὲ βε-
βρωκώς· ἀλλὰ ὅταν μὲν ἄσιτος ᾖ, τὰ σπλάγχνα
μύζει, καὶ καρδιώσσει, καὶ ἐμέει ἄλλοτε ἀλλοῖα,
καὶ χολὴν καὶ σίαλα καὶ λάππην καὶ δριμύ· καὶ
ἐπὴν ἐμέσῃ, ῥάων δοκέει εἶναι ἐπ' ὀλίγον. ἐπὴν
δὲ φάγῃ, ἐρυγμᾷ τε[2] καὶ φλογιᾷ, καὶ ἀποπατή-
σειν αἰεὶ οἴεται πολύ· ἐπὴν δὲ καθίζηται, φῦσα
ὑποχωρέει. καὶ τὴν κεφαλὴν ὀδύνη ἔχει, καὶ τὸ
σῶμα πᾶν ὥσπερ ῥαφὶς κεντέειν δοκέει ἄλλοτε
ἄλλῃ, καὶ τὰ σκέλεα βαρέα καὶ ἀσθενέα, καὶ
μινύθει καὶ ἀσθενὴς γίνεται.

Τοῦτον φάρμακα[3] πιπίσκειν, πρῶτον μὲν
κάτω, ἔπειτα ἄνω· καὶ τὴν κεφαλὴν καθαίρειν.
καὶ σιτίων ἀπέχεσθαι γλυκέων καὶ ἐλαιηρῶν καὶ
πιόνων, καὶ θωρηξίων. ἐμέειν δὲ[4] τοῖσι χυλοῖσι
καὶ ἀπὸ σιτίων, καὶ τὴν ὥρην ὄνου γάλα ἢ ὀρὸν
πιπίσκων, φάρμακον προσπῖσαι, ὁποτέρου ἄν[5] σοι
δοκέῃ μᾶλλον δεῖσθαι. ψυχρολουτέειν δὲ[6] τὸ θέρος
καὶ τὸ ἔαρ, τὸ φθινόπωρον δὲ καὶ τὸν χειμῶνα ἀλείμ-
ματι χρῆσθαι. καὶ περιπατέειν, καὶ γυμνάζεσθαι ὀλίγα· ἢν δ' ἀσθενέστερος ᾖ ὥστε γυμνάζε-
σθαι, ὁδοιπορίῃ χρῆσθαι. καὶ σιτίοισι ψυχροῖσι |
102 καὶ διαχωρητικοῖσι χρήσθω· καὶ ἢν ἡ γαστὴρ μὴ
ὑποχωρέῃ, ὑποκλύζειν κλύσματι μαλθακῷ. ἡ δὲ

---

[1] M: Λυαντή Θ.   [2] Θ: ἐρύγματα M.   [3] Θ: -κον M.
[4] Θ adds ἐν.   [5] ἄν om. M.   [6] δὲ om. M.

314

DISEASES II

66. Withering disease: the patient can tolerate neither fasting nor eating: when he does not eat, his inward parts rumble, he suffers pain in the cardia, and he vomits one time one thing, another time another thing: bile, sputum, scum and sharp substances; after he has vomited, for a short time he seems better. If he eats, he belches, becomes flushed, and continually has the feeling that he is about to pass copious stools, but, when he sits down, only wind passes. Pain occupies his head, and there seems to be a needle pricking him all through his body, sometimes here, sometimes there; his legs are heavy and weak, he wastes away, and he is powerless.

Have this patient drink medications, first those that act downwards, then ones that act upwards; also clean out his head. Let him refrain from foods that are sweet, oily and fat, and from drunkenness. Induce vomiting with fluids and foods; in season, have the patient drink ass's milk or whey, and also whichever medication you think he most needs. Have him bathe in cold water in summer and spring; in fall and winter he should be anointed. Have him take walks, and do a few exercises; if he is too weak for exercises, then let him take strolls. Have him eat cold laxative foods; if his belly does not pass anything, apply a gentle enema. The

νοῦσος χρονίη καὶ ἀπογηράσκοντας, ἢν μέλλῃ,
ἀπολείπει· ἢν δὲ μή, συναποθνήσκει.

67. Φονώδης· πυρετὸς ἴσχει καὶ ῥῖγος, καὶ αἱ
ὀφρύες ἐπικρέμασθαι δοκέουσι, καὶ τὴν κεφαλὴν
ἀλγέει· καὶ ἐμέει σίαλον θερμὸν καὶ χολὴν πολ-
λήν· ἐνίοτε καὶ κάτω ὑποχωρέει. καὶ τοὺς ὀφθαλμοὺς αἱ χῶραι οὐ χωρέουσι, καὶ ὀδύνη ἐς τὸν
αὐχένα καὶ ἐς τοὺς βουβῶνας[1] καὶ δυσθετεῖ καὶ
φλυηρεῖ. οὗτος ἑβδομαῖος ἢ πρότερον ἀπο-
θνήσκει· ἢν δὲ ταύτας ὑπερφύγῃ, τὰ πολλὰ ὑπεκ-
φυγγάνει·[2] ἡ δὲ νοῦσος θανατώδης.

Τούτῳ ψύγματα χρὴ προσίσχειν πρὸς τὰ
σπλάγχνα καὶ πρὸς τὴν κεφαλήν· καὶ πίνειν διδόναι ἐρίξαντα κάχρυς σὺν τοῖσιν ἀχύροισι [τὰ φύλ-
λα],[3] ἀποβρέχοντα, ἀπηθέοντα τὸ ὕδωρ, ἐν τούτῳ
μελίκρητον ποιέοντα· ὑδαρὲς τοῦτο διδόναι. σιτίον
δὲ μὴ προσφέρειν μηδὲ ῥύφημα ἑπτὰ ἡμερέων,
ἢν μὴ ἀσθενής τοι[4] δοκέῃ εἶναι· ἢν δ' ἀσθενὴς
ᾖ, χυλὸν πτισάνης ψυχρὸν καὶ λεπτὸν ὀλίγον
διδόναι δὶς τῆς ἡμέρης, καὶ[5] ἐπιπίνειν ὕδωρ.
ἐπὴν δὲ αἱ ἑπτὰ ἡμέραι παρέλθωσι καὶ τὸ πῦρ
μεθῇ, κέγχρον λείχειν· ἐς ἑσπέρην δὲ κολοκύντην
ἢ τεῦτλα ὀλίγα διδόναι, καὶ οἶνον λευκὸν ὑδαρέα
ἐπιπίνειν, ἔστ' ἂν γένηται ἐναταῖος. ἔπειτα σιτίῳ ὡς ἐλαχίστῳ διαχρήσθω, ἀριστιζόμενος κέγ-

---

[1] ἐς τ. β. Θ: τ. β. ἴσχει Μ.   [2] Θ: ἐκ- Μ.   [3] Deleted in the
printed editions except the Aldine.   [4] Θ: σοι Μ.   [5] καὶ om. Μ.

disease lasts a long time, and leaves, if at all, only when patients are growing old; otherwise, it continues on until their deaths.

67. Malignant disease: there are fever and chills, the eyebrows seem to overhang, and the patient suffers pain in his head; he vomits hot saliva and much bile; sometimes he also has a downward movement. The sockets do not have room for his eyes, and pain invades his neck and groins; he is greatly distressed and talks nonsense. This patient dies on the seventh day, or before; if he survives for that many, in most cases he escapes. The disease is mortal.

You must give this patient agents that cool the inward parts and head; have him drink the following: pound parched barley with its husks, steep it well, strain off the water, and make melicrat from this; give dilute. Do not administer food or gruel for seven days, unless you think the patient is weak; if he is, give him a little thin cold barley-water to drink twice a day, and afterwards water. When the seven days have passed and the fever goes away, have the patient take millet; towards evening give gourd or a few beets, and afterwards have him drink dilute white wine; do this until the ninth day. After that let him have as little food as

χρον· λουτρῷ δέ, ἔστ' ἂν μὲν ἡ ὀδύνη ἔχῃ καὶ ὁ
πυρετός, μὴ χρήσθω· ἐπὴν δὲ παύσηται, λούσθω[1]
πολλῷ. ἢν δὲ ἡ γαστὴρ μὴ ὑποχωρέῃ, ὑποκλύζειν κλύσματι μαλθακῷ, ἢ βαλάνους προστιθέναι.
ἐπὴν δὲ ἰσχύσῃ, προσθεὶς πρὸς τὰς ῥῖνας φάρμακον μαλθακόν, τὴν κοιλίην κάτω κάθηρον· ἔπειτα
γάλα ὄνου μετάπισον.

104   68. Πελίῃ πυρετὸς ἴσχει ξηρὸς καὶ φρὶξ ἄλλοτε καὶ ἄλλοτε, καὶ τὴν κεφαλὴν ἀλγέει, καὶ τὰ
σπλάγχνα ὀδύνη ἔχει, καὶ ἐμέει χολήν· καὶ[2] ὅταν
ἡ ὀδύνη ἔχῃ, οὐ δύναται ἀνορᾶν, ἀλλὰ βαρύνεται.
καὶ ἡ γαστὴρ σκληρὴ γίνεται, καὶ ἡ χροιὴ πελιδνή, καὶ τὰ χείλεα καὶ τῶν ὀφθαλμῶν πελιδνὰ
τὰ λευκά, καὶ ἐξορᾷ ὡς ἀγχόμενος. ἐνίοτε καὶ
τὴν χροιὴν μεταβάλλει, καὶ ἐκ πελιδνοῦ[3] ὑπόχλωρος γίνεται.

Τοῦτον φάρμακον πιπίσκειν καὶ κάτω καὶ
ἄνω, καὶ ὑποκλύζειν, καὶ ἀπὸ τῆς κεφαλῆς ἀπο-
καθαίρειν. καὶ θερμῷ ὡς ἥκιστα λούειν, ἀλλ'
ἐπὴν λούηται, ἐλειθερεῖν.[4] καὶ ὀρὸν τὴν ὥρην καὶ
γάλα ὄνου πιπίσκειν· καὶ σιτίοισιν ὡς μαλθακωτάτοισι χρῆσθαι καὶ ψυχροῖσιν, ἀπεχόμενον τῶν
δριμέων καὶ τῶν ἁλμυρῶν, λιπαρωτέροισι δὲ καὶ
γλυκυτέροισι καὶ πιοτέροισι χρήσθω. ἡ δὲ[5] νοῦσος
ὡς τὰ πολλὰ συναποθνήσκει.

---

[1] Θ adds μή.   [2] καὶ ἐμέει . . . καὶ om. Θ.   [3] Μ: ἐκπελιδνοῦ-
ται Θ.   [4] Potter (ἐλι- Μ[2]): ελειθερει Θ: ἐλιθέρει Μ.   [5] δὲ om. Μ.

DISEASES II

possible, and breakfast on millet; let him avoid the bath as long as pain and fever are present; when these stop, have him bathe in plenty of water. If his belly does not pass anything downwards, apply a gentle enema or suppositories. When the patient becomes strong, insert a mild medication into his nostrils, and clean out his lower cavity; after that have him drink ass's milk.

68. Livid disease: a dry fever comes on, occasional shivering, the patient suffers pain in his head and inward parts, and he vomits bile; when the pain is present, he cannot look up, but feels weighed down. His belly is costive, and his complexion, lips, and the whites of his eyes become livid; he stares as if he were being strangled. Sometimes his colour changes too, and turns from livid to yellow-green.

Give this patient potions that act both upwards and downwards, administer an enema, and clean out his head. Wash him in a very little hot water, and after he has been washed let him bask in the sun. In season, have him drink whey and ass's milk. Let him take foods that are as soft as possible and cold, avoid sharp and salty ones, but have those that are richer, sweeter and fatter. In most instances this disease continues until the person's death.

69. Έρυγματώδης· ὀδύνη λάζυται ὀξέη, καὶ
πονέει ἰσχυρῶς, καὶ ῥιπτάζει αὐτὸς ἑωυτόν, καὶ
βοᾷ· καὶ ἐρεύγεται θαμινά, καὶ ἐπὴν¹ ἀπερύγῃ,
δοκέει ῥάων εἶναι· πολλάκις δὲ καὶ χολὴν ἀπεμέει
ὀλίγην ὅσον βρόχθον. καὶ ὀδύνη λαμβάνει ἀπὸ
τῶν σπλάγχνων ἐς τὴν νείαιραν γαστέρα καὶ τὴν
λαπάρην, καὶ ἐπὴν τοῦτο γένηται, ῥάων δοκέει
εἶναι· καὶ ἡ γαστὴρ φυσᾶται καὶ σκληρὴ γίνεται
καὶ ψοφέει· καὶ ἡ φῦσα οὐ διαχωρέει οὐδὲ ὁ ἀπόπατος.

Τοῦτον ἐπὴν ἡ² ὀδύνη ἔχῃ, λούειν πολλῷ θερμῷ, καὶ χλιάσματα προστιθέναι. ὅταν δ᾽ ἐν τῇ
γαστρὶ ἡ ὀδύνη ᾖ καὶ ἡ φῦσα, ὑποκλύζειν· καὶ
τῆς λινοζώστιος ἑψῶν τὸν χυλὸν συμμίσγειν τῆς
πτισάνης τῷ χυλῷ, καὶ ἐπιπίνειν οἶνον | ὑδαρέα
γλυκύν· σιτίον δὲ μὴ προσφέρειν ἔστ᾽ ἂν ἡ ὀδύνη
χαλάσῃ. πινέτω δὲ ἓξ ἡμέρας, ἐκ νυκτὸς στέμφυλα βρέχων γλυκέα, τὸ ὕδωρ τὸ ἀπὸ τούτων· ἢν
δὲ μὴ ἔχῃ στέμφυλα, μέλι καὶ ὄξος ἑφθόν. ἐπὴν
δὲ τῆς ὀδύνης ἀποκινήσῃ, φαρμάκῳ τὴν κάτω
κοιλίην ἀποκαθαίρειν.³ σιτίοισι δὲ χρήσθω μαλθακοῖσι καὶ διαχωρητικοῖσι,⁴ καὶ θαλασσίοισι μᾶλλον
ἢ κρέασι, κρέασι δὲ ὀρνιθείοισι καὶ μηλείοισιν
ἑφθοῖσι· καὶ τεῦτλα καὶ κολοκύντην· τῶν δ᾽
ἄλλων ἀπέχεσθαι. ἡ δὲ νοῦσος ὅταν μὲν νέον

¹ Θ: ἐὰν Μ.   ² ἐπὴν ἡ Θ: ἢν Μ.   ³ Θ: καθ- Μ.   ⁴ Θ: ὑπο- Μ.

DISEASES II

69. Disease with belching: sharp pains afflict the patient, he is greatly distressed, he casts himself about, and he cries out. He belches frequently, and after belching he seems to be better; often he also vomits up a little bile—about a mouthful. Pain moves from the inward parts into the lower belly and flank, and when this happens the patient seems to be better; his belly puffs up, becomes costive, and makes sounds; but wind does not pass off below, nor do faeces.

When pain is present in this patient, wash him in copious hot water and apply fomentations. When there are pain and flatulence in his belly, administer an enema; boil the herb mercury, and mix this juice with barley-water for the patient to drink; afterwards let him drink dilute sweet wine; do not administer food until the pain slackens. For six days, let the patient soak sweet pressed grapes during the night, and drink the water from them; if pressed grapes are not available, then let him drink boiled honey and vinegar. When he has got over his pain, clean out his lower cavity with a medication. Let the patient have foods that are mild and laxative, sea-foods more than meats, of meats boiled fowl and mutton; also beets and gourd; have him abstain from the rest. When this disease

λάβῃ, χρόνῳ ἐξέρχεται· ἢν δὲ πρεσβύτερον, συναποθνήσκει.

70. Φλεγματώδης· λάζεται μὲν καὶ ἄνδρα, μᾶλλον δὲ γυναῖκα· καὶ παχέη μέν ἐστι καὶ εὔχρως, ὁδοιπορέουσα δὲ ἀσθενέει, μάλιστα δ᾽ ἐπὴν πρὸς αἶπος ἴῃ καὶ πυρετὸς λεπτὸς λαμβάνει, ἐνίοτε καὶ πνῖγμα. καὶ ἀπεμέει, ὅταν ἄσιτος ᾖ, χολὴν πολλὴν[1] καὶ σίαλα πολλά, πολλάκις δὲ καὶ ὅταν φάγῃ, τοῦ δὲ σιτίου οὐδέν. καὶ ὅταν πονήσῃ, ὀδυνᾶται ἄλλοτε ἄλλῃ τὸ στῆθος καὶ τὸ μετάφρενον, καὶ καταπίμπλαται πολφῶν ὡς ὑπὸ κνίδης.

Τοῦτον φάρμακον πιπίσκειν, καὶ ὀρὸν καὶ γάλα ὄνου πινέτω. ἢν δὲ ὀροπωτέῃ, προπῖσαι φάρμακον κάτω ὡς πλείστας ἡμέρας· καὶ ἢν ἀπολήγῃ τῆς ὀροπωτίης, μεταπιέτω γάλα ὄνειον. ἐπὴν δὲ πίνῃ, σιτίων μὲν ἀπεχέσθω· οἶνον δὲ πινέτω ὡς ἥδιστον, ἐπὴν παύσηται καθαιρόμενος. ἐπὴν δ᾽ ἀπολήξῃ τῆς πόσιος, ἀριστιζέσθω μὲν κέγχρον, ἐς[2] ἑσπέρην δὲ σιτίῳ ὡς μαλθακωτάτῳ χρήσθω καὶ ἐλαχίστῳ· ἀπεχέσθω δὲ πιόνων καὶ γλυκέων καὶ ἐλαιηρῶν.[3] καὶ ἄλλοτε καὶ ἄλλοτε, 108 τοῦ | χειμῶνος μάλιστα, ἀπεμείτω τῷ φακίῳ, λάχανα προτρώγων. καὶ θερμῷ ὡς ἥκιστα λούσθω, ἀλλὰ ἐλειθερείτω. ἡ δὲ νοῦσος συναποθνήσκει.

---

[1] πολλὴν om. M.   [2] ἐς om. M.   [3] M: ἐλατήρων Θ.

befalls a young person, in time it departs, but if it occurs in an older person, it remains until death.

70. Phlegmatic disease: it attacks men, but more often women. The patient is corpulent and has a good colour, but on walking is weak, especially when she walks against a grade; there is mild fever, sometimes also choking. When she goes without food, she vomits much bile and saliva, and often even when she eats, but never any of the food. When the patient exerts herself, she suffers pain at one time in one part of her chest and back, at another time in another part; she becomes covered with blisters as if from the stinging-nettle.

Have this patient drink a medication, and whey and ass's milk; if he is to drink whey, have him first drink medications that act downwards for several days; if he stops drinking whey, let him then drink ass's milk. As long as the patient is drinking, have him abstain from foods, but drink very sweet wine once he is no longer being cleaned. When he stops drinking, let him breakfast on millet, and towards evening have a very small amount of very soft food; let him avoid fat, sweet and oily foods. From time to time, especially in winter, have the patient vomit by employing a decoction of lentils after eating vegetables. Let him wash in as little hot water as possible, and then bask in the sun. The disease remains with the patient until his death.

ΠΕΡΙ ΝΟΥΣΩΝ Β

71. Φλέγμα λευκόν· οἰδέει ἅπαν τὸ σῶμα λευκῷ οἰδήματι, καὶ ἡ γαστὴρ παχέη ψαυομένη, καὶ οἱ πόδες καὶ οἱ μηροὶ οἰδέουσι καὶ αἱ κνῆμαι καὶ ἡ ὄσχη. καὶ ἀναπνεῖ ἀθρόον, καὶ τὸ πρόσωπον ἐνερευθές, καὶ τὸ στόμα ξηρόν, καὶ δίψα ἴσχει, καὶ ἐπὴν φάγῃ, τὸ πνεῦμα πυκινὸν ἐπιπίπτει. οὗτος τῆς αὐτῆς ἡμέρης τοτὲ μὲν ῥάων γίνεται, τοτὲ δὲ κάκιον ἴσχει.

Τούτῳ δὲ[1] ἢν μὲν ἡ γαστὴρ ταραχθῇ αὐτομάτη ἀρχομένης τῆς νούσου, ἐγγυτάτω ὑγιὴς γίνεται· ἢν δὲ μὴ ταραχθῇ, φάρμακον διδόναι κάτω, ὑφ' οὗ ὕδωρ καθαρεῖται. καὶ θερμῷ μὴ λούειν, καὶ πρὸς τὴν αἰθρίην κομίζειν, καὶ τὴν ὄσχην ἀποτύπτειν, ἐπὴν πιμπρῆται. σιτίοισι δὲ χρῆσθαι ἄρτῳ καθαρῷ ψυχρῷ καὶ τεύτλοισι καὶ σκορπίοις ἐφθοῖσι καὶ σελάχεσι καὶ κρέασι τετρυμένοισι μηλείοισιν ἐφθοῖσι·[2] τῷ δὲ ζωμῷ ὡς ἐλαχίστῳ καὶ ψυχρὰ πάντα, καὶ μὴ[3] γλυκέα μηδὲ λιπαρά, ἀλλὰ τετρυμένα καὶ ὀξέα καὶ δριμέα, πλὴν σκορόδου ἢ κρομμύου ἢ πράσου. ὀρίγανον δὲ καὶ θύμβραν πολλὴν ἐσθίειν· καὶ οἶνον ἐπιπίνειν οἰνώδεα, καὶ ὁδοιπορέειν πρὸ τοῦ σιτίου. ἢν δ' ὑπὸ τῶν φαρμάκων οἰδίσκηται, κλύζειν, καὶ τῷ σιτίῳ πιέζειν καὶ περιπάτοισι καὶ ἀλουσίῃσι[4] φάρμακα δὲ ὡς ἐλάχιστα δοῦναι, ἄνω δὲ μηδ' ἔμπροσθεν ἢ τὰ

---

[1] δὲ om. M.   [2] καὶ σελάχεσι ... ἐφθοῖσι om. M.   [3] μὴ om. Θ.
[4] Θ: -σίη M.

324

DISEASES II

71. White phlegm: the whole body swells up
with a white swelling, the belly feels stout to the
touch, and the feet and thighs swell, the legs below
the knees, and the scrotum. The patient breathes
rapidly, his face becomes flushed, and his mouth is
dry; he is thirsty, and when he eats he falls prey to
rapid breathing. On one and the same day this
patient is at one time better, at another time worse.

If the belly is set in motion spontaneously at the
beginning of the disease, the patient recovers very
soon; if it is not set in motion, give a medication
that will clean water downwards. Do not wash the
patient in hot water; expose him to the air; and
incise his scrotum, when it is distended. As food
give cold white bread, beets, boiled scorpion fish,
selachians, and boiled minced mutton—but as little
sauce as possible—all cold; give nothing sweet or
rich, but whatever is minced, acid and sharp,
except for garlic, onion and leek; have the patient
eat much marjoram and savory. After his meals let
him drink strong wine, and before them take walks.
If the patient swells up from the medications,
administer an enema, and squeeze him out by
means of food, walks, and abstinence from the bath;
give as few medications as possible, and none that
acts upwards, before the swellings have moved

οἰδήματα κατέλθῃ ἐς τὰ[1] κάτω. ἢν δὲ ἰσχνοῦ
ἤδη ἐόντος πνῖγμα ἐν τοῖσι στήθεσιν ἐγγίνηται,
ἐλλέβορον δὸς πιεῖν, καὶ τὴν κεφαλὴν καθῆραι,
κἄπειτα κάτω πῖσαι. ἡ δὲ νοῦσος μάλιστα δια-
κρίνει ἐν οὐδενί.

72. Φρενῖτις[2] δοκεῖ ἐν τοῖσι σπλάγχνοισιν
110   εἶναι | οἷον ἄκανθα καὶ κεντέειν, καὶ ἄση αὐτὸν
λάζυται· καὶ τὸ φῶς φεύγει καὶ τοὺς ἀνθρώπους,
καὶ τὸ σκότος φιλέει, καὶ φόβος λάζεται. καὶ αἱ
φρένες οἰδέουσιν ἐκτός, καὶ ἀλγέει ψαυόμενος.
καὶ φοβεῖται, καὶ δείματα ὁρᾷ καὶ ὀνείρατα φοβε-
ρὰ καὶ τοὺς τεθνηκότας ἐνίοτε. καὶ ἡ νοῦσος
[ἐνίοτε][3] λαμβάνει τοὺς πλείστους τοῦ ἦρος.

Τοῦτον πιπίσκειν ἐλλέβορον, καὶ τὴν κεφαλὴν
καθαίρειν· καὶ μετὰ τὴν κάθαρσιν τῆς κεφαλῆς
κάτω πῖσαι φάρμακον, καὶ μετὰ ταῦτα πίνειν
γάλα ὄνου. σιτίοισι δὲ χρῆσθαι ὡς ἐλαχίστοισιν,
ἢν μὴ ἀσθενὴς ᾖ, καὶ ψυχροῖσι[4] διαχωρητικοῖσι
καὶ μὴ δριμέσι μηδ' ἁλμυροῖσι μηδὲ λιπαροῖσι
μηδὲ γλυκέσι. μηδὲ θερμῷ λούσθω, μηδ' οἶνον
πινέτω, ἀλλὰ μάλιστα μὲν ὕδωρ· εἰ δὲ μή, οἶνον
ὑδαρέα· μηδὲ γυμναζέσθω, μηδὲ περιπατείτω.
ταῦτα ποιέων ἀπαλλάσσεται τῆς νούσου χρόνῳ·
ἢν δὲ μὴ ἐπιμελήσῃ, συναποθνήσκει.

---

[1] Θ: τὸ M.   [2] Potter: Φροντίς ΘM.   [3] Del. Ermerins
after Littré.   [4] M adds καὶ.

down to the lower parts. If, when the swelling has already gone down, choking occurs in the person's chest, give him hellebore to drink, clean out his head, and then have him drink a medication that acts downwards. This disease rarely has a crisis in a patient.

72. Phrenitis: something like a thorn seems to be in the inward parts and to prick them; loathing attacks the patient, he flees light and people, he loves the dark, and he is seized by fear. His diaphragm swells outwards, and is painful when touched. The patient is afraid, and he sees terrible things, frightful dreams, and sometimes the dead. This disease attacks most people in spring.

Give the patient hellebore to drink, and clean out his head; after you have cleaned the head, have him drink a medication to act downwards, and after that ass's milk. Give as few foods as possible, unless the patient is weak, and ones that are cold and laxative, but not sharp, salty, rich or sweet. He should not bathe in hot water, and he should not drink wine, but preferably water; if not water, then dilute wine. Let him not take exercises or walks. If the patient follows these instructions, in time he will recover from the disease; but if you do not take care of him, the disease continues until his death.

73. Μέλαινα· μέλαν ἐμέει οἷον τρύγα, τοτὲ δὲ αἱματῶδες, τοτὲ δὲ οἷον οἶνον τὸν δεύτερον, τοτὲ δὲ οἷον πωλύπου θολόν, τοτὲ δὲ δριμὺ οἷον ὄξος, τοτὲ δὲ σίαλον καὶ λάππην, τοτὲ δὲ χολὴν χλωρήν. καὶ ὅταν μὲν[1] μέλαν καὶ τὸ αἱματῶδες ἐμέῃ, δοκέει οἷον φόνου ὄζειν, καὶ ἡ φάρυγξ καὶ τὸ στόμα καίεται ὑπὸ τοῦ ἐμέσματος, καὶ τοὺς ὀδόντας αἱμωδιᾷ, καὶ τὸ ἔμεσμα τὴν γῆν αἴρει. καὶ ἐπὴν ἀπεμέσῃ, δοκέει ῥάων εἶναι ἐπ' ὀλίγον. καὶ οὐκ ἀνέχεται[2] οὔτ' ἄσιτος οὔθ' ὁπόταν πλέον βεβρώκῃ, ἀλλ' ὁπόταν μὲν ἄσιτος ᾖ, τὰ σπλάγχνα μύζει, καὶ τὰ σίαλα ὀξέα· ὅταν δέ τι φάγῃ, βάρος ἐπὶ τοῖσι σπλάγχνοισι, καὶ τὸ στῆθος καὶ τὸ μετάφρενον δοκέει οἷον γραφείοισι[3] κεντεῖσθαι. καὶ τὰ πλευρὰ ἔχει ὀδύνη, καὶ πυρετὸς βληχρός, καὶ τὴν κεφαλὴν ἀλγέει, | καὶ τοῖσιν ὀφθαλμοῖσιν οὐχ ὁρᾷ· καὶ τὰ σκέλεα βαρέα, καὶ ἡ χροιὴ μέλαινα, καὶ μινύθει.

Τοῦτον φάρμακον πιπίσκειν θαμὰ καὶ ὀρὸν καὶ γάλα τὴν ὥρην· καὶ σιτίων ἀπέχειν γλυκέων καὶ ἐλαιηρῶν καὶ πιόνων, καὶ χρῆσθαι ὡς ψυχροτάτοισι καὶ ὑποχωρητικωτάτοισι. καὶ τὴν κεφαλὴν καθαίρειν· καὶ μετὰ τὰς φαρμακοποσίας τὰς ἄνω ἀπὸ τῶν χειρῶν τοῦ αἵματος ἀφιέναι, ἢν μὴ ἀσθενὴς ᾖ. ἢν δ' ἡ κοιλίη μὴ ὑποχωρέῃ, ὑποκλύ-

---

[1] M adds τὸ.   [2] Θ: ἄχθεται M.   [3] M: γραφίοισι Θ: ῥαφίοισι later mss, Littré.

DISEASES II

73. Dark disease: the patient vomits up dark material that is like the lees of wine, sometimes like blood, sometimes sharp like vinegar, sometimes saliva and scum, sometimes yellow-green bile. When he vomits dark bloody material, it smells of gore, his throat and mouth are burned by the vomitus, his teeth are set on edge, and the vomitus raises the earth.[1] After he has vomited, for a short time the patient seems better. He can tolerate neither fasting nor eating too much; for when he does not eat, his inward parts rumble and his saliva is acid; but when he eats something, there is a heaviness in his inward parts, and his chest and back seem to be being pricked by styluses. Pains occupy his sides, there is a mild fever, he has a headache, and he is unable to see; his legs are heavy, his complexion is dark, and he wastes away.

Have this patient drink frequent medications, and in season whey and milk; let him refrain from foods that are sweet, oily and fat, but have as cold and laxative ones as possible. Clean out his head. After he has drunk medications to act upwards, draw blood from his arms, unless he is weak. If his cavity does not pass anything downwards,

[1] I take this statement literally in the sense "causes the earth to froth up or bubble".

ζειν μαλθακῷ κλύσματι. καὶ θωρηξίων ἀπέχεσθαι καὶ λαγνείης· ἢν δὲ λαγνεύῃ, νῆστις καὶ[1]
πυριᾶσθαι· καὶ τοῦ ἡλίου ἀπέχεσθαι, μηδὲ γυμνάζεσθαι πολλά, μηδὲ περιπατεῖν, μηδὲ θερμολουτέειν, μηδὲ δριμέα ἐσθίειν μηδὲ ἀλυκά. ταῦτα
ποιέων[2] ἅμα τῇ ἡλικίῃ ἀποφεύγει, καὶ ἡ νοῦσος
καταγηράσκει ἐν[3] τῷ σώματι· ἢν δὲ μὴ μελεδαν-
θῇ, συναποθνήσκει.

74. Ἄλλη μέλαινα· ὑπόπυρρος καὶ ἰσχνὸς καὶ
τοὺς ὀφθαλμοὺς ὑπόχλωρος γίνεται· καὶ λεπτό-
δερμος καὶ ἀσθενὴς τελέθει. ὅσῳ δ' ἂν χρόνος
πλείων ᾖ, ἡ νοῦσος μᾶλλον πονέει. καὶ ἐμέει
πᾶσαν ὥρην οἷον σταλαγμὸν ὀλίγον, κατὰ δύο
βρόχθους, καὶ τὸ σιτίον θαμινά, καὶ σὺν τῷ σιτίῳ
χολὴν καὶ φλέγμα· καὶ μετὰ τὴν ἔμεσιν ἀλγέει τὸ
σῶμα πᾶν, ἔστι δ' ὅτε καὶ πρὶν ἐμέσαι· καὶ φρῖκαι
λεπταὶ καὶ πυρετὸς ἴσχει· καὶ πρὸς τὰ γλυκέα καὶ
ἐλαιώδεα μάλιστα ἐμέει.

Τοῦτον καθαίρειν χρὴ φαρμάκοισι[4] κάτω καὶ
ἄνω, καὶ μεταπιπίσκειν γάλα ὄνου· καὶ σιτίοισι
χρῆσθαι ὡς μαλθακωτάτοισι καὶ ψυχροῖσιν, ἰχθύ-
σιν ἀκταίοις καὶ σελάχεσι καὶ τεύτλοις καὶ κολοκύντῃ καὶ κρέασι τετρυμένοις, οἶνον δὲ πίνειν
λευκὸν οἰνῶδες ὑδαρέστερον· ταλαιπωρίῃ δὲ περιπάτοισι χρῆσθαι, καὶ μὴ θερμολουτέειν, καὶ τοῦ

---

[1] καὶ om. M.   [2] Θ: ποιέειν M.   [3] Θ: σὺν M.   [4] M adds καὶ.

administer a gentle enema. The patient must
abstain from drunkenness and venery; however, if
he does engage in venery, let him take a vapour-
bath in the fasting state. He should also avoid the
sun, and not do too many exercises, go for walks,
take hot baths, or eat sharp or salty foods. If he follows this regimen, as he reaches the prime of his
life he escapes, and the disease grows old in his
body; but, if he is not cared for, the disease continues until his death.

74. Another dark disease: the patient becomes
reddish and lean, and his eyes are yellow-green; his
skin becomes thin, and he is weak. The more time
goes on, the more severe the disease becomes. The
patient continually vomits up a few drops, two
mouthfuls at a time, frequently food, and with the
food bile and phlegm; after vomiting, he suffers
pain through his whole body, sometimes even
before he vomits; there is mild shivering, and fever.
The patient vomits most from sweet and oily foods.

You must clean out this patient with medications that act downwards and upwards, and afterwards have him drink ass's milk. Have him eat
very mild cold foods: of fish those of the coast and
selachians; also beets, gourd, and minced meats.
As wine let him drink a strong white, quite dilute.
Have him exercise by taking walks, and let him go

ἡλίου ἀπέχεσθαι. ταῦτα ποιέειν, ἡ δὲ νοῦσος θανατώδης μὲν οὔ, ξυγκαταγηράσκει δέ.

114   75. Σφακελώδης· τὰ μὲν ἄλλα ταὐτὰ πάσχει, ἐμέει δὲ θρόμβους πεπηγότας χολῆς, καὶ κάτω ὁμοιοῦται ἐπὴν τὰ σιτία ἀποπατήσῃ. δρᾶν δὲ χρὴ τὰ αὐτὰ ἅπερ ἐπὶ τῆς προτέρης, καὶ ὑποκλύζειν.

without the hot bath and avoid the sun. Do these things; the disease is seldom mortal, but grows old together with the patient.

75. Sphacelous disease: what this patient suffers is the same, except that in his vomitus there are congealed clots of bile, and likewise below when he evacuates what he has eaten. You must do the same as in the preceding disease, and administer an enema.

THE LOEB CLASSICAL LIBRARY

VOLUMES ALREADY PUBLISHED

Latin Authors

AMMIANUS MARCELLINUS. J. C. Rolfe. 3 Vols.

APULEIUS: THE GOLDEN ASS (METAMORPHOSES). W. Adlington (1566). Revised by S. Gaselee.

ST. AUGUSTINE: CITY OF GOD. 7 Vols. Vol. I. G. E. McCracken. Vols. II and VII. W. M. Green. Vol. III. D. Wiesen. Vol. IV. P. Levine. Vol. V. E. M. Sanford and W. M. Green. Vol. VI. W. C. Greene.

ST. AUGUSTINE, CONFESSIONS. W. Watts (1631). 2 Vols.

ST. AUGUSTINE, SELECT LETTERS. J. H. Baxter.

AUSONIUS. H. G. Evelyn White. 2 Vols.

BEDE. J. E. King. 2 Vols.

BOETHIUS: TRACTS and DE CONSOLATIONE PHILOSOPHIAE. Rev. H. F. Stewart and E. K. Rand. Revised by S. J. Tester.

CAESAR: ALEXANDRIAN, AFRICAN and SPANISH WARS. A. G. Way.

CAESAR: CIVIL WARS. A. G. Peskett.

CAESAR: GALLIC WAR. H. J. Edwards.

CATO: DE RE RUSTICA. VARRO: DE RE RUSTICA. H. B. Ash and W. D. Hooper.

CATULLUS. F. W. Cornish. TIBULLUS. J. B. Postgate. PERVIGILIUM VENERIS. J. W. Mackail. Revised by G. P. Goold.

CELSUS: DE MEDICINA. W. G. Spencer. 3 Vols.

CICERO: BRUTUS and ORATOR. G. L. Hendrickson and H. M. Hubbell.

[CICERO]: AD HERENNIUM. H. Caplan.

CICERO: DE ORATORE, etc. 2 Vols. Vol. I. DE ORATORE, Books I and II. E. W. Sutton and H. Rackham. Vol. II. DE ORATORE, Book III. DE FATO; PARADOXA STOICORUM; DE PARTITIONE ORATORIA. H. Rackham.

CICERO: DE FINIBUS. H. Rackham.

CICERO: DE INVENTIONE, etc. H. M. Hubbell.

CICERO: DE NATURA DEORUM and ACADEMICA. H. Rackham.

CICERO: DE OFFICIIS. Walter Miller.

CICERO: DE RE PUBLICA and DE LEGIBUS. Clinton W. Keyes.

Cicero: De Senectute, De Amicitia, De Divinatione. W. A. Falconer.

Cicero: In Catilinam, Pro Flacco, Pro Murena, Pro Sulla. New version by C. Macdonald.

Cicero: Letters to Atticus. E. O. Winstedt. 3 Vols.

Cicero: Letters to His Friends. W. Glynn Williams, M. Cary, M. Henderson. 4 Vols.

Cicero: Philippics. W. C. A. Ker.

Cicero: Pro Archia, Post Reditum, De Domo, De Haruspicum Responsis, Pro Plancio. N. H. Watts.

Cicero: Pro Caecina, Pro Lege Manilia, Pro Cluentio, Pro Rabirio. H. Grose Hodge.

Cicero: Pro Caelio, De Provinciis Consularibus, Pro Balbo. R. Gardner.

Cicero: Pro Milone, In Pisonem, Pro Scauro, Pro Fonteio, Pro Rabirio Postumo, Pro Marcello, Pro Ligario, Pro Rege Deiotaro. N. H. Watts.

Cicero: Pro Quinctio, Pro Roscio Amerino, Pro Roscio Comoedo, Contra Rullum. J. H. Freese.

Cicero: Pro Sestio, In Vatinium. R. Gardner.

Cicero: Tusculan Disputations. J. E. King.

Cicero: Verrine Orations. L. H. G. Greenwood. 2 Vols.

Claudian. M. Platnauer. 2 Vols.

Columella: De Re Rustica. De Arboribus. H. B. Ash, E. S. Forster and E. Heffner. 3 Vols.

Curtius, Q.: History of Alexander. J. C. Rolfe. 2 Vols.

Florus. E. S. Forster.

Frontinus: Stratagems and Aqueducts. C. E. Bennett and M. B. McElwain.

Fronto: Correspondence. C. R. Haines. 2 Vols.

Gellius. J. C. Rolfe. 3 Vols.

Horace: Odes and Epodes. C. E. Bennett.

Horace: Satires, Epistles, Ars Poetica. H. R. Fairclough.

Jerome: Selected Letters. F. A. Wright.

Juvenal and Persius. G. G. Ramsay.

Livy. B. O. Foster, F. G. Moore, Evan T. Sage, and A. C. Schlesinger and R. M. Geer (General Index). 14 Vols.

Lucan. J. D. Duff.

Lucretius. W. H. D. Rouse. Revised by M. F. Smith.

Manilius. G. P. Goold.

Martial. W. C. A. Ker. 2 Vols. Revised by E. H. Warmington

Minor Latin poets: from Publilius Syrus to Rutilius Namatianus, including Grattius, Calpurnius Siculus, Nemesianus, Avianus and others, with "Aetna" and the "Phoenix." J. Wight Duff and Arnold M. Duff. 2 Vols.

Minucius Felix. Cf. Tertullian.

NEPOS, CORNELIUS. J. C. Rolfe.

OVID: THE ART OF LOVE and OTHER POEMS. J. H. Mozley. Revised by G. P. Goold.

OVID: FASTI. Sir James G. Frazer. Revised by G. P. Goold.

OVID: HEROIDES and AMORES. Grant Showerman. Revised by G. P. Goold.

OVID: METAMORPHOSES. F. J. Miller. 2 Vols. Revised by G. P. Goold.

OVID: TRISTIA and EX PONTO. A. L. Wheeler. Revised by G. P. Goold.

PERSIUS. Cf. JUVENAL.

PERVIGILIUM VENERIS. Cf. CATULLUS.

PETRONIUS. M. Heseltine. SENECA: APOCOLOCYNTOSIS. W. H. D. Rouse. Revised by E. H. Warmington.

PHAEDRUS and BABRIUS (Greek). B. E. Perry.

PLAUTUS. Paul Nixon. 5 Vols.

PLINY: LETTERS, PANEGYRICUS. Betty Radice. 2 Vols.

PLINY: NATURAL HISTORY. 10 Vols. Vols. I.–V. and IX. H. Rackham. VI.–VIII. W. H. S. Jones. X. D. E. Eichholz.

PROPERTIUS. H. E. Butler.

PRUDENTIUS. H. J. Thomson. 2 Vols.

QUINTILIAN. H. E. Butler. 4 Vols.

REMAINS OF OLD LATIN. E. H. Warmington. 4 Vols. Vol. I. (ENNIUS AND CAECILIUS) Vol. II. (LIVIUS, NAEVIUS PACUVIUS, ACCIUS) Vol. III. (LUCILIUS and LAWS OF XII TABLES) Vol. IV. (ARCHAIC INSCRIPTIONS).

RES GESTAE DIVI AUGUSTI. Cf. VELLEIUS PATERCULUS.

SALLUST. J. C. Rolfe.

SCRIPTORES HISTORIAE AUGUSTAE. D. Magie. 3 Vols.

SENECA, THE ELDER: CONTROVERSIAE, SUASORIAE. M. Winterbottom. 2 Vols.

SENECA: APOCOLOCYNTOSIS. Cf. PETRONIUS.

SENECA: EPISTULAE MORALES. R. M. Gummere. 3 Vols.

SENECA: MORAL ESSAYS. J. W. Basore. 3 Vols.

SENECA: TRAGEDIES. F. J. Miller. 2 Vols.

SENECA: NATURALES QUAESTIONES. T. H. CORCORAN. 2 VOLS.

SIDONIUS: POEMS and LETTERS. W. B. Anderson. 2 Vols.

SILIUS ITALICUS. J. D. Duff. 2 Vols.

STATIUS. J. H. Mozley. 2 Vols.

SUETONIUS. J. C. Rolfe. 2 Vols.

TACITUS: DIALOGUS. Sir Wm. Peterson. AGRICOLA and GERMANIA. Maurice Hutton. Revised by M. Winterbottom, R. M. Ogilvie, E. H. Warmington.

TACITUS: HISTORIES and ANNALS. C. H. Moore and J. Jackson. 4 Vols.

TERENCE. John Sargeaunt. 2 Vols.

TERTULLIAN: APOLOGIA and DE SPECTACULIS. T. R. Glover. MINUCIUS FELIX. G. H. Rendall.

3

TIBULLUS. Cf. CATULLUS.
VALERIUS FLACCUS. J. H. Mozley.
VARRO: DE LINGUA LATINA. R. G. Kent. 2 Vols.
VELLEIUS PATERCULUS and RES GESTAE DIVI AUGUSTI. F. W. SHIPLEY.
VIRGIL. H. R. Fairclough. 2 Vols.
VITRUVIUS: DE ARCHITECTURA. F. Granger. 2 Vols.

Greek Authors

ACHILLES TATIUS. S. Gaselee.
AELIAN: ON THE NATURE OF ANIMALS. A. F. Scholfield. 3 Vols.
AENEAS TACTICUS. ASCLEPIODOTUS and ONASANDER. The Illinois Greek
    Club.
AESCHINES. C. D. Adams.
AESCHYLUS. H. Weir Smyth. 2 Vols.
ALCIPHRON, AELIAN, PHILOSTRATUS: LETTERS. A. R. Benner and F. H.
    Fobes.
ANDOCIDES, ANTIPHON. Cf. MINOR ATTIC ORATORS Vol. I.
APOLLODORUS. Sir James G. Frazer. 2 Vols.
APOLLONIUS RHODIUS. R. C. Seaton.
APOSTOLIC FATHERS. Kirsopp Lake. 2 Vols.
APPIAN: ROMAN HISTORY. Horace White. 4 Vols.
ARATUS. Cf. CALLIMACHUS.
ARISTIDES: ORATIONS. C. A. Behr.
ARISTOPHANES. Benjamin Bickley Rogers. 3 Vols. Verse trans.
ARISTOTLE: ART OF RHETORIC. J. H. Freese.
ARISTOTLE: ATHENIAN CONSTITUTION, EUDEMIAN ETHICS, VICES AND
    VIRTUES. H. Rackham.
ARISTOTLE: GENERATION OF ANIMALS. A. L. Peck.
ARISTOTLE: HISTORIA ANIMALIUM. A. L. Peck. Vols. I.–II.
ARISTOTLE: METAPHYSICS. H. Tredennick. 2 Vols.
ARISTOTLE: METEOROLOGICA. H. D. P. Lee.
ARISTOTLE: MINOR WORKS. W. S. Hett. On Colours, On Things
    Heard, On Physiognomies, On Plants, On Marvellous Things
    Heard, Mechanical Problems, On Indivisible Lines, On Situations
    and Names of Winds, On Melissus, Xenophanes, and Gorgias.
ARISTOTLE: NICOMACHEAN ETHICS. H. Rackham.
ARISTOTLE: OECONOMICA and MAGNA MORALIA. G. C. Armstrong (with
    METAPHYSICS, Vol. II).
ARISTOTLE: ON THE HEAVENS. W. K. C. Guthrie.
ARISTOTLE: ON THE SOUL, PARVA NATURALIA, ON BREATH. W. S. Hett.
ARISTOTLE: CATEGORIES, ON INTERPRETATION, PRIOR ANALYTICS. H. P.
    Cooke and H. Tredennick.

ARISTOTLE: POSTERIOR ANALYTICS, TOPICS. H. Tredennick and E. S. Forster.

ARISTOTLE: ON SOPHISTICAL REFUTATIONS.
On Coming-to-be and Passing-Away, On the Cosmos. E. S. Forster and D. J. Furley.

ARISTOTLE: PARTS OF ANIMALS. A. L. Peck; MOTION AND PROGRESSION OF ANIMALS. E. S. Forster.

ARISTOTLE: PHYSICS. Rev. P. Wicksteed and F. M. Cornford. 2 Vols.

ARISTOTLE: POETICS and LONGINUS. W. Hamilton Fyfe; DEMETRIUS ON STYLE. W. Rhys Roberts.

ARISTOTLE: POLITICS. H. Rackham.

ARISTOTLE: PROBLEMS. W. S. Hett. 2 Vols.

ARISTOTLE: RHETORICA AD ALEXANDRUM (with PROBLEMS. Vol. II). H. Rackham.

ARRIAN: HISTORY OF ALEXANDER and INDICA. Rev. E. Iliffe Robson. 2 Vols. New version P. Brunt.

ATHENAEUS: DEIPNOSOPHISTAE. C. B. Gulick. 7 Vols.

BABRIUS and PHAEDRUS (Latin). B. E. Perry.

ST. BASIL: LETTERS. R. J. Deferrari. 4 Vols.

CALLIMACHUS: FRAGMENTS. C. A. Trypanis. MUSAEUS: HERO AND LEANDER. T. Gelzer and C. Whitman.

CALLIMACHUS, Hymns and Epigrams and LYCOPHRON. A. W. Mair; ARATUS. G. R. Mair.

CLEMENT OF ALEXANDRIA. Rev. G. W. Butterworth.

COLLUTHUS. Cf. OPPIAN.

DAPHNIS AND CHLOE. Thornley's translation revised by J. M. Edmonds: and PARTHENIUS. S. Gaselee.

DEMOSTHENES I.: OLYNTHIACS, PHILIPPICS and MINOR ORATIONS I.–XVII. and XX. J. H. Vince.

DEMOSTHENES II.: DE CORONA and DE FALSA LEGATIONE. C. A. Vince and J. H. Vince.

DEMOSTHENES III.: MEIDIAS, ANDROTION, ARISTOCRATES, TIMOCRATES and ARISTOGEITON I. and II. J. H. Vince.

DEMOSTHENES IV.–VI.: PRIVATE ORATIONS and IN NEAERAM. A. T. Murray.

DEMOSTHENES VII.: FUNERAL SPEECH, EROTIC ESSAY, EXORDIA and LETTERS. N. W. and N. J. DeWitt.

DIO CASSIUS: ROMAN HISTORY. E. Cary. 9 Vols.

DIO CHRYSOSTOM. J. W. Cohoon and H. Lamar Crosby. 5 Vols.

DIODORUS SICULUS. 12 Vols. Vols. I.–VI. C. H. Oldfather. Vol. VII. C. L. Sherman. Vol.VIII. C. B. Welles. Vols. IX. and X. R. M. Geer. Vol. XI. F. Walton. Vol. XII. F. Walton. General Index. R. M. Geer.

DIOGENES LAERTIUS. R. D. Hicks. 2 Vols. New Introduction by H. S. Long.

DIONYSIUS OF HALICARNASSUS: ROMAN ANTIQUITIES. Spelman's translation revised by E. Cary. 7 Vols.

DIONYSIUS OF HALICARNASSUS: CRITICAL ESSAYS. S. Usher. 2 Vols.
EPICTETUS. W. A. Oldfather. 2 Vols.
EURIPIDES. A. S. Way. 4 Vols. Verse trans.
EUSEBIUS: ECCLESIASTICAL HISTORY. Kirsopp Lake and J. E. L.
    Oulton. 2 Vols.
GALEN: ON THE NATURAL FACULTIES. A. J. Brock.
GREEK ANTHOLOGY. W. R. Paton. 5 Vols.
GREEK BUCOLIC POETS (THEOCRITUS, BION, MOSCHUS). J. M. Edmonds.
GREEK ELEGY AND IAMBUS with the ANACREONTEA. J. M. Edmonds. 2
    Vols.
GREEK LYRIC. D. A. Campbell. 4 Vols. Vols. I. and II.
GREEK MATHEMATICAL WORKS. Ivor Thomas. 2 Vols.
HERODAS. Cf. THEOPHRASTUS: CHARACTERS.
HERODIAN. C. R. Whittaker. 2 Vols.
HERODOTUS. A. D. Godley. 4 Vols.
HESIOD AND THE HOMERIC HYMNS. H. G. Evelyn White.
HIPPOCRATES and the FRAGMENTS OF HERACLEITUS. W. H. S. Jones and
    E. T. Withington. 7 Vols. Vols. I.–VI.
HOMER: ILIAD. A. T. Murray. 2 Vols.
HOMER: ODYSSEY. A. T. Murray. 2 Vols.
ISAEUS. E. W. Forster.
ISOCRATES. George Norlin and LaRue Van Hook. 3 Vols.
[ST. JOHN DAMASCENE]: BARLAAM AND IOASAPH. Rev. G. R. Wood-
    ward, Harold Mattingly and D. M. Lang.
JOSEPHUS. 10 Vols. Vols. I.–IV. H. Thackeray. Vol. V. H.
    Thackeray and R. Marcus. Vols. VI.–VII. R. Marcus. Vol.
    VIII. R. Marcus and Allen Wikgren. Vols. IX.–X. L. H.
    Feldman.
JULIAN. Wilmer Cave Wright. 3 Vols.
LIBANIUS. A. F. Norman. 2 Vols..
LUCIAN. 8 Vols. Vols. I.–V. A. M. Harmon. Vol. VI. K. Kilburn.
    Vols. VII.–VIII. M. D. Macleod.
LYCOPHRON. Cf. CALLIMACHUS.
LYRA GRAECA, III. J. M. Edmonds. (Vols. I.and II. have been re-
    placed by GREEK LYRIC I. and II.)
LYSIAS. W. R. M. Lamb.
MANETHO. W. G. Waddell.
MARCUS AURELIUS. C. R. Haines.
MENANDER. W. G. Arnott. 3 Vols. Vol. I.
MINOR ATTIC ORATORS (ANTIPHON, ANDOCIDES, LYCURGUS, DEMADES,
    DINARCHUS, HYPERIDES). K. J. Maidment and J. O. Burtt. 2 Vols.
MUSAEUS: HERO AND LEANDER. Cf. CALLIMACHUS.
NONNOS: DIONYSIACA. W. H. D. Rouse. 3 Vols.
OPPIAN, COLLUTHUS, TRYPHIODORUS. A. W. Mair.
PAPYRI. NON-LITERARY SELECTIONS. A. S. Hunt and C. C. Edgar. 2
    Vols. LITERARY SELECTIONS (Poetry). D. L. Page.

6

PARTHENIUS. Cf. DAPHNIS AND CHLOE.
PAUSANIAS: DESCRIPTION OF GREECE. W. H. S. Jones. 4 Vols. and Companion Vol. arranged by R. E. Wycherley.
PHILO. 10 Vols. Vols. I.–V. F. H. Colson and Rev. G. H. Whitaker. Vols. VI.–IX. F. H. Colson. Vol. X. F. H. Colson and the Rev. J. W. Earp.
PHILO: two supplementary Vols. (*Translation only*.) Ralph Marcus.
PHILOSTRATUS: THE LIFE OF APOLLONIUS OF TYANA. F. C. Conybeare. 2 Vols.
PHILOSTRATUS: IMAGINES; CALLISTRATUS: DESCRIPTIONS. A. Fairbanks.
PHILOSTRATUS and EUNAPIUS: LIVES OF THE SOPHISTS. Wilmer Cave Wright.
PINDAR. Sir J. E. Sandys.
PLATO: CHARMIDES, ALCIBIADES, HIPPARCHUS, THE LOVERS, THEAGES, MINOS and EPINOMIS. W. R. M. Lamb.
PLATO: CRATYLUS, PARMENIDES, GREATER HIPPIAS, LESSER HIPPIAS. H. N. Fowler.
PLATO: EUTHYPHRO, APOLOGY, CRITO, PHAEDO, PHAEDRUS. H. N. Fowler.
PLATO: LACHES, PROTAGORAS, MENO, EUTHYDEMUS. W. R. M. Lamb.
PLATO: LAWS. Rev. R. G. Bury. 2 Vols.
PLATO: LYSIS, SYMPOSIUM, GORGIAS. W. R. M. Lamb.
PLATO: REPUBLIC. Paul Shorey. 2 Vols.
PLATO: STATESMAN, PHILEBUS. H. N. Fowler; ION. W. R. M. Lamb.
PLATO: THEAETETUS and SOPHIST. H. N. Fowler.
PLATO: TIMAEUS, CRITIAS, CLEITOPHON, MENEXENUS, EPISTULAE. Rev. R. G. Bury.
PLOTINUS: A. H. Armstrong. 7 Vols.
PLUTARCH: MORALIA. 16 Vols. Vols. I.–V. F. C. Babbitt. Vol. VI. W. C. Helmbold. Vols. VII. and XIV. P. H. De Lacy and B. Einarson. Vol. VIII. P. A. Clement and H. B. Hoffleit. Vol. IX. E. L. Minar, Jr., F. H. Sandbach, W. C. Helmbold. Vol. X. H. N. Fowler. Vol. XI. L. Pearson and F. H. Sandbach. Vol. XII. H. Cherniss and W. C. Helmbold. Vol. XIII. 1–2. H. Cherniss. Vol. XV. F. H. Sandbach.
PLUTARCH: THE PARALLEL LIVES. B. Perrin. 11 Vols.
POLYBIUS. W. R. Paton. 6 Vols.
PROCOPIUS. H. B. Dewing. 7 Vols.
PTOLEMY: TETRABIBLOS. F. E. Robbins.
QUINTUS SMYRNAEUS. A. S. Way. Verse trans.
SEXTUS EMPIRICUS. Rev. R. G. Bury. 4 Vols.
SOPHOCLES. F. Storr. 2 Vols. Verse trans.
STRABO: GEOGRAPHY. Horace L. Jones. 8 Vols.
THEOCRITUS. Cf. GREEK BUCOLIC POETS.
THEOPHRASTUS: CHARACTERS. J. M. Edmonds. HERODAS, etc. A. D. Knox.

7

THEOPHRASTUS: ENQUIRY INTO PLANTS. Sir Arthur Hort, Bart. 2 Vols.
THEOPHRASTUS: DE CAUSIS PLANTARUM. G. K. K. Link and B. Einarson. 3 Vols. Vol. I.
THUCYDIDES. C. F. Smith. 4 Vols.
TRYPHIODORUS. Cf. OPPIAN.
XENOPHON: CYROPAEDIA. Walter Miller. 2 Vols.
XENOPHON: HELLENICA. C. L. Brownson. 2 Vols.
XENOPHON: ANABASIS. C. L. Brownson.
XENOPHON: MEMORABILIA and OECONOMICUS. E. C. Marchant. SYMPOSIUM and APOLOGY. O. J. Todd.
XENOPHON: SCRIPTA MINORA. E. C. Marchant. CONSTITUTION OF THE ATHENIANS. G. W. Bowersock.